C Stuart Houston

Birds of Yorkton-Duck Mountain

by

C. Stuart Houston and William Anaka

SASKATCHEWAN NATURAL HISTORY SOCIETY

2003

TABLE OF CONTENTS

LIST OF MAPS

LIST OF PHOTOGRAPHS

LIST OF APPENDICES

FOREWORD

By J. Frank Roy

The 50th anniversary issue of *Blue Jay* (September 1992) was dedicated to the memory of Isabel Priestly, its first editor. An inspirational woman, she co-authored a weekly nature column in the local newspaper and was recognized as the local expert in both botany and ornithology. It probably never entered her mind that when she offered to take Stuart Houston and several of his friends on weekly hikes around what was called "the muskeg" on the western outskirts of Yorkton, she was forming the nucleus of a group that would become the Yorkton Natural History Society. In July 1942 she and the students, now mostly in high school, produced a modest mimeographed list of 193 species of birds it was possible to see within a 30-mile radius of Yorkton. News of that list soon got around the province and by September, Mrs Priestly suggested founding a society which promised to issue a quarterly bulletin that could become "a worthwhile clearing house for nature records in Saskatchewan."

I was a 14-year-old farm boy at the time, living hundreds of miles distant in southwestern Saskatchewan. When I read about the formation of the Yorkton Society and realized that I could communicate with other people interested in nature, particularly in birds, I immediately sent in the 25¢ annual fee. Mrs Priestly wrote me a kind letter of encouragement and provided me with Stuart's address, suggesting that we share our experiences. For the next 13 years Stuart and I exchanged letters, but we didn't meet face to face until 1955, when Stuart came to Saskatoon to begin postgraduate study in radiology. We have remained close friends throughout the years, both of us profoundly influenced by Mrs Priestly, who died unexpectedly only five years after founding the Yorkton Society and *Blue Jay*.

Since the exciting days when Stuart began his study of birds and in summer months banded thousands of ducks for Ducks Unlimited, he has gone on to become one of North America's best known authorities on birds. He and his wife Mary have banded over 125,000 birds of 206 species, among them 7,204 Great Horned Owls, the North American banding record. A member of the American Ornithologists' Union since 1943, Stuart has written more than 240 papers on ornithological subjects.

Quite apart from a distinguished career in medicine, he is a noted historian, having authored five books for McGill-Queen's, three of them on the Franklin expeditions to the Arctic, another on the history or medicare in Saskatchewan and, most recently, the eighteenth-century naturalists of Hudson Bay. Stuart knows farmers and birders in every part of Saskatchewan. He likes people, remembers their names after having met them (not many can do that!), and maintains contacts even 50 years after an initial meeting. Although he hasn't lived in Yorkton for over 40 years, he keeps in touch with people there, particularly Bill and Joyce Anaka. Much of the data on breeding bird numbers and migration comes from them. So important is their contribution that Bill is listed as co-author and Joyce is recognized for "major assistance." Stuart's skills as writer, researcher, and historian have made him the inevitable choice to author *Birds of Yorkton-Duck Mountain*, one of the last in an extended series of regional works published by the Saskatchewan Natural History Society (SNHS). The much anticipated *Birds of Saskatchewan*, which he will co-author with Alan Smith, is already a work in progress.

INTRODUCTION

The late E. Manley Callin (1980) compiled one of the finest regional bird species accounts in Canada. Manley believed that his regional study was but one of a number of stepping stones to a more definitive publication, *Birds of Saskatchewan.* To make his desire possible, he left almost his entire estate to fund such a project. The raison d'être[1] of this publication, then, is to provide one more regional contribution towards the realization of Manley's dream.

Why should someone who left Yorkton in 1960 be a co-author? First, Yorkton was where I got my start. Second, I have been collecting the data for over 60 years, from a great many observers who deserve to have their observations credited. As a result of my failure to publish earlier, a number of relatively common species have no confirmed nest record whatever within the Yorkton area in Alan R. Smith's *Atlas of Saskatchewan Birds* (1996). Or, there are confirmed breeding dots for only one or two of his map sheet blocks. Third, the wetland complex south of Yorkton truly is a major component of "the great duck factory"—as amply demonstrated by the widely ranging ducks in Kathy Meeres' magnificent maps from the relatively small Yorkton-Duck Mountain (YDM) area. Fourth, the parkland country around Yorkton was one of the first areas in Saskatchewan to be studied at the end of the 19th century (the Baines family at Crescent Lake from 1883 and the Gunn family at Good Spirit Lake [GSL] from 1887); some of our information spans 120 years. Fifth, the Yorkton area was the birthplace or home of a disproportionate number of naturalists, and many aspects of birdlife have been documented more carefully than for most other parts of Saskatchewan. There is a mass of data to share, from museum collecting trips to Duck Mountain Provincial Park (DMPP) in 1926 and 1951, Breeding Bird Surveys, Christmas Bird Counts, seasonal and regional reports in *American Birds* and *Birders Journal*, the Prairie Nest Records Scheme (which had more junior contributors from the Yorkton area than any other region of Saskatchewan), from Douglas Flack's detailed counts of

[1] *Birds of Saskatchewan*, in compilation by CSH and Alan R. Smith, will assess birds within our province in relation to other provinces and states, whereas this more local publication makes no such attempt for most species.

breeding birds near Good Spirit Lake, from Al Smith's and Trevor Herriot's Point Counts in DMPP, from observations during two "summer meets" at DMPP and one at GSL, and from the meticulous Gunn-Anaka data set of migration dates and other observations spanning 48 years. Sixth, and most important, the Yorkton-Duck Mountain area, on the edge of the mixed forest at Duck Mountain Provincial Park, has one of the richest bird diversities in North America. The park has never had a resident birder, but over its lifetime has had many birding visitors (please see detailed listings under Abbreviations), and will no doubt become every more of a mecca for birders in years to come.

The Yorkton-Duck Mountain area, including Good Spirit Lake, is right at the junction between closely related species now recognized as separate: Western and Eastern Wood-Pewee; Spotted and Eastern Towhee. Yorkton is just east of the junction between the two forms of Marsh Wren. These geographic interfaces warrant further study to determine the extent of interbreeding between closely related taxa.

This book is a compendium of information, more for reference and for browsing than for reading. Since breeding data are more valuable than occasional wanderings of stray birds, we present information on nests and eggs whenever available. Because Fred and Frank Baines earned pocket money by collecting birds' eggs for Walter Raine and E. Rippon, and were twice visited by Raine, and because many Baines and Raine sets are still extant in major museums, we are sharing such information, tedious though some of it may perhaps seem.

The Area: This publication is designed to fill the geographic gap of 8,744 square miles between Callin's *Birds of the Qu'Appelle,* 1857-1979, and the late Donald F. Hooper's *Birds of East-Central Saskatchewan: Kelvington to Kelsey Trail,* published in 1992 (see map, p. 41).

Thus the south boundary meets Callin's area along Highway 22. For banding purposes, to conform to 10-minute blocks of latitude and longitude (these are not metric), from the Manitoba boundary the south border follows 50° 40' west to 102° 30', jogs north 10 or 11 miles and follows 50° 50' west to 103° 30', and then jogs north to follow 51° west to 104°.

The west boundary is approximately the same as Callin's and exactly the same as Hooper's, the 104th meridian. The north boundary, the 52nd parallel of latitude, is identical with Hooper's south boundary. As was the case with both lists, the eastern limit is the Saskatchewan-Manitoba boundary.

Since this is a largely an historic study, we have chosen to use the units of measure in which the land was surveyed, and in which all heights and distances were recorded. Units of measurement are presented in Arabic numerals in Imperial units. A conversion chart follows:

METRIC CONVERSION

1 inch = 2.54 cm	**1 acre = 0.4 ha**
1 foot = 30.5 cm	**1 ounce = 28.35 g**
1 mile = 1.6 km	**1 pound = 454 g**

Agriculture and Water: Most of the YDM area is now a fertile agricultural district, devoted largely to grain, beef cattle, and dairy farming, although the latter has declined since the 1950s. The terrain is undulating, with elevation varying from 1,445 feet above sea level at Kamsack to over 2,200 feet in the Beaver Hills. Most of the area drains into the Assiniboine River, via the Whitesand and other tributaries. South of Melville and Bredenbury, drainage is south into the Qu'Appelle River.

Wooded areas consist almost entirely of poplars--Black (Balsam) Poplar in lower, damper locations and Trembling Aspen (White Poplar)

Aspen Woods, 1977 SNHS Meet, Good Spirit Lake Dunes

Aspen Bluffs, Good Spirit Lake Community Pasture

Interior Basin, Sand Dune Area

Theodore Dam on Whitesand River, downstream

on higher ground. Willows, Chokecherry, Saskatoons, Pincherry, and other shrubs are found in profusion. The main tree planted on city boulevards of Yorkton and Melville, the Manitoba Maple (Box Elder), also occurs in a few small natural groves.

The district is dotted with sloughs of every size, many of which dry up as summer advances. With numerous permanent small lakes and deeper ponds and sloughs, and since the 1940s, the ubiquitous farm dugouts, 10 to 12 feet deep, most duck broods are within reach of water in all but the driest of years.

Upper and Lower Rousay lakes are normally shallow (not over five feet deep) with wide margins of bulrush, grasses, and some cattail affording excellent cover along the edges of open water. Crescent and York lakes are deeper, but rarely more than 10 or 15 feet. The Crescent (or "Maddaford" and "Maple Island") marshes and Rokeby marsh possess dense, almost impenetrable stands of high grasses and rushes, with little open water. Good Spirit Lake is a large lake with extremely wide sandy beaches. Ravine Reserve, on the western boundary of Yorkton, is bisected by the CPR, to form "muskeg" on the south and "Hopkins' Lake" on the north.

The Yorkton area has been subject to vast fluctuations in moisture levels. Water levels were extremely high from 1880 to 1882, with the Whitesand River a major obstacle to settlers and their horse-drawn carts. Then a dry cycle began. In 1889, the snow was gone by February and grass fires occurred in July. In 1896, another very dry year, Leech and Crescent lakes were dry. Wet years returned in 1901, and lake levels were back to normal in 1902. There were severe blizzards in 1903 and 1904, with trains unable to reach Yorkton for days at a time. The years 1902 through 1904 were wet, with Good Spirit Lake overflowing during those years and again in the 1920s, stopping in 1928 (Anaka 2000). Precipitation remained above average through 1910. Several bridges across the Whitesand River were washed out in 1913. In 1916, water levels were so high that GSL overflowed into Horseshoe Lake. Over 25 inches of precipitation fell each year in 1921 and 1923; fish ascended ditches to reach Gladstone Avenue in downtown Yorkton. The driest year was 1931; that year precipitation was only 9.14 inches. At GSL, the beach of fine white sand was 400 yards wide, and the water shallow far into the lake, perfect to allow children and even toddlers to play safely with little supervision. In the 1930s, water levels in York Lake became so low that

Rolling Topography, Good Spirit Lake Community Pasture

cottages were deserted. It was extremely dry in 1937. In 1942, water levels recovered, and once-dry sloughs filled again (Swallow 1946).

Water levels rose again in 1955; in 1956 and 1957, GSL flooded and overflowed via the creek at the south end of the lake, for the first time in 27 years. The Gunns' yard was partially flooded and the road from the west to Gunn's Beach was under water. Subsequent dry years included 1960-61, 1988, and 2002. The spring of 1980 was dry at Sheho (WN). The creek overflowed again in the wet years of 1985 and 1986.

From 1898 to 1906, and again in 1942, a diversion ditch from Willow Brook Creek, with strategically located dams, augmented the water supply to the Rousay lakes and York Lake. In 1976, a comprehensive proposal for water management went forward, but was not instituted, and Upper Rousay Lake went dry in 1989. The level of GSL is maintained by structures at its outlet.

Duck Mountain Provincial Park: (abridged from Wright et al. 1995) The park (map, p. 10) is situated along the Manitoba boundary in the Duck Mountain uplands, classified as the Duck Mountain landscape area in the Mid-Boreal Upland ecoregion by Padbury and Acton (1994). It rises 330 feet above the Swan River Plain to the north and about 660 feet above the Quill Lake-Yorkton Plain to the west. Most of the park is over 2,000 feet above sea level. The highest elevation, just north of the Little Boggy Creek valley, is 2,250 feet. As a result the park normally has higher precipitation and lower mean temperatures than the surrounding plains. The total area of the park is 101 square miles (66,575 acres). Most of it is characterized by hummocky knob and kettle topography with slopes of 10 to 30% on over 40% of the area. Irregularly-shaped sloughs and small lakes are common. The glacial till of ground moraines is from 300 to 600 feet thick. No bedrock exposures occur within the park. The most striking topographic feature, and arguably the spot with the greatest bird diversity, is the Little Boggy Creek valley, a 400-foot-deep glacial meltwater channel in the southeast corner of the park. Native grassland remnants in this valley are of provincial and national significance.

Water bodies cover 13.4% of the park area. Most of the watershed drains into and out of the largest lake, Madge Lake. When water levels are high, Clearwater Creek exits from the north end of Madge Lake, flows north into Bear Head Lake, and then down Bear Head Creek to the Swan River. The lake surface varies from 1,995 to 2,005 feet above sea level.

Moist soils, which develop beneath forest vegetation, are *luvisols* or

BOUNDARY OF THE DUCK MOUNTAIN 1931 FEDERAL FOREST RESERVE AND 1994 PROVINCIAL PARK

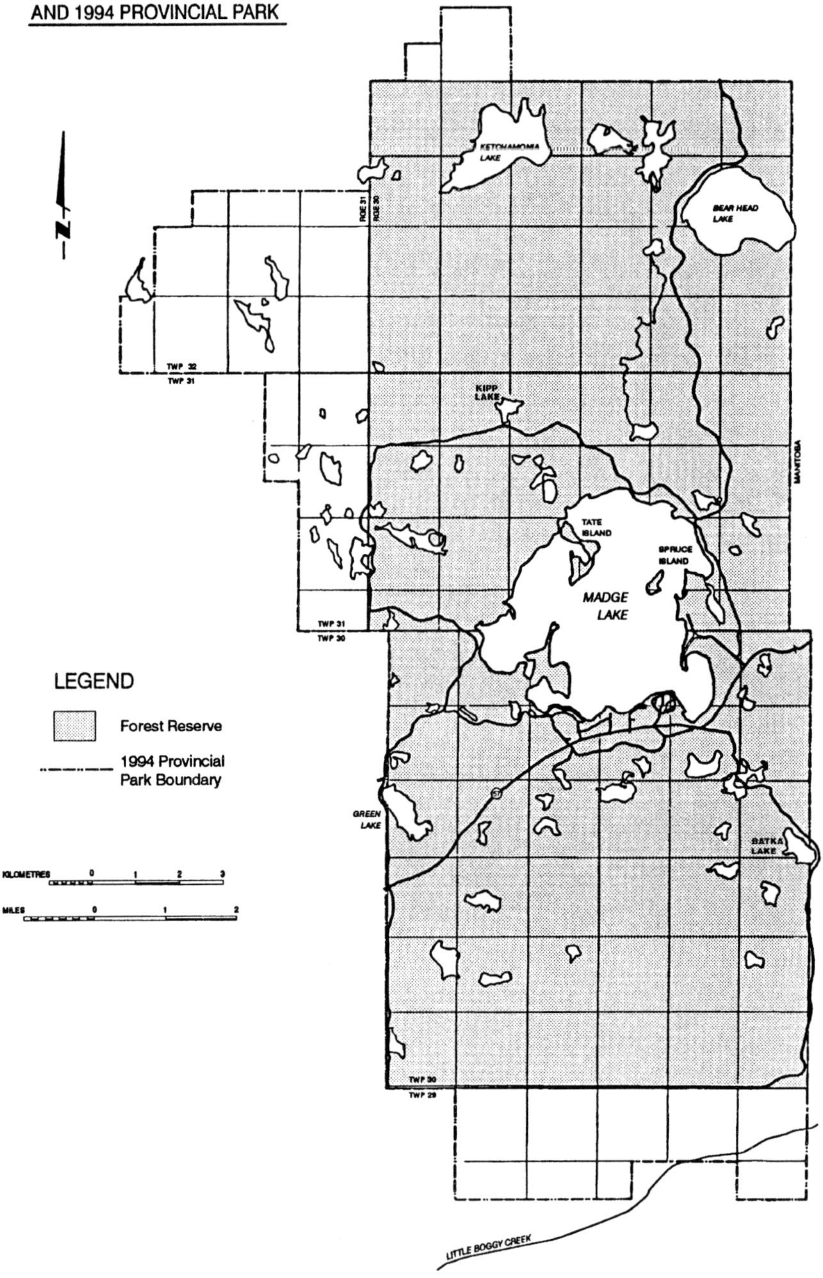

gray wooded soils. Depressional areas contain organic or *gleysolic* soils. In the Little Boggy Creek valley, calcium-rich groundwaters support calcareous fens, which, along with Trembling Aspen-Mountain Maple forest areas, are considered rare ecosystems in the province and contribute to the unique importance of DMPP in Saskatchewan's park system.

Vegetation in Duck Mountain Provincial Park has been strongly influenced by both fire and logging. More than 75% of the park's aspen stands, which in turn occupy 77.5% of the forest land, are over 90 years old. From 1883, consecutive wildfires through most of the mixedwood region resulted in the virtual elimination of White Spruce from the northern half of the park. The last major fire was before 1890, but a few smaller fires occurred as late as 1896. The 20th century has seen concerted suppression of all fire, resulting in an unnatural predominance of old (90+ yr) aspen stands throughout the park, but particularly north of Madge Lake. South of Madge Lake, selective logging for White Spruce between 1890 and 1914 resulted in a forest with an unnaturally high ratio of aspen and a low proportion of pure coniferous and mixedwood ecosystems. Fuelwood harvesting peaked in the 1930s and 1940s; in the 1930s there were six sawmills in the park. During World War II, 1939-1945, conscientious objectors built logging roads and cut much aspen for cordwood on the west side of the park, north of Green Lake. The mixedwoods consist of an admixture of White Spruce, Trembling Aspen, Balsam Poplar, and White Birch. Jack Pine is absent.

Minor components of the vegetation are the pure Black Spruce and Black Spruce-Tamarack stands in poorly drained areas. Balsam Fir is found almost exclusively on islands or in other "fireproof" settings. True bogs with scattered and stunted Black Spruce and Tamarack are rare.

In 1931, the 81-square-mile Duck Mountain Forest Reserve became Duck Mountain Provincial Park. The boundary was later extended to the north and west. The central and southern thirds of the eastern boundary, 20% of the total boundary, adjoin the forest landscapes of Manitoba's similarly named Duck Mountain Provincial Park. Another 47% of park boundary is buffered from adjacent agricultural land by only one mile of privately owned, highly fragmented forest, and 33% of the boundary fronts directly onto cleared agricultural land.

Because of striking ecological and ornithological differences, Duck Mountain observations for each species are under a separate heading, with the exception of Seton's early historical observations that also apply to adjacent areas.

Museum Collecting at Madge Lake, DMPP, 6 May-19 June 1926

There were two museum collecting camps at Madge Lake in May and June, the first in 1926 (Mitchell 1927, Bradshaw 1927), and the second in 1951 (Bard 1952).

Towards the end of the first museum trip, on 14 June 1926, H. Hedley Mitchell, "Provincial Naturalist," sent a list of the 101 species seen to the *Kamsack Times*, with an accompanying letter. The annual report of the museum and specimen labels show that an additional five species were collected: Western Wood-Pewee, Bay-breasted Warbler, American Redstart, Swamp Sparrow, and Dark-eyed Junco; perhaps one or more of these were not keyed out and positively identified until the party had returned to the museum. A California Gull may also have been collected at Madge Lake. Thus the total list is actually either 106 or 107 species. Mitchell's letter was printed with the list on page 1 of the 17 June 1926 newspaper as follows:

"In the many districts that I have worked in collecting museum specimens in this province, I can recall no more interesting or beautiful locality than this Madge Lake; it is indeed a great asset to the citizens of Kamsack and the surrounding country as a summer resort and I might say I

Madge Lake, DMPP

was struck by the orderly arrangements about the numerous cottages and general tidiness of the grounds, the well-kept roads, etc., due mainly to the energy and efficiency of the Forest Ranger Mr. Dell. But from a naturalist's point of view I have been especially delighted and surprised; I cannot record here all the various finds and observations, but might say that while the water-birds are somewhat limited in species, or varieties, the woodland birds are exceedingly numerous and interesting. Of course it is difficult to find any given district an ideal ground for all species of birds, etc., at the same time, but here is a "ground" not only interesting to the Ornithologist, but also to the Botanist and Entomologist. I have noted some new butterflies and moths as well as flowers and ferns, the latter plentiful and of at least four varieties. The list [of] birds included is a conservative one; any species not properly or satisfactorily identified has been omitted. I trust it may be of some interest to your publication. Most of these birds listed are already represented by specimens in the museum, but my assistant, Mr. F[red] Bard and myself have been successful in securing several new specimens of birds and eggs for that institution, notably the Ring-necked Duck, the Philadelphia Vireo, Blackburnian Warbler, and Connecticut Warbler."

Museum Collecting at Madge Lake, DMPP, 15 May-22 June 1951

"Some 120 varieties of birds were reported in the area, the most interesting being a male Yellow-throated Vireo, the first provincial specimen. Pileated Woodpeckers and a pair of Bald Eagles were some of the more interesting and beautiful birds....Typical birds from the region include Ring-necked Ducks, Common Loons, Red-breasted Mergansers, Connecticut Warblers, Blue-headed Vireos, and Blackburnian Warblers." (Bard 1952). Other egg sets included those of the Swamp Sparrow and Swainson's Thrush.

SNHS Summer Meets

Lists of species seen were kept during SNHS summer meets at DMPP, 9-19 June 1956 and 9-11 June 1989. Margaret Belcher then added nine more species to the 1989 list when she visited DMPP at the same time in 1990.

Hatch and Weidl Survey of DMPP, 1981

This intensive study continued from May through November 1981, with data on 181 species. An air survey of the park was carried out on 17 June. Kelvin Krieger and Louis L'Arrivee assisted Donald A. Weidl and David

R.M. Hatch (20 of the more interesting observations were published in Weidl 1983). "Hatch and Weidl" always refers to this 1981 study.

GENERAL SOURCES OF DATA

Blue Jay

Sixty volumes of *Blue Jay* were abstracted, 1942 through 2002 inclusive. Note that *Blue Jay* had seven years, under Lloyd Carmichael's editorship, when pagination began anew at page 1 with each issue; hence, for volumes 7 through 13, the issue number is provided, along with the volume number.

"Chickadee Notes," *Winnipeg Free Press*

CSH and MIH copied all Saskatchewan entries in Canada's most popular nature column of its time, nos. 1 through 666 (31 Mar 1921 through 29 Dec 1933), as part of research into the spread of Gray Partridge, Western Kingbird, and Black-billed Magpie across the prairies. Nine items of interest to the YDM area have been cited.

Christmas Bird Counts (CBC)

Bill Anaka has compiled the 241 CBCs for the entire YDM area: GSL 53 yrs, 1943-2002; Yorkton 50 yrs, 1941-2001; Bangor 36 yrs, 1956-2001; Kamsack-Togo-Duck Mountain 36 yrs, 1947-2001; Melville 20 yrs, 1951-2001; Saltcoats 13 yrs, 1942-1997; Sheho 12 yrs, 1942-1997; Preeceville 10 yrs, 1986-2001; Gerald 3 yrs, 1947-1951; Canora 2 yrs, 1977-1978; Ituna 2 yrs, 1946, 1977; Bredenbury 1952; Pelly 1962; Stornoway 1960; Tonkin 1983.

Breeding

Prairie Nest Records Scheme (PNRS)

The late 1950s were years of ferment, excitement, and enthusiasm for Saskatchewan ornithology. Premier T.C. Douglas had responded to requests from the Saskatchewan Natural History Society by establishing a magnificent new Saskatchewan Museum of Natural History (SMNH) to honour the pioneers and thus celebrate Saskatchewan's jubilee in 1955. The museum in turn brought Bob Nero to Regina. Bob had what seemed a presumptuous idea, to bring the American Ornithologists' Union (AOU) to Regina for its annual meeting in 1959. This required two years of preparation and attracted an unprecedented number of local people and general public (112 of the 391 registrants were from Saskatchewan), and these included four young people from the Yorkton area and two others: Gary Anweiler of Melville, Frank Switzer from Rokeby, Bill Horseman from Saltcoats, Lawrence Ostoforoff of Kamsack, Glen Fox from Kindersley, and Spencer Sealy from Battleford (Deignan 1960). Ripple effects from the museum, Nero, and the AOU spread over the province. A Prairie Nest Records Scheme was begun in 1958 by the SNHS and for some years the submitted nest cards were housed in the museum.

The first annual report of the PNRS, for 1958, singled out its four largest contributors: Bill Horseman, with 153 cards, Bill Anaka with 92, Stuart Houston with 79, and Glen Fox with 44 (Fox 1959); three of these four were from the Yorkton area. Joyce Gunn sent in cards from Spirit Lake, as did Jacob H. Jmaeff from Kamsack. In the second year, 1959, Spencer Sealy had jumped into the lead with 181 cards, but the Yorkton area was well represented by Horseman, who had improved his total to 155. Bill Anaka had another 84, Stanley Zazelenchuk sent in 65 from Stornoway, and other cards came from Walter Chudzik near Canora (Nero 1960). In 1960, they were joined by Zazelenchuk's neighbour, Ross Derkatch, and by Victor Schmidt, from a farm just south of Melville. In 1964, Larry Morgotch of Yorkton contributed a few cards, as did Maurice Maurer, from near Duff, in 1966-69, Ernie Kozak and Ivan Yaholnitsky from Mikado, 1966-69, and Jim Jowsey in 1971.

Never, before or since, have so many lads of high school age (plus a 20-year-old schoolteacher) spent so much time and effort in contributing to the science of ornithology. The Yorkton area led the rest of the prairies in this respect.

For preparation of this manuscript, Alan R. Smith made available to me the printouts of computerized nest records available on cards in the Prairie Nest Records Scheme through 1976. In addition, Danielle Cyr Pearson at the office of the Saskatchewan Nest Records Scheme in the Royal Saskatchewan Museum in Regina, compiled information on Brown-headed Cowbirds before 1976 (the computerized list had no column to give the species parasitized), and since 1976 for highly selected species where Smith's *Atlas* listed definite breeding records for any map sheets within the YDM area.

Following the PNRS section in species accounts, I have named those map sheets shown in Smith's 1996 *Atlas* as having definite breeding records, yet for which specific data are not cited.

Jowsey Oology Collection

Only one oology collection from the YDM area has been saved, though egg-collecting 100 years ago was a more popular hobby than stamp collecting. Norman Jowsey collected birds' eggs from his arrival in 1905 to about 1915 (hence all but two of his records are listed in the historical paragraph), and added a few more from 1935 to 1945. Most eggs were collected near his farm, 7 miles north of Saltcoats. In 1989, his nephew, Dr J.R. Jowsey, donated eggs of 38 species to augment the fine oology collection at the Provincial Museum of Alberta (PMA).

Cymbalisty Oology Collection

In the mid-1960s, Harvey Cymbalisty made a small egg collection, which he turned over to the SMNH before he completed high school at Canora.

Breeding Bird Surveys (BBS)

Bill Anaka has compiled data from all 33 Breeding Bird Surveys at Theodore (WA,FAS, Lionel Bonneville), and Alan R. Smith and Connie Downes have provided printouts of the other 20 BBS surveys in the YDM area: nine at Churchbridge/Langenburg (JRJ,CSH), one at Calder/Wroxton (CSH), three at Ituna/Hubbard (Wayne R. Gemmell), and seven at Willowbrook (WA; Appendix 2).

We consider all five BBS survey routes to be reasonably representative of the "parkland" habitat, which predominates in the YDM area, but it excludes DMPP, a very different habitat. During the Theodore counts a remarkable 142 species have been recorded; 25 species were identified during all 33 counts and 59 during 20 or more counts. Four more species have been added on the other four survey routes, for a grand total of 146 BBS species.

Species decreasing on the Theodore route are: Killdeer, Black Tern, Horned Lark, Bank Swallow, European Starling, Western Meadowlark, Yellow-headed Blackbird, and House Sparrow. Species increasing are: Least Flycatcher; Cliff Swallow at two bridges (over a thousand during one count!); Mountain Bluebird, American Robin, Yellow Warbler, and Clay-colored Sparrow.

Other information from the Theodore BBS includes an excellent indication of when the Common Raven first became a resident species (one was seen in 1980 and one in 1989, then in most years since).

Bird Banding Data

The numbers involved are impressive and worth sharing. The total of 266,644 individual birds of 174 species banded within the YDM area (Appendix 3) nicely surpasses the 73,619 individual birds of 168 species banded in the Saskatoon Bird Area--confirming that Yorkton is in the middle of what Ducks Unlimited (DU) has long termed the "duck factory."

A Word of Explanation

For the first 30 or so years of banding, "return" was the term used for a bird reported later, usually at a distance. This term should now be restricted to a bird that returns in a subsequent season to the location where it was banded. When such a record occurs with code "99," to indicate a bird retrapped or renetted in a subsequent year at the initial site of banding, it has been identified as a "return" and has been deleted from the number of "recoveries."

Next, "recovery" became the acceptable term for birds reported by someone else. Now that some bird bands are read by telescope, and other birds are caught in a network of distant banding stations (and released alive), experts restrict use of "recovery" to a bird that is dead, and use "encounter" as an all-encompassing term.

In this publication, because 99% of "encounters" are birds found dead, we cheat a little and use "recovery" as a term more widely understood over a much longer period than "encounter."

Incidentally, the term "repeat" is used for a bird that is caught again at the same banding station less than three months after banding. Repeats have not been entered into the banding computer, are not included in printouts, and are not mentioned in this monograph.

The number of waterfowl banded within the YDM area compares favourably with any similar-sized region in North America, and surpasses most.[2] Most of the waterfowl banding was done by men then known as United States Game Management Agents (Appendix 4), most of whom banded ducks in Canada for many years. It is interesting to note that the

[2] No YDM area bands have yet been recovered in Hawaii, the District of Columbia, or the Yukon Territory, but all other states, provinces, and territories are represented, most by rather substantial numbers. States and provinces that barely made the YDM list are as follows: A Mallard banded in Rhode Island and two banded in Vermont were shot in the YDM area; three Mallards banded in the YDM area were shot in Prince Edward Island and one in Newfoundland; one Blue-winged Teal was shot in Vermont, one in New Hampshire, and four in Maine; one Mallard, two Northern Pintails, one Green-winged Teal, and one Redhead were shot in Arizona.

number of ducks banded in the Yorkton area far exceeds the number ringed by DU banders (of which CSH was one) throughout the Prairie Provinces through the end of 1954 (Cartwright and Shortt 1956). That total was 77,068 with 8,435 recoveries; unlike in this report, a number of the northeastern states and Maritime Provinces were still missing from the 1956 booklet that summarized DU band recoveries to that date.

Langenburg

To fill in one obvious gap for Smith's *Atlas,* and for this publication, Trevor Herriot and Bob Luterbach made a two-day study of the Langenburg map sheet, 14-15 June 1990 (these dates are omitted from each species account). Dale Hjertaas, in the spring of 1976, recorded another 15 species from the Langenburg area.

Duck Mountain Provincial Park

"**Point Counts**" were undertaken by Alan R. Smith at DMPP during five years: Ten five-minute stops on the East Bay Nature Trail, 24 Jun 1990 (34 species detected); 15 stops on Core Loop, 25 Jun 1991 (41 species); 20 stops on Core Loop, 18 Jun 1992 (40 species); 20 stops on Core Loop, 17 Jun 1993 (45 species); 20 stops on Core Loop, 20 Jun 1994 (37 species), for a total of 85 stops, with 66 species detected. Smith was assisted by Rob W. Wapple in 1992.

"**Aspen counts**" were undertaken by Trevor Herriot on 15-19 June 1993, as part of a vegetation survey (Wright et al. 1995). He did 14 stops in old aspen 100+ yr forest (OA)— with lots of deadfall and openings in the canopy (mean of 9.6 species per sample site); 11 stops in young aspen <26 yr (YA) with densely packed small trees and a lot of crown closure (7.1 species/site); and 12 stops in mature but not dying 50-75 yr aspen (MA) with a fairly high crown closure (5.4 species/site).

At each stop, which lasted 10 minutes, following a five- minute wait to allow full resumption of singing after their entry, Rob Wright recorded the dominant species of trees, shrubs, and herbs, with percentage estimates of cover. Each stand was at least 150 yards from the previous. Trevor counted 168, 103, and 97 individual birds, respectively, in the three age classes of aspen, old to young.

Compared to the other two age classes, Herriot and Wright found a statistically significant increase in species diversity in the old aspen forest, with a range of 7 to 15 species, and a mean of 10.4 species per stop. Next was young aspen forest, with a range of 4 to 10 species, and a mean of 7.3 species per stop. Least, by a small margin, was mature, middle-aged aspen with 4 to 10 species, with a mean of 6.1 species per stop. The difference between young and mature aspen was not statistically significant (Wright et al. 1995).

The presence of Brown-headed Cowbirds in each aspen age class is worthy of note, because of the impact this parasite may have on nesting success of passerine birds.

ORGANIZATION OF SPECIES ACCOUNTS

Species name

Common and scientific names are derived from the American Ornithologists' Union *Check-List of North American Birds, Seventh Edition* (1998), and the supplements issued in 2000 and 2002. Although used for only a few recognizable subspecies, subspecific names are based on the Fifth Edition (1957).

Status Line

COSEWIC status: is given for endangered and threatened species.
Abundance: The 53 YDM BBS counts have allowed us to rank 146 species in numerical sequence (Appendix 2): 16 **abundant**, *in large numbers, found without any particular search,* with a mean of 20 or more sightings per BBS count; 15 **very common** (10 to 19.9); 45 **common,** *in substantial numbers with minimum search* (mean of 1.0 to 9.9); 33 **fairly common,** *in small numbers with some searching* (mean of 0.2 to 0.9); 38 **uncommon,** *in small numbers or alone, missed on some trips* (0.1 or less). This latter "uncommon" category means that only 1 to 7 individuals were sighted in total at 53 x 50 (= 2,650) three-minute BBS stops. Summer residents not sighted at a single stop are considered **rare,** *in extremely limited numbers.* (The italicized descriptions above are modified from Leighton et al. 2002, *Birds of the Saskatoon Area,* except that we have added the category, "Very Common").

The above stratagem uses an objective measure based on statistical cut-offs (admittedly chosen AFTER the fact, not in advance), replacing the traditional subjective estimate which may be more susceptible to personal bias. We admit that BBS totals likely underestimate the nocturnal Great Horned Owl, the crepuscular Common Nighthawk, and lake and river species far from designated roads. Of necessity, the status line resulting from this statistical compilation lumps results over 35 years, thus masking population trends; if we used only the third decade for the Theodore route, we would give a higher status for the Common Raven and House Finch, and a lower status for a declining species such as the Horned Lark.

Terms

A **resident** is *a breeding species, either a year-round resident or a summer resident.*

Vagrant refers to a species *occurring outside of its usual range and not normally to be expected.*

Visitor refers to a species *near the edge of its expected range and not ordinarily seen.*

A **transient** is a species *passing through both spring and fall en route to and from northern breeding grounds* (Roy 1996).

A **wanderer** is a species, *not known to breed in the area, seen on an irregular basis.*

A **hypothetical** species is one of three within the YDM area that *as yet lacks, anywhere within Saskatchewan, an identifiable specimen, photograph, or voice recording.*

Seasons follow Roy (1996): **Spring:** March, April, May. **Summer:** June, July. **Fall:** August, September, October, November. **Winter:** December, January, February.

History

Historical observations, when available, come immediately after the status line.

Observations and Comments

Migration Dates

Except when initials indicate someone else, all dates are from the 48-year data set of William and Joyce (Gunn) Anaka.

ESA: *Earliest Spring Arrival*
LSD: *Latest Spring Departure*
EFA: *Earliest Fall Arrival*
LFD: *Latest Fall Departure*

When there is another record earlier or later than Bill or Joyce Anaka's, it takes priority, and is included along with the Anakas' best date. Only the Anakas' dates have been used to compile averages.

Breeding

This section includes nests, eggs, and other evidence of breeding. When data are available, the first paragraph deals with early and late breeding dates, Mrs Priestly's observations, and the waterfowl nest study by the author in 1946. Sometimes a paragraph deals with one locality, such as Sheho or GSL, and rarely a paragraph deals with results of a special study. Within a given paragraph, nests are listed in chronological order by year. Egg numbers, because they are often numerous and repetitive, are given in Arabic numerals, although numbers of young below 10 are spelled out.

Prairie Nest Records Scheme/*Atlas* records

All records for each species are given in a separate paragraph. Breeding records for Smith's *Atlas* map sheets follow.

Jowsey Oology Collection

The eggs donated by J.R. Jowsey to the Provincial Museum of Alberta are noted as: "(NJ,PMA)."

Breeding Bird Survey

Whenever a species is recorded on BBS surveys other than Theodore (when the Theodore mean often suffices), each BBS paragraph begins with "overall mean," the average number of individuals sighted per count, for all 53 counts. As mentioned, abundance status is based on this mean. For a small number of species wherein the total number (not in this exception, the mean) of individuals seen at Calder, Churchbridge, Ituna, and Willowbrook, might be of interest, these numbers are provided.

Increasing and decreasing trends on the Theodore counts are noted when appropriate, and for less frequently seen species, the number of years seen, the highest number in a given year, the year, and the mean number of individuals of each species counted per year on the Theodore count.

For the 59 species seen on 20 or more Theodore counts, mean numbers are given for Theodore for the 1970s (1971-80), 1980s, and 1990s, followed by "n=" to indicate the number of years (of 33) each species was seen or heard. This provides a numerical index of abundance and consistency of presence. The Theodore mean can be compared to the overall mean, for an indication of similarity or difference between Theodore and the other four counts.

Summer
Any unusual non-breeding records

Christmas Bird Counts
Within the YDM area over 60 years, 1943-2001, 83 species have been reported on count day; 57 species have totals of three individuals or more, and are ranked in descending order. For each species, the summary is as follows:

CBC: species rank; number of counts; total number of individual birds; highest count, giving location, year, and number. For example, Snowy Owl. CBC: Rank #29. 51 counts, 86 birds. Highest: 9, Yorkton 1999. This is shorthand for "This species ranks 29th in frequency, has been found in 51 different counts with an overall total of 86 birds. The count with the highest total was in Yorkton in 1999, with nine birds on that one count."

Winter
Any unusual sightings

Bird Banding Data
Distribution of recoveries of selected species is illustrated by the 12 maps prepared by Kathy Meeres.

Banding (Appendix 3).[3]
Because only game bird banding (waterfowl and Mourning Doves in our case) has been computerized through 1954, pre-1955 banding requires separate totals for non-game species for the Yorkton area.

Banding results from Blakeley of Tadmore, Mrs Wickencamp of Stenen, and for J.A. Briggs of Regina have been summarized and published (Houston 2003c).

The number of individuals of each species banded by each waterfowl bander is given in Appendix 4.

Banding within the Yorkton-Duck Mountain Bird Area
Following each species account the number of individuals of that species banded is given.

[3] The single-spaced printout occupied 107 pages for each batch (of more than three batches) of 8192 lines of data.

Flightless young

For waterfowl, we provide the total of 10-minute blocks of latitude and longitude (there are 108, seven of them shared with Manitoba, within our YDM area) in which flightless young were banded.

Bird Banding Recoveries

Recoveries, banded in YDM area

Number and per cent of individuals banded and recovered in the YDM area (Saskatchewan first, then other Canadian provinces, then American states in alphabetic sequence [see abbrevations]; then Central and South American and Caribbean countries, with names spelled out in full.)

Recoveries, banded outside YDM area

Number and per cent of individuals banded outside but recovered within the YDM area.

Returns

Available for eight species only

Langenburg

Numbers from the Herriot-Luterbach and Hjertaas studies.

Duck Mountain Provincial Park

"Point Counts": In each species account, a rank between 1 and 48 is given. Where only a single individual was seen, no ranking is provided. Total individuals and total stops are given.

"Aspen Counts": In each species account, the number of total individuals seen in OA (old aspen), YA (young aspen), and MA (mature aspen) is given.

ABBREVIATIONS USED IN THIS BOOK

# indicates brief biography in following pages

** indicates deceased

AB - Alberta

AFN - *Audubon Field Notes* (through volume 24)

AK - Alaska

AL - Alabama

AmB - *American Birds* (begins with volume 25)
AMNH - American Museum of Natural History, New York City
AOU - American Ornithologists' Union
AR - Arkansas
ARS - Alan R. Smith#
AS - Adam Schmidt#
AZ - Arizona
BBS - Breeding Bird Survey
BC - British Columbia
BH - Bill Horseman#**
BL - Bob Luterbach#
CA - California
CBC - Christmas Bird Count
CCS - C.C. Shaw#**
CNR - Canadian National Railway
CO - Colorado
COSEWIC - Committee on the Status of Endangered Wildlife in Canada
CP - Curtis Pollock#
CPR - Canadian Pacific Railway
CSH - C. Stuart Houston
CT - Connecticut
CWS - Canadian Wildlife Service
DAW - Donald A. Weidl#
DE - Delaware
DF - Dallas Fairburn#
DH - Dale Hjertaas#
DGJ - Don G. Johnston#
DMPP - Duck Mountain Provincial Park
DV - Dan Vetter (Regina)
EET - Ernest E. Thompson#** [after 1902 known as Ernest Thompson Seton]
EFA - Earliest Fall Arrival
EK - Ernie Kozak#
ESA - Earliest Spring Arrival
ETS - Ernest Thompson Seton#**

FAS - Frank A. Switzer#
FB - Frank Baines#**
FL - Florida
FWL - Fred W. Lahrman#**
GA - Georgia
GAA - Gary A. Anweiler#
GSL - Good Spirit Lake
HWRC - Herbert W.R. Copland#
IA - Iowa
ID - Idaho
IL - Illinois
IMP - Isabel M. Priestly#**
IN - Indiana
IY - Ivan Yaholnitsky#
JA - Joyce Anaka#
JADF - J.A. Douglas Flack#
JAG - John Alexander Gunn#**
JAS - James A. Slimmon#
JDS - J. Dewey Soper#**
JFR - J. Frank Roy#
JG - Joyce Gunn#
JHJ - Jacob H. Jmaeff#
JHW - J.H. Wilson#
JP - John Pollock (Whitewood)
JRJ - James R. Jowsey#
DRMH - David R.M. Hatch#
KS - Kansas
KY - Kentucky
LA - Louisiana
LFD - Latest Fall Departure
LM - Larry Morgotch#**
LP - Leona Pollock
LSD - Latest Spring Departure
LTMcK - L. T. McKim#**
MA - Massachusetts

MB - Margaret Belcher#**
MB - Manitoba
MD - Maryland
MDG - Mary D. Gilliland#
ME - Maine
MI - Michigan
MIC - Muriel I. Carlson#
MIH - Mary I. Houston#
MJB - Marc J. Bechard (Boise, ID)
MK - Mark Kornder#
MM - Maurice Maurer#
MN - Minnesota
MO - Missouri
MS - Mississippi
MT - Montana
NB - New Brunswick
NC - North Carolina
ND - North Dakota
NE - Nebraska
NF - Newfoundland
NH - New Hampshire
NJ, PMA - Norman Jowsey, Provincial Museum of Alberta
NJ - New Jersey
NM - New Mexico
NS - Nova Scotia
NU - Nunavut
NV - Nevada
NWR - National Wildlife Refuge (in US)
NWT - Northwest Territories
NY - New York
OH - Ohio
OK - Oklahoma
ON - Ontario
OR - Oregon
PA - Pennsylvania

PC - Paul Chytyk#
PEI - Prince Edward Island
PFRA - Prairie Farm Rehabilitation Administration
PMA - Provincial Museum of Alberta
PNRS - Prairie Nest Records Scheme
PP - Provincial Park
PPa - Phil Pawluck#
PSa - Pat Sargent (Regina)
QU - Quebec
RB - Robert Barnhart#
RD - Ross Derkatch#
RE - Robert (Bob) Ewart (Regina)
RI - Rhode Island
RJL - Robert J. Long#
RK - Robert Kreba#**
RLS - R. Lorne Scott#
ROMZ - Royal Ontario Museum of Zoology, Toronto
SAJ - Shirley A. Johnston#
SC - South Carolina
SD - South Dakota
SJ - Shirley Jowsey#
SJS - Stanley J. Shadick (Saskatoon)
SK - Saskatchewan
SL - Sheila Lamont#
SMNH 1926 - field camp at Madge Lake, 6 May-19 June
SMNH 1951 - field camp at Madge Lake, 15 May-22 June
SMNH 1956 - field camp at Good Spirit Lake, 9-10 June 1956
SMNH - Saskatchewan Museum of Natural History, Regina [now the Royal Saskatchewan Museum]
SNHS - Saskatchewan Natural History Society
SNHS 1956 - SNHS "summer meet" at DMPP on 9-10 June 1956
SNHS 1977 - SNHS "summer meet" at GSL on 10-12 June 1977
SNHS 1989 - SNHS "summer meet" at DMPP on 9-11 June 1989
SZ - Stanley Zazelenchuk#
TH - Trevor Herriot#

TN - Tennessee
TX - Texas
USNM - United States National Museum
UT - Utah
VA - Virginia
VS - Victor Schmidt (Melville)
VT - Vermont
WA - William Anaka#
WA - Washington
WC - Walter Chudzik#**
WCH - Wayne C. Harris#**
WHC - W.H. Carrick**
WN - William Niven#**
WFVZ - Western Foundation of Vertebrate Zoology
WH - Warren Hjertaas#**
WI - Wisconsin
WRG - Wayne R. Gemmell#
WV - West Virginia
WY - Wyoming
YNHS - Yorkton Natural History Society

SELECTED BIOGRAPHIES

Joyce Gunn Anaka was born at Gunn's Beach, Good Spirit Lake, into a family always interested in the natural world. She attended the Yorkton Collegiate Institute, and then joined the Canadian Women's Army Corps (January 1944 through September 1946). Otherwise, she has lived in the same yard in which she was born. Until 1969, she was a postmistress, storekeeper, and resort operator. The Gunn property was sold to the park in 1970, retaining a 10-acre lease for the summer home. Joyce married Bill Anaka in 1971 and worked at Good Spirit Provincial Park as an entry clerk and later as park naturalist, retiring in 1985.

William Anaka grew up on an isolated farm near the southeast corner of GSL. With no siblings for eight years, he made the outdoors and nature his companions, and birds were the jewels he searched for. The weekly *Yorkton Enterprise* brought news of the very active Yorkton Natural

History Society and whetted his appetite. When aged 12, he decided to write the quintessential field guide--only to learn that Roger Tory Peterson had beaten him to it. In 1953, with testimonials from Stuart Houston and Cliff Shaw, he obtained a CWS bird banding permit. He put a band on "everything that had feathers," but later concentrated his efforts on raptor and corvid nestlings and birds visiting his winter feeding station. He abandoned his banding activities before taking employment with Good Spirit Provincial Park from 1968 through 1987. From 1973 to 1977, he and Joyce were co-editors of *Blue Jay News* for the Saskatchewan Natural History Society (Houston 1989). They retain their summer home in the park, but reside in Yorkton in winter, still interested in birds and nature.

Gary A. Anweiler attended high school in Melville, where he collected birds' eggs and submitted nest record cards. After a career as a wildlife consultant for Blood and Associates, and later as a building manager, he is now a keen member of the lepidopterists' group in Edmonton.

Frank Baines was born in Manchester, England, 1 October 1877. In 1883 his family settled at a cluster of tents known as "Crescent City," Assiniboia, 18 miles south of the later site of Yorkton. From 1887 to 1890, he collected, prepared, and sold skins and birds' eggs to E. Rippon, then until 1894 to E.J. Cousins, and subsequently to Walter Raine, all three commercial oologists from Toronto. Raine stayed at the Baines farm for nearly a week in 1901. Frank was one of the first 75 charter members of the fledgling Yorkton Natural History Society in 1942, participated in CBCs, and was president of the Saltcoats and District Conservation Society in 1954. He died on 19 October 1968 (Houston 1969b).

Robert Barnhart kept bird migration dates on his farm, 5 miles east of Saltcoats, from 1943 to 1954. He recently retired from farming at age 89, after converting his cropland to grassland as a conservation measure in perpetuity.

Margaret Belcher was born at Dilke, SK, on 19 December 1920. She taught French at the Regina College, later transformed into the University of Regina. She was author of two editions of *Birds of Regina*, and *The Isabel Priestly Legacy: Saskatchewan Natural History Society, 1949-1990.* She contributed to sightings at the SNHS meets at Good Spirit Lake in 1977 and at DMPP in 1989. She made a follow-up trip to Madge Lake in 1990, adding another nine species to the list made by the SNHS the previous year.

Robert M. Blakeley banded birds at Tadmore in 1929 (Houston 2003a).

Muriel I. Carlson, of Turtle Lake and Vancouver, helped Stuart and Mary Houston with two days of birding, including a walk into the north end of DMPP, 24-25 June 1991.

Walter Chudzik submitted nest record cards to PNRS. He grew up on a farm 2 miles west and 2 miles north of Canora, and attended high school in Canora. He died of a heart attack; his funeral was held on what should have been his 30th birthday.

Paul Chytyk attended Thom Collegiate in Regina. He worked with Al Smith and CWS on wader surveys at Little Quill Lake and general studies at Preeceville and Sturgis. He is now the manager of YUNI Environmental Consulting in Victoria, BC.

Herbert W.R. Copland worked as a curator at the Manitoba Museum of Man and Nature in Winnipeg. He used his in-laws' cottage at Madge Lake as a base for each BBS in the adjacent Duck Mountains within Manitoba.

Ross Derkatch grew up on a farm 8 miles north of Stornoway. He contributed PNRS nest record cards, 1960-1963. He is now a businessman in Calgary.

Loney Dickson held the CWS banding permit under which the extensive wader banding was done at Little Quill Lake, 1988-1991.

Dallas Fairburn runs an auto body shop at Ketchen, SK, and is one of the founding members of the Kelsey Ecological Society. He has arguably the largest bird feeding structure in Saskatchewan, 4 feet square and 5 feet high, that attracts hundreds of birds each winter.

J.A. Douglas Flack is the author of *Bird Populations of Aspen Forests in Western North America.* Of his 41 study sites across North America, four were in Saskatchewan: near Paynton, in Moose Mountain, and two near Good Spirit Lake. He kindly shared his detailed, unpublished field notes with CSH.

Wayne R. Gemmell undertook the Ituna BBS in 1973, 1974, and 1975.

Mary D. Gilliland, a retired university teacher of languages, has been chair of special publications for the SNHS since 1982.

John Alexander Gunn was born at Fort Qu'Appelle on 21 August 1883. He was the youngest son of Donald Gunn II, who in turn was the son of

Donald Gunn, a member of Manitoba's first legislature, co-author of *The History of Manitoba* (1880), and collector of specimens for the Smithsonian Institution in Washington, DC (Hunt 1880). Donald Gunn II settled at the south end of Good Spirit Lake in June 1887. Donald's second wife and children joined him in September. He operated a trading post for more than 10 years. When the land was first surveyed in 1894, he applied for his homestead under "squatter's rights," listing 80 cattle and 26 horses, a log house, and two stables. To provide extra hay and grazing he had leased all of Section 34, 2 miles to the south (Anaka 1996).

John volunteered for the Boer War but it was over by the time he arrived in South Africa in June 1902. In the winter of 1906-07 he ran a trading post near the Pewei River, 5 miles south of the present location of Weekes. He then homesteaded in the Peace River country of northwestern Alberta, near Grande Prairie, for three years. He joined the army again in 1914, was taken prisoner in 1916, and returned home to ranch on the south shore of Good Spirit Lake only in May 1919. He was also postmaster. Like his grandfather, John was a competent naturalist, the first to reside in the area covered by this book. He died at Fort San on 12 August 1947 (Anaka 1996).

Wayne C. Harris made important bird observations during visits to GSL and Duck Mountain Provincial Park. He participated in more Christmas Bird Counts than anyone else in Canada and contributed about 20% of the total breeding records in Smith's *Atlas of Saskatchewan Birds.* He died at Raymore in 2002 (Houston 2002b).

David R.M. Hatch attended school in Oak Lake, Manitoba. His life-long passion for nature, especially birds, led to formation of his own ecological consulting company, based in Winnipeg.

Trevor Herriot, naturalist, artist, and author of the prize-winning *River In a Dry Land* (2000), did the aspen point counts at DMPP and did the Langenburg survey with Bob Luterbach in 1990.

Dale Hjertaas made a survey of Horseshoe Lake and its environs, the **Horseshoe Lake Survey**, in 1975. He recorded 78 bird species (Hjertaas 1975). He also carried out the **Rokeby Marsh Survey** on 10-11 May 1993 and 17-23 July 1994, preliminary to more detailed studies of Sandhill Cranes there by Wendy Lyon and Ann Burke. He has shared seven sets of computerized field observations from other trips to the Yorkton-Duck Mountain area, 1975-2002.

Warren Hjertaas was born at Wauchope on 31 January 1921, and farmed there until he moved to Yorkton in 1977. He was president of the Yorkton Natural History Society from 1981 until 1997. The Ravine Ecological Reserve project began under his guidance and that of Lorne Lepp in 1982, Yorkton's centennial year, with a 1.5-mile guided nature trail, with 20 interpretive stations, including a boardwalk extending into the marsh. In 1985, Yorkton City council set aside the Ravine and associated uplands as a reserve. Warren died in Yorkton on 12 July 1999.

Bill Horseman was born on 21 January 1942. While in high school in Saltcoats, Bill contributed many PNRS cards in 1958 and 1959. The many Great Horned Owl nests he found while on his trapline opened my eyes to the high density of this species in parkland areas. Bill made his living by trapping muskrats in winter and shingling roofs in summer. He died on 19 June 2003.

Mary I. Houston has banded more Bohemian Waxwings (4,835) than anyone else in North America, as well as nearly 7,500 Mountain Bluebirds and over 18,000 Tree Swallows. She is an author or co-author of 84 ornithology papers and of a book, *Eighteenth-Century Naturalists of Hudson Bay,* due in print about the same time as this book.

Jacob H. Jmaeff grew up on a farm 5 miles south of Kamsack. While in school, he contributed bird observations. He now has a lawn care business in Saskatoon and mows my lawn each week.

Don Johnston and **Shirley A. Johnston** until recently were apiarists near Lintlaw in summer, who spent part of each winter on southern wildlife tours. They are founding members of the Kelsey Ecological Society.

James R. Jowsey, co-author of two editions of *Wild Flowers Across the Prairies,* did the BBS at Churchbridge for seven years, 1970-74, and 1976-77. He did other BBS surveys near Radville.

Norman Jowsey farmed 7 miles north of Saltcoats. His egg collection was donated to the Provincial Museum of Alberta.

Mark Kornder, an employee of DU, with assistance from Ken Belcher, Jim Jowsey, and Warren Hjertaas, made a **Barvas Marsh Survey** in 1996. They listed 72 species of birds. Kornder succeeded Warren Hjertaas as president of the YNHS, but he was later transferred to Saskatoon.

Ernie Kozak submitted PNRS cards from Mikado, where he taught school. A part-time taxidermist, he continues in retirement as a substitute teacher.

Robert H. Kreba was born in Yorkton on 7 February 1954. An employee of the Saskatchewan Museum of Natural History, Bob made important observations in DMPP, especially detailed observations concerning the habitat of each species of warbler, reproduced (through the kindness of Bob Luterbach), in this book. He died in 2001 (Chytyk et al. 2002).

Fred W. Lahrman was born at Mortlach on 4 September 1921, and raised on a farm south of there. He painted the background dioramas for many artistic museum displays, participated in the 1951 SMNH collecting trip at DMPP, and assisted Fred G. Bard in the tremendously successful re-introduction of the Canada Goose into southern Saskatchewan. He developed asbestosis from working with paper-maché mounts for large mammals but carried on courageously in spite of reduced lung function until his death on 29 June 2003.

Sheila Lamont obtained her education at Waseca and the University of Saskatchewan. An expert in botany, she accompanied her husband, Wayne C. Harris, on trips to the YDM area.

Robert J. Long is a naturalist and cinematographer who lives in Qu'Appelle. He studied and filmed wildlife in DMPP in July and August 1970.

Bob Luterbach was a social worker based in Yorkton in 1972 and 1973. He retired to Regina, but still makes occasional visits to DMPP.

Maurice Maurer grew up on a farm near the extreme southwest corner of the YDM area, and sent in PNRS record cards. He is now a taxidermist and mountain guide in Cranbrook, BC.

Louis T. McKim was born in Napanee County, Ontario, on 19 July 1879. When he was four, his father moved to Deloraine, Manitoba, as the owner of a furniture store. Louis obtained his A.B. and LL.B. degrees from the University of Manitoba, the latter in 1908. He practised law in Melville from 1909 to 1945, when he was appointed district judge at Melfort. He contributed migration dates to "Chickadee Notes." When the Yorkton Natural History Society was formed in 1942, he was its first honorary president. He died in May 1955 (Ledingham 1956; Hawkes 1924).

Larry Morgotch was born at Hyas on 19 September 1929 and raised on the family farm there. He worked for various automobile shops in Yorkton from 1955 and operated his own service station from 1967 until his death on 28 January 1980. He was a skilled nature photographer. The annual Larry Morgotch Award for the best slide presentation at a Saskatchewan Natural History Society meet is given in his memory (Switzer 1980).

Farley Mowat visited CSH for three days in August 1946. He had an admiring audience each evening as he prepared study skins for the Royal Ontario Museum in the Houston basement.

William Niven and his brother George were four and two years old when their Scottish parents settled on a farm near the southeast corner of Salt Lake, 5 miles north of Sheho, in 1910. Will became interested in birds while attending Newburn School, to the north of their farm. By 1924 he had recorded 84 species, and by 1973, 181 species. His 1973 article is invaluable in understanding the many changes in birdlife during the better part of his lifetime. In 1975, Will and George bought a house in the town of Foam Lake, and moved back to the farm for the summer only at the end of May, and thus missed most of the spring migrants. Summer residence on the farm ceased in 1989, his final year of migration records. He died on 1 February 2001 (Houston 2001).

Curtis Pollock was a newspaperman for several years at Canora, when he contributed important observations to *Blue Jay*. He is now a tax consultant in Lumsden. His mother, **Leona Pollock**, resides in Preeceville.

J.A.M. Patrick, a Yorkton lawyer, banded birds at Yorkton, Foam Lake, and Big and Little Quill lakes, 1929-1933 (Houston 1967c).

Phil Pawluck worked for Canada Post in Yorkton. He was the fourth president of the Yorkton Natural History Society.

Isabel M. Priestly was the founder of the Yorkton Natural History Society and first editor of *Blue Jay*. Born in Newbury, Berks., England, on 23 July 1893, she married a Canadian soldier, Robert J. Priestly, on New Year's Eve, 1918, and emigrated to Victoria, British Columbia. Bob's newspaper career took them to Winnipeg in 1929 and to Yorkton in July 1935. Mrs Priestly died suddenly in Yorkton on 23 April 1946 (Mayhew 1992).

J. Frank Roy, raised on a farm near Tullis (see Foreword), was one of Saskatchewan's best high school English teachers, and a superb nature photographer and lecturer. His 1996 book, *Birds of the Elbow*, is recognized as one of the best regional bird books on this continent.

Adam Schmidt, now Regional Biologist for the Parkland Ecoregion for

Saskatchewan Environment, based in Melville, carried out a valuable survey of the wetlands around and south of Yorkton, the **Yorkton Wetlands Survey**, from 25 April through 16 August 1973 (Schmidt 1973). He recorded 130 species, 44 of which had definite evidence of breeding. In 1974 he made many visits to the Foam Lake marsh. Adam owns 320 acres of largely aspen forest, with a beaver dam and large beaver pond, 3 miles east of Parkerview. He has banded many Merlin nestlings throughout the entire YDM area.

R. Lorne Scott grew up on a farm south of Indian Head. He built over 2,000 nest boxes for bluebirds and tree swallows, joining his "bluebird house trail" with that of Jack Lane on the east and the Saskatoon Junior Natural History Society's on the west. As an MLA for two terms, he was Minister of the Environment. He banded bluebirds, swallows, Great Horned Owls, and Red-tailed Hawks in the southwest corner of the YDM area.

Ernest Thompson Seton (Ernest Evan Thompson) was born in South Shields, England, on 14 August 1860. His parents immigrated to Canada in 1870 and settled near Lindsay, Ontario. After studying art in England for more than two years, Ernest returned to Canada in the fall of 1881, having assumed the family name of Seton on his 21st birthday. Only in 1902 did his use of the surname Seton become consistent (Houston 1980: xiv).

He arrived with his friend William Brodie at his brother Arthur's farm near present Carberry, Manitoba, on 4 April 1882 and began his serious study of birds and mammals. (Willie visited the upper Assiniboine River and unfortunately drowned when his canoe overturned south of present Kamsack).

Seton's first visit to the Duck Mountain Area, with his friends George Richardson and John Duff, began on 10 October 1883; they stayed until 24 October. He and his two friends each having selected a homestead, they filed on each at the Birtle Land Office on the return trip to Carberry. Seton's land was the south half of section 36, township 28, range 31 west of the first meridian.

Seton returned to his homestead on 8 June 1884 with his brother. He built an 8- by 12-foot log claim shanty, but spent most of his time until 22 June studying the birds. He found the world's first nest of a Philadelphia Vireo and discovered that both species of wood-pewee were summer residents, the Eastern in spruce and the Western in more open aspen groves (Houston 1980). Seton's observations constitute historical

Seton Coulee and Runnymede

Seton Log Claim Shanty

sections for 33 species in this monograph.

Seton performed another valuable service. When he was preparing his 1891 annotated check-list of Manitoba birds, he borrowed John Macoun's field notes from his trip on the upper Assiniboine, from July to mid-September 1881. Seton included Macoun's notes on 39 species, of which only three had been published by Macoun, who strangely omitted them from both editions of his *Catalogue of Canadian Birds*, although the trip itself is mentioned in his autobiography (Macoun 1922). Macoun had entered the Swan River on 1 September 1881, was at Fort Livingstone from 4 through 13 September, at Fort Pelly from 15-19 September, at Cote's Reserve near present Kamsack on 21-23 September, and was below the Shell River in present Manitoba by 24 September. Macoun reached Fort Ellice at the junction of the Qu'Appelle River on 30 September (Houston 1980).

Clifford Carlyle Shaw was born at Elgin, Manitoba, on 13 September 1907 and raised at Bulyea, where his father was the blacksmith. An "all-round naturalist," as well as a newspaperman, Cliff was the second editor of *Blue Jay*, 1946-1948, and was the second president of the Saskatchewan Natural History Society. A Cliff Shaw Award is presented annually for the best article in *Blue Jay*. He died on 13 July 1959 (Belcher 1959).

James A. Slimmon, an executive with Saskatchewan Federated Cooperatives, was a keen birder, photographer, and oologist. He spent several weeks of holidays with his family each summer at Madge Lake, 1974 through 1977.

Alan R. Smith, author of *Atlas of Saskatchewan Birds,* works for the Canadian Wildlife Service in Saskatoon and in spring and fall operates the bird banding station at the upper end of Last Mountain Lake. For demonstration purposes, Al banded a few birds in DMPP.

J. Dewey Soper, then Chief Federal Migratory Bird Officer for the Prairie Provinces, observed birds at GSL on 23 June 1936 (Soper 1936). He also visited Mery Creek and Lac La Course, southwest of Pelly, near the northwest edge of DMPP, on 26-28 May 1942 (Soper 1942). He died on 2 November 1982 (Houston 1997).

Frank A. Switzer grew up on a farm near Rokeby, where he attended Armstrong and Meadowvale rural schools and then the Yorkton Collegiate Institute. He worked for SaskPower, latterly as environmental consultant, until his retirement in September 1987. He participated in the Theodore BBS with Bill Anaka, first as the recorder and for three years as the observer.

Robert Wapple, schooled in Biggar, and a co-author of the forthcoming revised *Birds of Rosetown-Biggar*, prepared a study of the Assiniboine River drainage area in 1998 and 1999 (Wapple 1999).

Donald A. Weidl was born and schooled in Broadview, where he contributed 13 years of bird observations to Callin's *Birds of the Qu'Appelle.* He played a key role in the lengthy study of DMPP (Hatch and Weidl 1981). Don is an ecological consultant.

Emma Wickencamp banded birds near Stenen from 1933 to 1939 (Houston 2003a).

J.H. Wilson, from Indian Head, banded birds for Ducks Unlimited at Leech Lake in 1944 and 1945 (Houston 1946a).

Ivan Yaholnitsky, a student of Kozak's in the Mikado school, contributed catbird and robin nest records in 1969.

Stanley Zazelenchuk grew up on a farm 8 miles north of Stornoway. He was a keen nest-finder while in high school; he sent many nest record cards to PNRS, 1959-1963. Most of his career was spent as a teacher in Arctic schools. After retirement he set up a business selling Inuit art, in St Andrew's, NB.

ACKNOWLEDGMENTS

I owe a personal debt to the late Isabel M. Priestly. It was she who initiated me into the fellowship of the Great Outdoors. She recommended me to Bertram W. Cartwright, chief naturalist of DU, to gain me special dispensation for an underage bird banding permit. This provided gainful and interesting summer employment, 1943 through 1948, banding ducks for DU (Canada). These summers changed my life (Houston 1992).

From weekly hikes around the "muskeg" to later collaboration on *Blue Jay*, it was Mrs Priestly who guided my way, directed my energies, corrected my mistakes, and encouraged any small successes. During my high school years, I was secretary-treasurer of the Yorkton Natural History Society, typed the stencils, and mimeographed *Blue Jay*.

Carol Beaulieu of the Saskatchewan Research Council prepared the magnificent overall map of the area. Kathy Meeres prepared 12 exquisite maps showing the migration pathways of waterfowl banded in the YDM area. Kelwin A. Wylie of Ottawa, who in years past helped me band owls in the YDM area, wrote a computer program to produce maps of the flightless young banded in each 10-minute block of latitude and longitude. Robert A. Wright generously allowed us to include large sections of text from his 1995 DMPP Vegetation Management Plan. Rick Shussel, DMPP superintendent, allowed use of the park map and a photograph (p. 12). Glen Sutter and the late Fred W. Lahrman checked data on museum labels in the Royal Saskatchewan Museum, Regina.

Alan R. Smith kindly loaned me the Canadian Wildlife Service four-volume copy of printouts of computerized nest records available on cards in the Prairie Nest Records Scheme through 1976. From his files, he also provided additional details about rare bird observations.

Many museum curators provided information about egg sets, especially Lloyd F. Kiff, who hosted CSH and MIH at the Western Foundation of Vertebrate Zoology (WFVZ) in Los Angeles for a full week of inventory, 12-19 November 1984; Paul James of the Saskatchewan Museum of Natural History in Regina, allowed me access to its egg collection, catalogue, and correspondence; more recently, Keith Roney and Glenn Sutter have allowed Danielle Pearson to search the Nest Record cards for specific species since 1976. Others who supplied egg data were Thomas Erdman at University of Wisconsin-Green Bay; Wayne Campbell at the British Columbia Provincial Museum, Victoria; Alison Victor Andors of the American Museum of Natural History in New York; James Dean of the United States National Museum, Washington; Ross James of the Royal Ontario Museum of Zoology. Chuck Deschamps of Wadena and Lois Wilson of Langenburg added local information.

Brian W. Johns provided recent confirmed sightings of the Whooping Crane in the YDM area.

Trevor Herriot, Dale Hjertaas, James R. Jowsey, Bob Luterbach, and Alan R. Smith criticized an earlier version of the manuscript, offering helpful suggestions, additions, and corrections. Mary Gilliland has been a constant source of advice and encouragement, and has checked four versions of this manuscript with her legendary diligence, detecting discrepancies that are somehow invisible to the author.

Anyone with additional information is encouraged to contact: C. Stuart Houston, 863 University Drive, Saskatoon, SK, S7N OJ8.

FOR THE READER: A few items may cause confusion and require explanation.

Spirit Lake: The post office at GSL, in the J.A. Gunn store for some years, was named simply Spirit Lake, hence many observations gave that short form as the locality.

Foam Lake: is both a town and a "lake," the latter of which we have more accurately designated as Foam Lake marsh.

24 vs 108 place names: In the *Atlas of Saskatchewan Birds*, Alan Smith used 24 rectangles of 15 minutes latitude north-south and 30 minutes longitude, east-west, each corresponding to a 1:50,000 map,[4] covering roughly 21 by 17 miles. The banding office, for all bandings and all recoveries, uses 10-minute blocks, of which there are 108 (Appendix 5); 511-1022 for Yorkton means 51° 10' north latitude and 102° 20' west longitude). Each 10-minute block at this latitude is roughly 11 miles north-south and 7 miles east-west.

[4] Map sheets 1:50,000: 62K/13 Langenburg; 62L/16 Bredenbury; 62L/15 Melville; 62M/3 Goodeve; 62M/4 Jasmin; 62M/2 Willowbrook; 62M/1 Yorkton; 62N/4 Calder; 62N/5 Togo; 62M/8 Rhein; 62M/7 Springside; 62M/6 Parkerview; 62M/5 Kelliher; 62M/12 Foam Lake; 62M/11 Sheho; 62M/10 Buchanan; 62M/9 Canora; 62N/12 Kamsack.

SPECIES ACCOUNTS

As we go to press, we list 305 species, plus three species still on the Saskatchewan hypothetical list. For 161 species we have evidence of breeding on at least one occasion.

This represents a great deal of additional study by a great many people since the initial mimeographed list in 1942 of 198 species (Priestly 1942a) and since "Birds of the Yorkton District, Saskatchewan" (Houston 1949) listed 231 species plus eight hypotheticals.

SEQUENCE WITHIN SPECIES ACCOUNTS:

Status
History
Observations and comments
Migration dates, **ESA, LSD, EFA, LFD**
Breeding
  Nests, eggs, or flightless young
  PNRS
  "Also breeds at ..." (ARS, *Atlas*)
  Oology: Jowsey and Cymbalisty Oology Collections
  BBS
Summer (where relevant)
CBC [Note: 1966 was the first year for counting Rock Doves in CBCs.]
Winter (where relevant)
Banding
Recoveries, banded inside YDM area
Recoveries, banded outside YDM area
Returns
Langenburg
Duck Mountain

PACIFIC LOON *Gavia pacifica*

Vagrant.

An adult in basic plumage was sighted at the south end of GSL on 28 Sep 1997 (CP, RK and S. Benoit; Bain and Holder 1997d).

COMMON LOON *Gavia immer*

Uncommon summer resident. Breeds only on large water bodies.

Was more common at Crescent Lake when fish were plentiful there during the 1920s. Originally present on Salt Lake near Sheho, but since 1930s seen only as a transient. Fairly common on the larger lakes, with summer sightings at GSL during 16 of 48 years. Rarely resides on a large slough, such as one near Beaver Dale in the 1940s. In 1973, resident pairs on Upper and Lower Rousay lakes, York Lake, and Crescent Lake (AS). One pair calling on Lake of the Prairies, 21 Jun 1989 (CSH,MIH). On 15 Sep 1998, Geoff Smith saw 20 loons eating Tiger Salamanders at York Lake (Bain and Shanahan 1998c).

ESA(25): 12 Apr 2000, Av 2 May. **LFD**(16): 16 Nov 1983, Av 31 Oct.

Breeding: At Ravine Reserve, two newly hatched young in nest 10 Jun 1947. The frantic mother put on an interesting aquabatic display. Mother and two week-old young at Ravine Reserve on 19 Jun 1948 (CSH). A large beaver pond 3 miles east of Parkerview has been occupied fairly regularly since at least 1993, when a pair had two large downy young on 26 Jun 1993, two young on 4 Jun 1994, and again on 23 Jun and 4 Aug 2002, with sightings of adults in the intervening years (AS).

Breeds at Pelly (ARS, *Atlas*).

BBS: overall mean, 0.1. Theodore: 4 years; highest, 3, 1979.

Banded 1 (CSH).

Langenburg: One in Cutarm valley, 11 Jun 1989 (CSH,MIH). One on lake 10 miles northwest of Churchbridge, 1990 (TH,BL).

DUCK MOUNTAIN: SMNH 1926: "Very numerous, unusually tame, always interesting, with their evident sense of curiosity and weird notes and calls which they kept up day and night." Several nests; three sets of 2 eggs each collected on 25 May, 1 and 3 Jun, #1951, #1952, #1953, were "replaced within a few days" by the adults. SMNH 1951: 2 eggs on 30 Jun, #2728, and 2 on 1 Jul, #5507. SNHS 1956. Nest, 2 eggs, in the west bay of Madge Lake, 1973 (BL). In 1974, five broods with one young, two with two; also broods in 1975, 1976, 1977 (JAS). Breeding on Madge, Batka, Ketchamonia and Jackfish lakes; three pairs with young, 18 Jun 1981 (Hatch and Weidl). SNHS 1989.

PNRS: Nest, 2 eggs, 22 May 1982, hatched successfully (B. De Vries).

Point counts: Rank #20, 11 individuals at 10 stops (ARS).

PIED-BILLED GREBE *Podilymbus podiceps*

Common summer resident on marshy ponds and larger sloughs. Breeds.

Recorded only in wet years at Sheho. Resident at Anderson Lake, Adams Lake and Barvas marsh (JRJ).

ESA(40): 15 Apr 1958, Av 1 May. **LFD**(21): 25 Oct 1995, Av 7 Oct.

Breeding: GSL: Earliest, 20 June 1987, adult with young. Adult with two young, 18 July 1955. Latest, 20+ adults with young, 28 Aug (WA).

Sheho: Only brood at Sheho was on a large slough which bordered the Niven farm, 27 Jul 1947 (Niven 1973). Yorkton: five young, 3 miles east of Rokeby, 25 Jun 1952 (CSH). In 1973: nest, Upper Rousay Lake, 3 eggs, 10 Jun; nest with only 1 egg, hatching, at Lower Rousay Lake on 22 Jun; at least six broods at Upper Rousay Lake, five at Lower Rousay, two at Crescent marsh, two at Little Leech, two at York, one at Crescent, three at Little Bone Lake (AS). In 1975: on nest at Foam Lake marsh on 30 May (AS); at Horseshoe Lake, 20 adults and 21 chicks, with broods of one to eight young, 9-12 Jul (DH). Adult with young at Ketchen on 17 Jun 2001 (DF).

PNRS: Nest, Saltcoats, 2 eggs, 3 Jun 1958 (BH).

Also breeds west of Canora (ARS, *Atlas*).

BBS: overall mean, 2.6; Theodore, mean 3.2 (1970s, 5.1; 1980s, 2.1; 1990s, 3.0; n=27).

Banded 1 (CSH) + pre-1955, 3 (CSH 2, Wotherspoon 1).

Langenburg: Fairly common on medium to large sloughs; one on nest, 1990 (TH,BL).

DUCK MOUNTAIN: SMNH 1926. SMNH 1951, nest, eggs. Seen in 1977 (JAS). Fairly common at Batka, Green, and Kipp lakes in June 1981 (Hatch and Weidl). SNHS 1989.

HORNED GREBE *Podiceps auritus*

Common summer resident. Breeds.

Common northward to Duck Mountain, 1884 (Seton 1891). Full-page photo of Crescent Lake nest, 8 eggs, on floating platform of bulrushes, 6 June 1901 (Reed 1904:6).

A pair on most potholes, two or more pairs on larger sloughs; 16 at Foam Lake marsh on 28-31 May 1974 (AS). Resident at Anderson Lake, Adams Lake, and Barvas marsh (JRJ).

ESA(47): 21 Apr 1958, 1981, and 1991, Av 30 Apr. **LFD**(30): 19 Nov 1983, Av 13 Oct.

Breeding: In 1973, early nest, 1 egg, 23 May; broods observed on both Rousay, York and Crescent lakes and Crescent marsh (AS).

Nests on small sloughs near Sheho most years (WN). New nest at Hilltop School on 8 Jun 1943 (IMP). Nest, 2 eggs, 1 Jun 1946; early hatching, 20 Jun 1946; at Saltcoats, 2 eggs, 2 and 6 Jul 1961 (CSH). Adult with two downy young near GSL on 15 July 1956 (WA). Nests with 2 and 4 eggs, Melville, 25 May 1957 (GAA). Two pairs with nest at Ravine Reserve, 30 May 1987, and adult with three young on 23 Jul 1987 (DH).

PNRS: at Melville, 4, 4, and 6 eggs, 10 Jun to 4 Jul 1960 (VS).

BBS: overall mean, 1.9. Theodore, mean 1.8 (1970s, 2.7; 1980s, 1.7; 1990s, 1.9; n=20).

Banded 1 (WA) + pre-1955, 2 (CSH).

Langenburg: Common, but less numerous than Eared Grebe, 1990 (TH,BL).

DUCK MOUNTAIN: SMNH 1926. SNHS 1956. Uncommon, in June seen only at Green Lake and a small slough along west park boundary (Hatch and Weidl 1981).

RED-NECKED GREBE *Podiceps grisegena*

Fairly common summer resident on larger sloughs and lakes. Breeds.

Nest, 3 eggs, "floating mass of wet rushes, moored by a few growing rushes," in 2 feet of water, 18 June 1884, Duck Mountain above Boggy Creek (Seton 1891).

Has decreased since the 1920s, when fish were plentiful at Crescent Lake, and before some lakes and large marshes were drained. Was once a common resident of the larger sloughs and lakes near Sheho (Niven 1973). Increasing since 1987 (WA). Became a regular summer resident at GSL in 1989; 15 on Whitesand Lake, 17 Jun 1989 (WA,FAS). Resident at Anderson Lake, Adams Lake, and Barvas marsh (JRJ,SJ).

ESA(29): 21 Apr 1998, Av 4 May. **LFD**(8): 8 Oct 1996, Av 13 Oct.

Breeding: Early nest with 1 egg, east of Parkerview on 26 May 2001, later destroyed by strong wind; nest in same area 23 Jun 2002, with young on adult's back on 21 July (AS). In 1973, six nests, the earliest on Lower

Rousay Lake, 4 eggs, 27 May; later six broods on this lake, plus other broods on Upper Rousay and York lakes (AS).

Nest with 2 eggs near Saltcoats on 19 Jun 1938 and another at Hilltop School on 8 Jun 1943 (IMP). Nest near Crescent Lake, 8 Jun 1943 (IMP). Nest, 4 eggs, in 17 inches of water on 29 Jun 1946; 1 egg rolled into the water but 3 hatched (CSH). Two adults with three downy young, 3 Aug 1956 (WA). Nest, 2 eggs, 29 May 1958, CNR Dam, Melville (GAA). Two adults and one young at Horseshoe Lake, 12 Jul 1975 (DH). Nest in large slough near Sheho in 1979 (WN). Adult on nests at Sturgis and Preeceville, 7 Jul 1989 (ARS,PC), and near Crescent Lake, 20 Jun 1989 (CSH,MIH). Two pairs on slough near GSL on 29 July 1990; one young was downy and one was almost full-sized. One pair with one downy young riding on adult's back, 7 Jul 1991 (WA).

PNRS: Saltcoats, 4 eggs, 13 Jun 1958 (BH); Arran, 2 and 3 eggs, 14 and 28 Jun 1958 (LM).

BBS: overall mean, 0.8. Theodore, 11 years; highest, 8, 1998.

Banded pre-1955, 2 (CSH).

Langenburg: Pair on large slough, one on nest, 1990 (TH,BL).

DUCK MOUNTAIN: SMNH 1926: #1872; young, #1886; 4 eggs on 14 Jun, #1950. SMNH 1951: nest, 5 eggs on 27 Jun, #5508. Two broods with one, two with three young, 1975, 1976, 1977 (JAS). Common on Madge Lake, two nests on Ketchamonia Lake, 11 nests on Bear Head Lake, 17 Jun; also observed on Batka Lake and a small slough along west park boundary (Hatch and Weidl 1981). Nest, SNHS 1989. Two nests found June 1989 and adults swimming with young, 3 Jul 1989 (TH). Adult on early nest, 25 May 1993 (DH).

Point counts: Rank #20, 11 individuals at 6 stops (ARS).

EARED GREBE *Podiceps nigricollis*

Common summer resident on most larger bodies of water. Breeds.

At Sheho, recorded in only 5 of 54 years, all between 1947 and 1954 (Niven 1973); over 50 seen on Salt Lake on 17 Jun 1989 (WA,FAS). In 1973, present on Upper Rousay and Leech lakes; maximum count of 182 adults on Crescent marsh on 26 May (AS). Regular on Horseshoe Lake (WA). Resident at Anderson Lake, Adams Lake, and Barvas marsh (JRJ). Colony in large slough southwest of Preeceville in wet years (DF).

ESA(16): 28 Apr 1987, Av 9 May. **LFD**(2): 17 Sep 1954 and 26 Sep 1987 (WA).

Breeding: Early nesting, 7 Jun 1946, when a colony of eight nests in 15 inches of water in a bay on Upper Rousay Lake had 4 to 9 eggs, averaging 5.9 eggs. Each nest was built of wet and decayed vegetation, with an outside diameter averaging 11.7 inches, inside diameter 4.5 inches, height 3 inches and depth of cup 2 inches (Houston 1946b).

On 17 Jun 1948 at Leech Lake there were 32 nests with 1 to 6 eggs, and other nests under construction (CSH). In 1973, 55 broods on Crescent marsh, 35 on Little Leech, 6 on Crescent Lake, 18 Jul (AS). In 1974, counted 277 adults at Foam Lake marsh, 23-25 May (AS). At Horseshoe Lake in 1975, a colony of 70 nests, some still empty, had been built by 3 Jun; 139 adults and 119 chicks on 9 Jul; 148 adults and 48 chicks on 11 Jul (DH). Adults with young in July and August (WA).

PNRS: Nests, each with 3 eggs, Gorlitz, 19 Jun 1960 (SZ) and Kamsack, 10 Jun 1961 (SZ,RD); Mikado, four newly-hatched young on 21 Jun 1966 (EK).

BBS: overall mean, 1.7. Theodore, 7 years; highest, 15, 1991, mean 1.7.

Langenburg: Common on larger sloughs, largest colony 22 birds, 1990 (TH,BL).

DUCK MOUNTAIN: Uncommon, three on Green Lake on 3 Jun 1981 (Hatch and Weidl).

WESTERN GREBE *Aechmophorus occidentalis*

Uncommon summer resident, noted chiefly during migration. Breeds.

A single bird at Upper Rousay Lake, 5 Jul 1943 (IMP,JRF), and a pair at Ravine Reserve, 10 Jun 1947 (CSH). Individuals were forced down in Yorkton after heavy snowstorms, on 15 Sep 1944, and 13 and 26 Oct 1959. Because their feet are placed so far back on their bodies, they are unable to take off from land, so these birds were taken to water and released. At Sheho, only two observations in 54 years, including 12 May 1950 (Niven 1973). There were 20 on GSL on 20 Jun 1962 (Dorothy E. Wade, Nero 1962c). In 1973, seen all summer on the Rousay and Leech lakes, with a flock of 77 on York Lake on 13 May, and 62 and 67 counted on Leech Lake on 16 Jul and 13 Aug, respectively; no nests were found (AS). Pair in courtship display on Foam Lake marsh, 25 May 1974 (AS); two adults present at Horseshoe Lake, 2 July 1975 (DH). Present on Lake of the Prairies, 21 Jun 1989 (CSH,MIH).

ESA(27): 26 Apr 1977 and 1981, Av 8 May. **LFD**(27): 17 Nov 1983, Av 18 Oct. Migration peak of over 1,000 scattered all over GSL, 28 Sep

1988 (WA).

Breeding: Two adults with two young on Horseshoe Lake, 10 July 1976 (DH). In the 1980s, high water levels at GSL created good breeding habitat in a large marsh at the north end of the lake, where there were five adults with three young, 2 Jul 1987, and two pairs, each with one begging young, 31 Aug 1990 (WA).

BBS: Theodore, two years; 1 in 2000 and 4 in 2001, mean 0.1.

Banded pre-1955 1, forced down in storm unharmed, 15 Sep 1944 (CSH).

DUCK MOUNTAIN: SMNH 1951. Seen 1977 (JAS). Rare in summer, one on 4 Jun, three on 15 Jul 1981 (Hatch and Weidl).

AMERICAN WHITE PELICAN *Pelecanus erythrorhynchos*

Fairly common transient, uncommon in summer. Breeds at Little Quill Lake some years.

A large spring flock of 100 flew over Sheho on 16 Apr 1972, and again on 26 Apr 1989 (WN), but even larger flocks of up to 300 birds passed overhead prior to 1950 (Niven 1973). Flocks of non-breeding birds make sporadic visits, often of one or two days duration, in any week during the summer. There were 100 at Leech Lake, 16 Aug 1971 (FAS). In 1973, pelicans were present all summer, with the largest flocks of 120 on 9 Jul, 220 on 20 Jul and 310 on 27 Jul at Big Leech Lake; many were grey-headed immature birds raised on another lake that year (AS). Three on Fishing Lake on 6 Jun 1974 (AS). A flock of 12 lit on Salt Lake on 25 Jul 1986 (WN). One hundred at York Lake on 11 Aug 1991 (DH). Nine on beaver pond 3 miles east of Parkerview on 26 May 2002 (AS).

ESA(33): a flock of 25, 6 Apr 1991, Av 30 Apr. **LFD**(26): 26 Oct 1942 (CSH) and 18 Oct 1983 (WA), Av 21 Sep.

Breeding: A colony on the south shore of Little Quill Lake, barely within 104° west longitude, the western boundary of the Yorkton area, raised only 19 young in 1956, but over 300 (of which 219 were banded) in 1957 (Houston 1957e). [The colony subsequently moved less than 20 miles northwest, to Middle Quill or "Muddy Quill" Lake.]

BBS: Theodore, 1 year; 23, 1977, mean 0.1.

Banded 157 (CSH) + pre-1955, 38 (Patrick), at Little Quill Lake.

Recoveries, banded in area, 14 [7.2%] (SK 2, AR 1, MN 1, MO 1, ND 1, OK 1, SD 4, Mexico 3).

Recoveries, banded outside area, 6 (SK 3, MB 1, ND 1 [Chase L,

Stutsman Co.], WY 1 [Yellowstone National Park].

DUCK MOUNTAIN: SMNH 1926: pelicans inexplicably absent. Present but not breeding, SMNH 1951. Fairly common, groups of 50, 10, and 30 were seen on Ketchamonia Lake, 17 Jun, and 20 on Batka Lake on 19 Jun 1981 (Hatch and Weidl, Weidl 1983). SNHS 1989. Large flock of 90 on upper Assiniboine River, 7 Sep 2002 (TH).

Point counts: 1 individual at 1 stop (ARS).

DOUBLE-CRESTED CORMORANT *Phalacrocorax auritus*

Fairly common in summer. Has bred at Little Quill Lake.

One or two individuals summered at GSL until about 1922 (John A. Gunn). Uncommon during spring migration in the 1950s and 1960s. Non-breeding flocks present on GSL and McOuat Lake [spelled McQuat on some maps] every summer since 1973. Three on Fishing Lake on 6 Jun 1974 (AS). Stops in migration at Anderson, Adams, and Soda lakes near Saltcoats (JRJ).

ESA(32): 15 Apr 1978, Av 1 May. **LFD**(18): 30 Oct 1998, Av 7 Oct.

Breeding: Two deserted nests with eggs on island in Crescent Lake, July 1953 (CSH).

BBS: Theodore, 6 years; highest, 8, 1981, mean 0.5.

Banded pre-1955, 60 (Patrick, Little Quill Lake).

Recoveries, banded in area, 4 [6.7%] (SK 1, AR 1, MS 1, TN 1).

Recoveries, banded outside area, 11 (SK 11 [8 at Big or Middle Quill Lake, 2 at Last Mountain and 1 at Redberry Lake]; one banded at Big Quill Lake on 16 Jul 1935 by F.G. Bard, picked up dead at GSL by John A. Gunn, 15 Aug 1938).

DUCK MOUNTAIN: SNHS 1956. Uncommon, 40 resting on dead trees on an island in Ketchamonia Lake, but no nests, 17 Jun 1981 (Weidl 1983). One on 13-15 Jun 1990 (MB).

AMERICAN BITTERN *Botaurus lentiginosus*

Common summer resident, now declining. Breeds.

After a noticeable decrease during the dry years of the 1930s, numbers improved during the wetter years of the late 1940s and 1950s, Another downturn was evident in 1978 when Niven, for the first time in his life, did not see or hear a bittern; this decline persisted at least into the late 1980s. Two at Foam Lake marsh on 23 and 30 May and 24 Jun 1974

(AS). Five sightings at Horseshoe Lake from 9 to 12 Jul 1975 (DH). Resident at Anderson Lake, Adams Lake, and Barvas marsh (JRJ).

ESA(30): 25 Apr 1955 and 2000, Av 9 May. **LFD**(12): 21 Oct 1956 and 1979 (except for CBC below), Av 22 Sep.

Breeding: At GSL: Early nest, 6 eggs, 6 Jun 1955, young hatching 12 Jun 1955; early hatch, four young, 26 Jun 1955; nest with 2 eggs on 2 Jun and 4 eggs on 21 Jun 1958 was predated; latest and most unusual nest, 4 eggs, in sweet clover field over 150 yards from water, 10 Jul 1956; nest with 4 eggs, 8 Jun 1960; nest with 1 egg on 12 Jun 1960 had 2 eggs on Jun 19, but only one hatched by 17 Jun; nest with 3 eggs on 3 and 10 Jul 1960, was the last nest found at GSL (WA).

Nest at Upper Rousay Lake, 5 eggs, 24 Jun 1946, hatched 1 Jul (CSH). In 1973, an adult flushed from a completed but empty nest, Leech Lake, 28 Jun (AS).

PNRS: at Stornoway, nest, 5 eggs, 8 Jun 1960 (SZ).

BBS: overall mean, 3.4. Theodore, mean 2.3 (1970s, 3.1; 1980s, 3.4; 1990s, 2.2; n=20).

Banded 24 (CSH 1, WA 23) + pre-1955, 9 (Briggs 1, Houston 8, including a flightless immature caught at Upper Rousay Lake, 12 Sep 1943).

Recoveries, banded in area, 2 [6.3%] (SK 1, AL 1). One Anaka bird was found dead at Goodwater, Alabama, the same fall, while another flew into a power line at Canwood, Saskatchewan, four years after banding.

CBC: 1 count, 1 bird, Yorkton 1953, but this individual died a few days later.

Langenburg: Eight seen or heard, 31 May 1976, and one on 12 May 1980 near Churchbridge (DH). Only two seen and none heard, 1990 (TH,BL).

DUCK MOUNTAIN: SMNH 1926. SMNH 1951: 5 eggs on 31 May, #5506. Seen in 1977 (JAS). Fairly common, calling from marsh areas in late evening and early morning through June and July (Hatch and Weidl 1981). SNHS 1989. Resident in marsh north of Pickerel Point, 1987, 1989, 1990 (TH).

GREAT BLUE HERON *Ardea herodias*

Common summer resident. Breeds.

Rarely seen at Sheho, with early sightings only on 28 May 1943 and 20 Jun 1952. For the next 20 years, seen nearly every spring, with an occasional solitary bird during summer. Fall sightings in 1947, 1949, and almost yearly from 1952 to 1972 (WN). In 1973, one or two seen almost daily south of Yorkton throughout the summer (AS). In 1974, four on Fishing Lake on 6 Jun and three on Foam Lake on 20 Jul (AS).

ESA(45): 4 Apr 1991 and 1992, Av 15 Apr. **LFD**(34): 7 Nov 1977, Av 12 Oct.

Breeding: Breeds sparingly in any given year. 1. Two nests were located in a dead Black Poplar at Leech Lake on 25 Jul 1944. 2. At Horseshoe Lake in 1961, there were 15 nests; one accessible nest had parents present on 23 Apr, 3 eggs on 13 May, three young and an unhatched egg on 9 Jun, only two young in the nest on 2 Jul, a fully-grown young perched nearby on 23 Jul, and all young able to fly on 19 Aug (Morgotch 1961b). The colony had 24 active nests in 1965 (Hatch 1965; Houston 1989), peaked at 50 nests in 1970 (Vermeer and Anweiler 1970); had 20 nests in 1975; 25 nests were counted, unimpeded by trees leafing, on 20 Apr 1976 (DH); 35 nests in 1979; 10 to 15 nests in 1984 (Smith in Wapple 1999); and 6 nests in 1989, all in spindly aspen that were unsafe to climb (WA,CSH). Bill Anaka took over the banding of this colony, with 16, 16, 22, and five young banded yearly in 1962 through 1965, consecutively, and 24 in 1967. Not all nests could be reached safely. After all the trees died, the colony moved to the northeast side of Horseshoe Lake, where suitable trees were limited, so that there were only five to eight nests each year for about six years until the site was abandoned in 1992 (WA). 3. A colony of 50 to 60 nests was present at Fishing Lake in 1979 and a colony of 8 to 10 nests at Ebel Lake, southeast of Fishing Lake, in 1980 (Smith in Wapple 1999). 4. A colony between the two Rousay lakes had five nests in 1985 (DH, Smith in Wapple 1999). 5. A colony at “Little Bone Lake,” the southwest extension of Leech Lake, had 31 nests in 1994 (DH). 6. A colony of 25 nests, not all occupied, was present in the Cutarm Creek valley, 2 miles west of Gerald, in 1996 (Hruska 1997). 7. A colony with 12 nests was present along “Blackbird Creek,” 3 miles south and 2 miles east of MacNutt in 1996, 9 nests in 1998 and 12 nests in 1999 (Hanson and Smith 2000). 8. A colony of 29 nests in aspen, 10 miles south and 2 miles west of Rama, was active in 2001 (Thom Carnahan, via ARS).

BBS: overall mean, 2.3. Theodore, mean 3.3; decreasing (1970s, 7.9; 1980s, 1.1; 1990s, 1.2; n=26).

Banded 95 (CSH 12, WA 83) + pre-1955, 19 (Patrick 11 at Little Quill Lake, CSH 8 [2 at Leech and 6 at Horseshoe Lake]).

Recoveries, banded in area, 7 [6.1%] (SK 4, LA 1 [Shreveport], TX

1 [Dallas], Mexico 1 [Fresnillo, Zacatecas]). One banded 10 Jul 1954 at Horseshoe Lake was shot 4 miles south of Melville in late October 1954 (CSH).

Recoveries, banded outside area, 1 (MT, near Malta).

Langenburg: One in Cutarm valley, 11 Jun 1989 (CSH,MIH). Two singles seen, in coulees, 1990 (TH,BL).

DUCK MOUNTAIN: SMNH 1926. SMNH 1951. SNHS 1956. Seen in 1974, 1975, 1976, 1977 (JAS). Thirty nests in park in 1970 (Vermeer and Anweiler 1970), and 22 nests with young on Watt Island, 1981 (HWRC, Weidl 1983). Common, about 20 young in 12 nests on Watt Island, 21 Jun 1988 (TH). SNHS 1989. Large group, 16, in migration, 17 Sep 2002, upper Assiniboine River (TH).

GREAT EGRET *Ardea alba*

Occasional visitor.

An emaciated bird appeared in the yard of Gilbert Hulston near Crescent Lake on 27 Oct 1941 and was collected by Frank Baines (Houston 1949). One sighting at Stenen on 16 Aug 1972 (DRMH, Houston 1972a). One on island between Upper and Lower Rousay lakes on 11 Jun 1973 (AS). Two were seen from 24 Sep to 15 Oct 1992 beside Highway 9, 1 mile north of the Yorkton airport (photographed by WH; Hombeck 1993; Koes and Taylor 1993a). Two along southwest side of GSL, 9 and 10 Sep 1994. One at north side of GSL, 23 Apr 1995 (WA).

DUCK MOUNTAIN: Very rare, a single bird at north end of Ketchamonia Lake, 17 Jun 1981 (Weidl 1983).

SNOWY EGRET *Egretta thula*

Occasional visitor.

One at GSL, 23-29 May 1956 (Anaka 1956a; Gammell 1956a), and one at Little Quill Lake, 30 May 1985 (Michael Gollop, Gollop 1985).

LITTLE BLUE HERON *Egretta caerulea*

Vagrant.

Four seen at GSL on 24 May 1956 by Joyce Gunn (Anaka 1956b). One immature sighted at Crescent Lake by Betty McDonald and Mr and Mrs W.J. McDonald on 9, 14, and 15 Sep 1956, verified by Cliff Shaw and B.W. Cartwright on 14 Sep and by CSH and MIH on 15 and 16 Sep. Smith says it was present until 21 Sep (Houston 1957c; Gammell 1957a).

CATTLE EGRET *Bubulcus ibis*

Vagrant.

One seen by John Tautin [misspelled "Totten" in Gollop 1978] at southeast corner of Little Quill Lake: five on 2 Aug, one on 11 Aug and seven on 31 Aug 1977 (Gollop 1978; ARS, *Atlas*).

GREEN HERON *Butorides striatus*

Vagrant.

One sighted by CSH and MIH, Lower Rousay Lake, from a canoe, at 10 feet distance, on 10 Aug 1977.

BLACK-CROWNED NIGHT-HERON *Nycticorax nycticorax*

Rare summer resident, declining in numbers. Breeds.

Twelve seen on Assiniboine River near mouth of Shell River, 25 Sep 1881 (Macoun in Seton 1891). Uncommon summer resident of marshes. Several hundred rested at GSL in the fall of 1922, the only time that John A. Gunn saw this species.

In 1929, J.A.M. Patrick banded 68 nestlings at Foam Lake marsh, 4.5 miles north of the town of the same name (Houston 1967c).

Regular until about 1960, now rarely seen. Thirty at Little Quill Lake on 10 Jul 1976 and six at York Lake, 31 May 1986 (DH). One standing in marsh beside Highway 10 at Sturdee, 13 Jun 1989 (CSH,MIH).

ESA(12): 15 Apr 1942 (WN) and 30 Apr 1939 (WN) and 1987 (WA), Av 15 May. **LFD**(7): 6 Oct 1957, Av 6 Sep.

Breeding: A large colony of "thousands" nested in trees between the Upper and Lower Rousay lakes in 1934 (E.S. Wood); every tree was occupied and many contained three, four, and five nests. A colony of 26 nests in willows, but chiefly Black Poplar, 5 to 20 feet above ground, was present on the island in Upper Rousay Lake in 1944 (31 banded). The mystery as to where they moved in 1945 became understandable in 1946, when they were found nesting in three different sites at Lower Rousay Lake, with 38 banded in 21 floating marsh nests between 27 Jun and 25 Jul. Early nests with 3 eggs, 2 eggs and 1 egg, on 26 May 1946, had four young, three young, and one young and 3 eggs respectively on 19 June; the incubation period was thus 22 to 24 days. In 1947, they were flooded out and returned to build 13 nests in Chokecherry bushes on the island in Upper Rousay Lake, where, due to late nesting, seven were banded on 15 Jul and eight on 3 Aug. In 1948, nests were again located in the marsh, but all but two fledglings had already left their nests when discovered on 10 Jul. Only two were banded, while another two wandered into a funnel-opening duck trap on 31 Aug. At Crescent marsh on 19 Jun 1955, 24 banded in nests in the bulrushes. Although night-herons are known to nest in trees or in marsh nests, I know of no other locality that has documented the tree-marsh and marsh-tree switch on two occasions. Present in 1973 on all the lakes, but no nests found (AS). Nesting in bulrush and in small willows at Foam Lake marsh, 6-8 Jun 1974 (AS). A group of 25 adults beginning to build nests in bulrushes at Rokeby marsh, 10 May 1993 (DH), abandoned later that year, but they nested successfully in 1995 in a year of high water levels at the marsh (Rob Kirkness).

BBS: Theodore, 1 year; 1, 2000, mean 0.03.

Banded 24 (CSH) + pre-1955, 156 (Patrick 68, CSH 88).

Recoveries, banded in area, 6 [3.3%] (SK 1, IL 1, KS 2, OK 1, TX 1). Of the five out-of-province recoveries: two in Kansas at three and eight years; one in Illinois at one year; one in Oklahoma at one year; one in Texas the same fall (Houston 1967b).

Langenburg: Two seen along Cutarm Creek, 1990 (TH,BL).

YELLOW-CROWNED NIGHT-HERON *Nyctanassa violacea*

Vagrant.

One sighting on Highway 10, along the channel between Lower Rousay Lake and York Lake [halfway between Yorkton and Otthon], on 3 and 7 Sep 1985 by Maurice Lindgren (1985). Another was seen south of Leross by Ian Lochtie, Phyllis Bordass, Ruth Englund, and Terry Haynes on 8 Jun 1990; it was present from 8:30 a.m. to 3:30 p.m. (Smith 1996).

WHITE-FACED IBIS *Plegadis chihi*

Vagrant.

One seen at southeast corner of Little Quill Lake, 31 Jul through 1 Sep 1977 by John Tautin [misspelled "Totten" in Gollop 1978]; there were two birds on 13 Aug and 4 on 17 Aug (Gollop 1978; ARS, *Atlas*).

TURKEY VULTURE *Cathartes aura*

Uncommon transient. Uncommon summer resident at Sturgis and Duck Mountain, where it breeds.

One shot on the upper Assiniboine, near 52° [Fort Pelly] (Bell 1882). Present at Cote's Reserve, 1883-84 (Seton 1891).

Present in plains and parkland in buffalo days. Fred Baines saw four or five at a dead cow near Crescent Lake in 1892 and Frank Baines saw one that had been shot at Leech Lake.

Fairly common all summer from 1888 to about 1896, soaring above the margins of GSL. From 1883 through 1947, John A. Gunn saw one or two birds each spring, but none in midsummer.

One sighting at Sheho on 10 May 1949 (WN). One soaring over Lake of the Prairies, 21 Jun 1989 (CSH,MIH). Two vultures sat on fence posts 8 miles east of Yorkton on 16 Jul 1991; one was present in the same place four days later (Jim Smart, *fide* CSH). Seen at GSL on 22 Jul 1990 and 12 Aug 1991, then three on 9 Jun 2000, two on 13 Aug 2001, seven on 26 May 2002, five sightings of two to six in June 2002, and three sightings of one to four in July 2002, at roadkills, coyote kills, fish kills, and other carrion (WA). One should now look for a nest in the attic of an abandoned house somewhere near GSL! (See Houston et al. 2002).

ESA(15): 21 Apr 1985, Av 6 May. Two fall dates: 10 Sep 1997 and 12 Sep 1998.

Breeding: Nest, two young, in manger of deserted barn, Boggy Creek valley near Runnymede, 24 Jul 1984 (B. DeVries). Nest with two half-grown young in attic of deserted small log house east of Parkerview on 18 Jul 1998, at least the fourth year of occupancy there; the adult was perched on the roof on 3 May 2003 (Terry Flesjer, AS). Nest in attic of barn between Canora and Stenen in 2001; the barn was demolished in 2002 (Fran and Peter Buryk). Two nestlings, nearly ready to fly, found by Fran and Peter Buryk in the attic of a deserted house near Sturgis, had an official, numbered wing tag applied by Brenton Terry on 8 August 2003.

CBC: 1 count, 1 bird, Yorkton 1996. Turkey Vultures rarely winter anywhere on the prairies, but Kenn Woods saw one on 3 Jan 1997; it later died (Bain and Holder 1997a).

DUCK MOUNTAIN: SMNH 1926: "Always to be seen in their grand soaring flight, in unusual numbers." Thirty in sight overhead at once, as late as 1945 (CSH). SMNH 1951. SNHS 1956. Robert J. Long had excellent opportunities to photograph them, 21 Jul to 29 Aug 1970. Once he counted 24 birds, usually perched in trees near the dump or soaring overhead; no nest located (Long 1971). The only possible nest record is of an adult vulture which came out of a hollow log on the forest floor of an island in DMPP in 1942 (F.J.H. Fredeen in Houston 1969a). One to eight seen in 1974, 1975, 1976, 1977 (JAS). Common summer resident; at garbage dump, 20 adults sunning on fence on 13 Aug 1981; 13 adults with four immatures, 9 Sep; also seen over Pelly Point and Tate Island (Hatch and Weidl 1981, Weidl 1983).

GREATER WHITE-FRONTED GOOSE *Anser albifrons*

Fairly common to common spring and fall transient.

About 1900, there were thousands at Crescent marsh each year (FB). Rarely noted in the 1930s, except for a flock that rested at GSL in the fall of 1938 (JAG). Hundreds at Leech Lake, fall 1944 and October 1945, and during last week of Apr 1945 (CSH). Now regular but uncommon and in recent years outnumbered by two new arrivals, the Blue Goose and the Ross's Goose.

None seen in migration at Sheho until 1939, but present during spring migration yearly thereafter. For example: many flocks, 10-24 Apr 1971 at Sheho (WN). In 1973, seven records, 26 Apr to 21 May 1973, with largest flocks of 150 and 50 on 6 and 10 May (AS). For the die-off of 130 birds at Soda Lake on 28 Sep 1954, see insert box in Snow Goose account and Harris 1985a. Flock of 150 with 100 Canada and 20 Snow geese on 13 May 1974 (AS).

ESA(35): 1 Apr 1991, Av 18 Apr. **LSD**: 26 May 1992.

EFA(20): 2 Sep 1974 (WA, Houston and Houston 1975), Av 12 Sep. **LFD**(10): 21 Oct 1984 (WA), Av 27 Sep.

Recoveries, banded outside area, 19 (NU 11, NWT 1, LA 3, ND 3, NE 1; 19 recoveries scattered between 1961 and 1996, with none before or since; some years no banded geese of this species are shot).

SNOW GOOSE *Chen caerulescens*

Abundant spring and fall transient.

Large flocks flew over Rothbury settlement each spring in the 1890s (R.P. Rooke). Frank Baines saw small flocks in migration each year until 1942, then none for six years. At GSL, John A. Gunn saw them only once, two or three in a flock of small Canada Geese, in the early 1900s.

At Yorkton, a noteworthy flock flew over the city 27 Apr 1939 (Mrs

M.J. Culver). At Sheho, seen during spring migration in 1919, 1923, 1926, 1927, then none until 1947 and 1954, then sighted almost yearly for next 20 years (Niven 1973), building up to flocks of 500 containing both colour phases by 1984-1988 (WN).

Seen more commonly in recent years, with 200 in mixed flock with White-fronts on 6 May 1973 (AS). Numbers increased steadily during the 1980s and especially the 1990s, until the fall migration of 18 Oct 1992, when 200,000 were counted (JP, FAS, Koes and Taylor 1993a). On 10 Apr 1999 there were 100,000 on GSL (WA,JA, Bain and Shanahan 1999c), and 200,000 at GSL on 18-20 Apr 2000 (WA,JA, Bain and Shanahan 2000a). In the fall of 2001 they peaked at 100,000 (WA, Koes and Taylor 2002a). In recent years, 160-acre fields have been blanketed with this species, and on GSL they stretch from shore to shore, hundreds of yards wide. About 25% of them are in the blue phase.

ESA(35): 29 Mar 2000, Av 17 Apr. **LSD**(26): 20 May 1995, Av 11 May. **EFA**(26): 30 Aug 2001, Av 10 Sep. **LFD**(24): 11 Nov 1992, Av 19 Oct.

Recoveries, banded outside area, 112 (MB 70, NU 28, ON 2, KS 1, KY 1, NM 1, TX 9; over half of the Manitoba geese were banded on Fred Cooke's long-term study area at La Pérouse Bay, east of Churchill).

Necrotizing enteritis has caused death of moderately large numbers of geese on three occasions:

First, at Soda Lake on 28 Sep 1984: 128 Snow Geese, including 86 adults and 12 juveniles in white phase and 27 adults and 3 juveniles in blue phase; Ross's Geese, three adults and four juveniles; White-fronted Geese, two adults.

Second, at Potts Lake on 4 Oct 1984: 147 Snow Geese including 100 adults and 26 juveniles in white phase and 17 adults and four juveniles in blue phase; Ross's Geese, four adults and two juveniles; White-fronted Geese, two adults and two juveniles; 53 Canada Geese, including a seven-year-old small Canada banded near Butler, OK, on 16 Feb 1977.

Third, at Potts Lake on 28 Oct 1990: 307 white-phase Snow Geese, including 157 adults and 35 juveniles; 102 adults and 3 juveniles in blue phase; ten "partial carcasses"; Ross's Geese, two juveniles; Canada Geese, two (AS, pathology by Dr Gary Wobeser).

GREATER SNOW GOOSE *Chen caerulescens atlantica*

This larger subspecies, from the eastern Arctic, particularly Greenland and Baffin Island, has one recovery.

Recoveries, banded outside area, 1 (NU, banded north of Pond Inlet, far northern Baffin Island, 12 Aug 2001, shot at Calder, 13 Oct 2001).

BLUE GOOSE, the blue morph of the Snow Goose

Initially this morph was a rare transient. C. Stuart Francis saw a pair with a flock of small Canada Geese, in a stubble field 8 miles west of Saltcoats, in 1923. Five were noted at Crescent Lake by C.H. Maddaford on 7 May 1938. John A. Gunn also saw one flock at GSL in the 1930s. Sightings were extremely rare before 1960, when I left Yorkton (CSH).

The largest flock of the blue morph seen was over 500 in a mixed flock of about 1,000 geese in the spring of 1953 (C.H. Maddaford). One hundred Blues stopped over at GSL on 22 Apr 1962 (WA, Nero 1962b). Since then, most flocks of Snow Geese have had a larger and appreciable percentage of the blue morph.

Recoveries, banded outside area, 44 (MB 19, NU 20, ON 3, LA 2, demonstrating the increase and westward spread from the long-known main flyway through eastern Manitoba; the first to be shot in the YDM area was banded in Louisiana in December 1970 and shot south of Pelly on 31 Oct 1973.) Only 3 of the 44 recoveries were shot in the 1970s and 7 in the 1980s, whereas 17 were shot in 2000 and 2001.

ROSS'S GOOSE *Chen rossii*

Rare transient until 1994; now uncommon; shot each year only since 1995.

Adam Schmidt saw none in the YDM area between 1978 and 1983. Two dead at Soda Lake on 29 Sep 1984 (see Snow Goose) (WA, Harris 1985a). Four flew over GSL, 29 Sep 1996 (WA).

Recoveries, banded outside area, 45 (NU, all banded in Nunavut, demonstrating the extremely recent increase and eastward spread from the long-known main flyway through the Kindersley region. There were only two Yorkton area recoveries prior to 1995: one in 1976 and one in 1990. The first had been banded as a flightless young, north of Lake Garry in what is now Nunavut, on 28 Jul 1965; it was shot near Tuffnell in Sep 1976, when over 11 years old. The 1990 goose was a bird of the year, shot near Elfros. Fifteen of the recoveries occurred in 2000 or 2001).

CANADA GOOSE *Branta canadensis*

Uncommon in 1930s. Now a very common to abundant summer resident, greatly increased in numbers. Breeds.

They bred on the upper Assiniboine River above Fort Ellice, at least until 1883 (Macoun 1883:365). Common in the early days of settlement. Eggs collected north of Saltcoats (NJ,PMA).

By the early 1940s, there were only a few resident pairs left, two or three per lake, at both Rousay lakes, and Crescent marsh and Rokeby marshes; three pairs at Horseshoe Lake; and one or two pairs at several lakes north of Sheho. Data on 10 breeding records of the giant subspecies, *B.c. maxima*, were deemed publishable due to their scarcity (Houston 1966b). Each sighting and each hunter kill in those years was a major event.

By 1973 there were at least 20 broods on the lakes south of Yorkton (AS, Houston 1973c). Several broods of 10 young were raised near Sheho in 1976 (WN). In the late 1970s and 1980s, 60 to 80 young were raised each year by wild pairs at Horseshoe Lake alone. At Foam Lake marsh, 27 nests found in 1974 and 23 nests in 1995 (AS). In the dry year, 1989, the geese left Horseshoe Lake from mid-June until 28 Aug, but there were two pairs with three young each at Wilson Lake, southwest of Springside, on 23 Jun (WA).

Similar reintroduction efforts throughout the northern United States were derived from a private flock of decoy geese, donated by Henry Wallace to the Seney National Wildlife Refuge in the upper peninsula of Michigan (Nelson 1963).

A partial albino goose, first seen at McOuat Lake, 14 miles north of Yorkton, photographed on 28 May, had a normally coloured mate and five young on 18 Jun (Morgotch 1966).

ESA(47): 26 Feb 1981 at Sheho (WN) and 29 Feb 1988 at GSL (WA), Av 25 Mar. **LFD**(47): 30 Nov 1981, Av 10 Nov. In 1991, a heavy snowfall on 21 October was followed by severe cold that crusted snow and froze even the larger lakes; the geese left by 24 October (WA).

For 41 years at Sheho, Niven not only kept dates for the first large Canada Goose each spring, but also for the smaller subspecies that include Hutchins' Goose. The large geese arrived between 26 Feb and 16 Apr, mean 27 Mar. The smaller geese arrived 4 to 48 days (mean of 20.2 days) later, from 28 Mar to 8 May, mean 16 Apr.

Breeding: Earliest nest, 40 feet above ground in a former Red-tailed Hawk nest, 1 mile west of Burgis, 21 Apr 1994 (WH, Harold

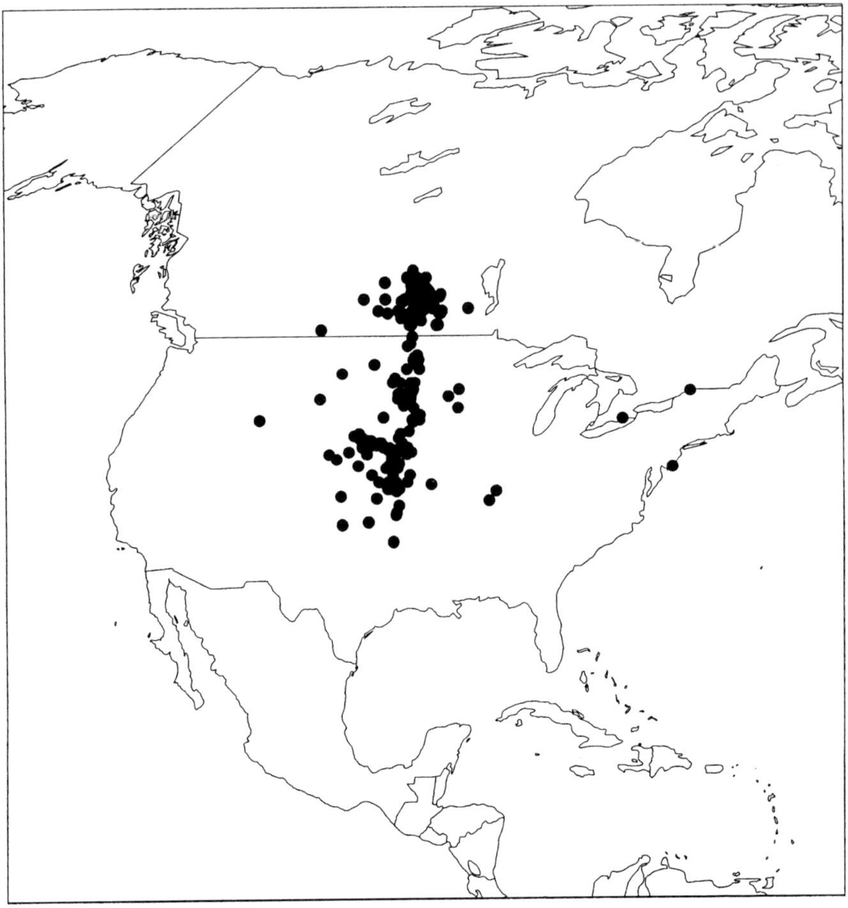

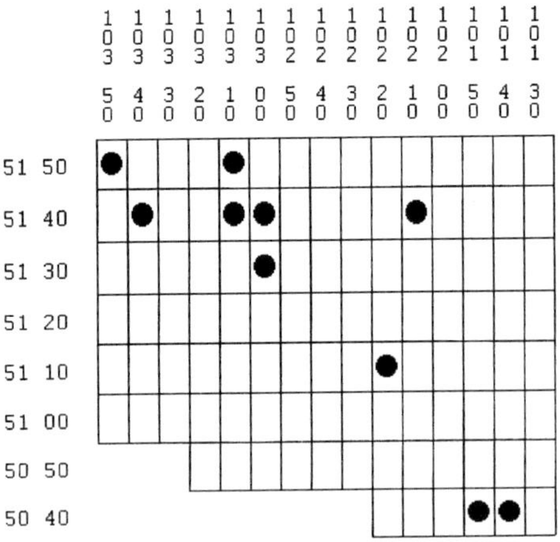

CANADA GOOSE

Wilkinson). Another early nest, on top of a haystack near GSL, contained 6 eggs on 7 May 1967 (WA). Early hatching, 14 and 15 May 1946 at Upper Rousay Lake and Rokeby marsh (five nests, one with 6 eggs and an infertile egg, and one with 8 eggs); goslings took their first short flights almost exactly two months later, on 15 Jul. Late nest, 6 eggs, 22 Jun 1947 (CSH). Nests, 7 eggs, Leech Lake, and 5 eggs, Rokeby marsh, 1 Jun 1948; 7 eggs, Upper Rousay Lake, 10 Jun 1948 (CSH). At Sheho, geese had a successful year in 1972 and again in 1975 when there were eight successful nests at Salt Lake (WN). In 1973, 28 pairs with 98 goslings in the Rousay lakes-Leech Lake area (AS). Twenty-five breeding pairs at Horseshoe Lake in June 1975, with at least 50 young in July (DH), and 60 to 80 in most years since (WA). However, in two years of below normal water levels, 1989 and 2002, there were no young Canada Geese

In the 1950s and 1960s a gradual buildup of numbers occurred, as in other prairie and parkland localities, spreading out from a single domesticated pair, named Hiawatha and Queenie, supplied to Fred Bard by Ralph Stueck of Abernethy in 1953. Stueck, president of the Saskatchewan Natural History Society in 1948-49, obtained a federal permit to raise geese. After losing his first flock of 12 (perhaps by theft), and his famous talking goose, Jonathon, and his brood to lightning, Stueck's third attempt at raising geese began with a pair named Maggie and Jiggs, received from J.A.M. Patrick in Yorkton (Motherwell 1990, Joyce Stueck McKen, pers. comm.). At Wascana marsh in Regina, the first set of eggs from this and subsequent pairs was hatched artificially, and each pair then allowed to raise their second clutch naturally. By 29 Jul 1957, Bard's flock had increased to 36 adults and 64 young (Bard 1957), and by the next summer, after 19 pairs brought off 46 young from 11 successful and six predated nests, there were 134 geese (Bard and Lahrman 1958). In 1959, Bard released about 500 goslings to launch the Nis'ku project near Eyebrow (Gray 1969). By 1970 or 1971, Bard began reintroducing geese throughout southern Saskatchewan from his ever-growing Regina flock.

As early as 1963, Hamilton M. Laing claimed that the rescue and return of the subspecies listed as "extinct" for 50 years was "the brightest star in conservation's crown" in the 20th century to that date (Nero's addendum to Houston 1966b).

raised at Horseshoe Lake. Most years there are 40 to 50 young fledged at Burgis Beach, GSL. Brood at Invermay on 9 Jul 1989 (ARS,PC). Nests with 4 eggs, with 2 eggs and two freshly hatched young, and broods of two and four at Rokeby marsh, 11 May 1993 (DH). Flightless young banded in only 10 of the 108 10-minute blocks.

BBS: overall mean, 10.9. Theodore, 17 years; highest, 85, 1978, mean 10.2. Willowbrook, mean 39.5.

Banded 4,563 (incl. 254 by D. Sherrat; computerized pre-1955 banding: Patrick 14, Bard 5, Colls 7, CSH 1). A three-quarters-grown young captured after a canoe chase on 13 Jul 1946 at Upper Rousay Lake, was shot at Bredenbury, Saskatchewan, on 15 Oct 1947 by George R. Morrison. It weighed 12 pounds when shot, evidently the large subspecies, *B.C. maxima*.

Recoveries, banded in area, 1,472 [32.3%] (Map; SK 826, AB 1, MB 6, ON 2, CO 5, ID 1, IL 2, KS 43, MN 2, MT 2, ND 29, NE 261, NJ 1, OK 8, SD 278, TX 1, WY 1, no state given 3).

Recoveries, banded outside area, 745 (SK 217, AB 19, MB 7, NWT 2, ON 1, AR 2, CA 1, CO 11, IA 24, ID 2, IL 4, KS 31, MN 27, MO 14, MT 9, ND 36, NE 183, NM 1, NV 1, OK 38, OR 1, SD 80, TX 2, UT 10, WA 1, WY 21).

Langenburg: Pairs, some with young, largest group 35, on major sloughs and a small lake, 1990 (TH,BL)

DUCK MOUNTAIN: SMNH 1926. SMNH 1951: "Captive" bird. Uncommon (Hatch and Weidl 1981). SNHS 1989.

SMALL GOOSE *Branta canadensis hutchinsii* and *B.c. parvipes*

Recoveries, banded outside area, 112 (CO 4, KS 11, OK 37, SD 3, TX 57).

TRUMPETER SWAN *Cygnus buccinator*

Rare summer resident, probably breeding, until 1850s, then absent for about a century; now expected as a rare transient.

From at least 1821 through 1842, there was a brisk trade in swan skins at the Hudson's Bay Company fur trade posts within what is now Manitoba and Saskatchewan. The Swan River district, which included Fort Pelly, sent 6,960 skins in 21 years, an average of 348 in the 20 years reporting, with a maximum of 900 skins in 1825. The trade in swan skins and quills remained large until 1890, and contributed greatly to the almost complete elimination of the Trumpeter Swan as a breeding species (Houston et al. 1997).

A pair of "large swans" spent the summer on a small island in Rokeby marsh, half a mile west of John Domon's farmyard, sometime in the 1940s. The birds acted in a territorial manner, one of the adults attacking Domon's daughter with its wings. (Letter from Frank Switzer to C.S. Houston, 25 Nov 1988).

Breeding: John Makowetski had a pair of swans that built a large nest, a mound of vegetation, on top of some willow regeneration, on an elongated 60-acre wetland, SE16-26-7w2, 1 mile south of Beaver Dale, in 1993. Adam Schmidt considered this to be typical Trumpeter nest construction, whereas a Tundra Swan typically builds a less ambitious structure on a dryland shore or island. Video footage and still films were obtained, but have only recently been relocated and have not as yet been checked with a swan specialist for absolute confirmation as to species.

Three young swans were raised but the weakest was killed by an eagle in the fall; the adult and remaining two young stayed almost to freeze-up. The swans did not return. This marsh had been dry in 1961 and was almost dry in 2002 (AS).

> Swans from restored flocks on wildlife refuges in the Dakotas began nesting in Saskatchewan's Porcupine Forest, immediately to the north of the YDM area, in 1986, with 11 pairs nesting at 11 sites by 1995 (Beaulieu 1999). Large swans in migration over Preeceville, Pelly, and Duck Mountain may include Trumpeters en route to wintering areas on wildlife refuges in the northern states.

TUNDRA SWAN *Cygnus columbianus*

Fairly common spring and fall transient.

Stops over on the larger bodies of water. Two birds spent a summer in Crescent marsh, north of Frank Baines' farm, in the 1920s. In 1973, the largest spring flock contained 62 birds on 8 May (AS). In fall, flocks of 35 on 8 Nov 1960 (WA, Krause 1961), 35 on 8 Nov 1970 (WA, Houston 1971a), and 45 on Leech Lake on 6 Oct 1993 (DH). Over 400 were present on GSL on 9 Nov 1998 (Bain and Shanahan 1998d).

ESA(46): 29 Mar 2000, Av 20 Apr. **LSD**(43): 27 May 1945 (five birds at Leech Lake, WHC) and 25 May 1993 (WA), Av 11 May. **EFA**(18): 500 at Leech Lake, 18 Aug 1971 (F. Bucsis, F.A. Switzer), and 3 Sep 1982 (WA), Av 28 Sep. **LFD**(47): 1 Dec 2001 over Parkerview (AS), 27 Nov 1981 at Sheho (WN) and 24 Nov 1981 at GSL (WA), Av 6 Nov.

Recoveries, banded outside area, 3 (MD 2, NC 1).

Langenburg: 2 on 26 Apr 1976 (DH).

WOOD DUCK *Aix sponsa*

Rare but increasing summer resident since the 1970s. No breeding records.

The first sight record in the area was a pair near Ituna, 27 May 1955 (Brennan 1965b), followed 19 years later by an adult male at Foam Lake marsh on 9 Jul 1974 (AS).

Banding records demonstrated the gradual influx into the area: an adult male trapped by J.D. Smith on south shore of Little Quill Lake, 9 Aug 1965 (Shaver 1966; an immature male was caught 10 days later); an adult male 2 miles east of Amsterdam by E. Ferguson on 12 Aug 1978 (Serr 1979); an immature female west of Wadena in 1989; at Leech Lake, an adult male in 1989, three in 1990, one in 1991, and an immature male in 1989; an adult male west of Springside in 1992; an adult female at the Rousay lakes in 1995; one west of Saltcoats in 1997 and three adult males near Otthon in 1998. Five adult males were banded north of Foam Lake, one in 1986 and four in 1992.

One flushed by canoeists on the Whitesand River near Highway 9 north of Yorkton, 11 Aug 1988 (DH). Only one sighting on south shore GSL with 30 other ducks--on 12 May 1998 (WA). Seen on the Assiniboine west of Preeceville in recent years; male in eclipse plumage on Robinson Creek near Loch Lomond, 5 Aug 2002 (DF).

Present on various wetlands east and southeast of Parkerview: a pair on 11 May 1997; a male in full breeding plumage and two males in partial moult in early Sep 1997; a possible brood on 19 Aug 2001, all full-grown and fully feathered with two females, the two males in identical stages of juvenal plumage, and an agitated female quite vocal as if protecting the other four; one on 15 Sep and three on 29 Sep 2001; three males on 12 May 2002, two on 9 June; at least 10 moulting males on 23 Jun and five on 1 Jul 2002; a male on 21 and 28 Jul, three on 10 Aug and two on 6 Sept 2002, the latter two males in full breeding plumage. Another four southwest of Jedburgh on 11 Jul 2002 (AS).

The Kamsack branch of the Saskatchewan Wildlife Federation purchased day-old ducklings from a hatchery in Illinois, took delivery of them at the Minot, ND, airport, kept them in a pen with a pool on a farm near Kamsack and then released them as flightless young later in the summer, 27 in 1984 and another 93 in 1986 (Richard Dix, pers. comm.); prior to release they were banded by Dan Nieman of CWS.

Breeding is to be expected within the YDM area, since a female with two flightless, half-grown young was seen on the Assiniboine River immediately north of 52 degrees, on 10 August 2003. [An erroneous GPS reading gave the location as immediately south of 52 degrees](DF,DHJ,SJ).

Banded 161.

Recoveries, banded in area, 3 [1.9%] (SK 2, TX 1).

Langenburg: Seven males and two females on a bare prairie pothole east of Langenburg, 1990 (TH,BL).

DUCK MOUNTAIN: One male, 30 May 1986, near west boundary of the park (MDG,SJS). Four males on a small boggy pond on road to Batka Lake, 20 May 1989. Five males along road to Kipp Lake, 17 Aug 1990 (TH). Seen at Green Lake (RK). Fall record, three on 7 Sep 2002 (TH).

GADWALL *Anas strepera*

Common summer resident, especially in the southern part of the YDM area. Breeds.

Only one specimen shot on the upper Assiniboine in September 1881 (Macoun in Seton 1891). Ten eggs, Yorkton, 2 Jun 1901, W. Raine, AMNH.

It was common in the 1920s, decreased in the 1930s, now again common and visibly increasing since 1985.

ESA(41): 31 Mar 1997, Av 30 Apr. **LFD**(21): 11 Nov 1983, Av 15 Oct.

Breeding: Female with five downy young, 10 Jul 1962 (WA). In 1973, nest with 2 eggs on 28 Jun and 7 eggs on 7 Jul; first brood of eight, 10 days old, on 9 Jul (AS).

Breeds also at Wadena, Margo, Preeceville, Willowbrook (ARS, *Atlas*). Flightless young banded in 21 of the 108 10-minute blocks within the area.

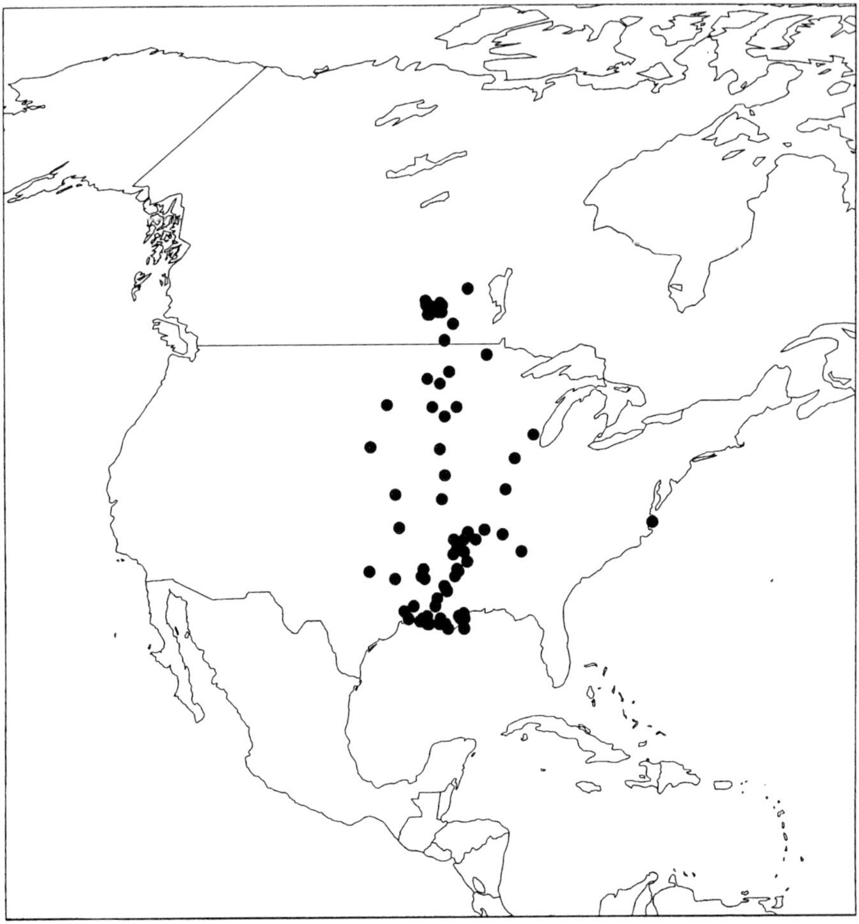

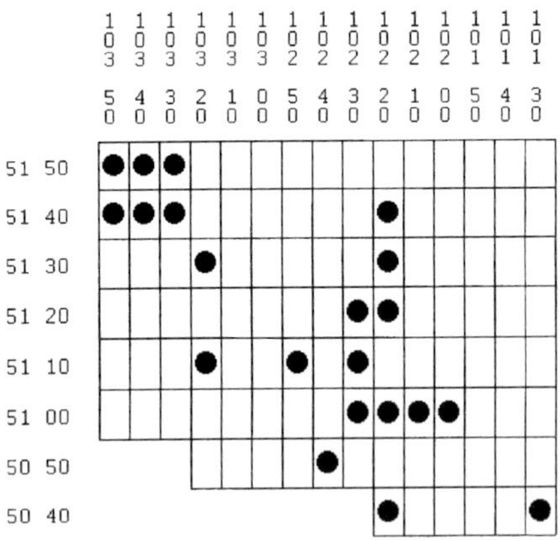
103 50
103 40
103 30
103 20
103 10
103 00
102 50
102 40
102 30
102 20
102 10
102 00
101 50
101 40
101 30
51 50
51 40
51 30
51 20
51 10
51 00
50 50
50 40

GADWALL

BBS: overall mean, 1.7. Theodore, none sighted until 6 recorded in 1985, then yearly with one exception from 1988 through 1997.

Banded 828 (CSH 1).

Recoveries, banded in area, 82 [9.9%] (Map; SK 12, MB 3, AL 1, AR 11, IL 2, KS 1, LA 25, MN 2, MS 5, NE 2, NC 2, ND 3, OK 1, SD 3, TN 1, TX 5, WI 1, no state given 2).

Recoveries, banded outside area, 4 (SK 2, AB 1, MB 1).

Langenburg: Uncommon, at only a few locations, 1990 (TH,BL).

DUCK MOUNTAIN: Brood of five in 1976, seen in 1977 (JAS). Uncommon, nests with 5, 6, 7, 7, 9, 10, 10, and 13 eggs, also 8 Gadwall and 1 scaup egg, and 13 Gadwall and 4 scaup eggs (Hatch and Weidl 1981). One seen 21 May 1989 (TH). SNHS 1989.

AMERICAN WIGEON *Anas americana*

Common summer resident. Breeds.

Frequent on the Assiniboine, 1881 (Macoun in Seton 1891).

ESA(47): 8 Apr 1958, 1977 and 1991, Av 19 Apr. **LFD**(23): 15 Nov 1983, Av 6 Oct.

Breeding: Early nest had 3 eggs on 17 May and 7 eggs on 21 May 1967 (WA). Nest, 11 eggs, 8 miles north of Melville, 20 Jun 1958. Nest, 9 eggs at Saltcoats, 27 Jun 1957 (BH). Brood of six day-old downy ducklings at York Lake, 6 Jul 1973 (AS).

PNRS: 8 and 7 eggs at Saltcoats on 6 and 28 Jun 1958 (BH); 9 eggs at Melville, 8 Jul 1960 (VS).

Also breeds at Wadena, Buchanan, Goodeve, Willowbrook (ARS, *Atlas*). Flightless young banded in 42 of the 108 10-minute blocks.

BBS: overall mean, 1.8. Theodore, 12 years; highest, 17, 1975, mean 1.8.

Banded 1,538.

Recoveries, banded in area, 184 [11.6%] (Map; SK 17, MB 4, ON 3, QU 1, AK 1, AL 3, AR 9, CA 6, CO 1, FL 5, GA 1, IA 4, ID 1, IL 2, KS 2, KY 2, LA 26, MD 3, MI 4, MO 3, MS 18, NC 6, ND 5, NE 4, NJ 1, OH 1, OK 3, SC 2, SD 4, TN 3, TX 16, VA 2, WI 8, no state given 2, Colombia 1, Cuba 1, Dominican Republic 1, Haiti 1, Mexico 7). An immature female banded at Upper Rousay Lake on 28 Aug 1943 was shot near Navajoa, Sonora, Mexico on 18 Feb 1955, 11.5 years after banding; the band was so thin that the band number could be deciphered only by chemical etching (Houston 1960a).

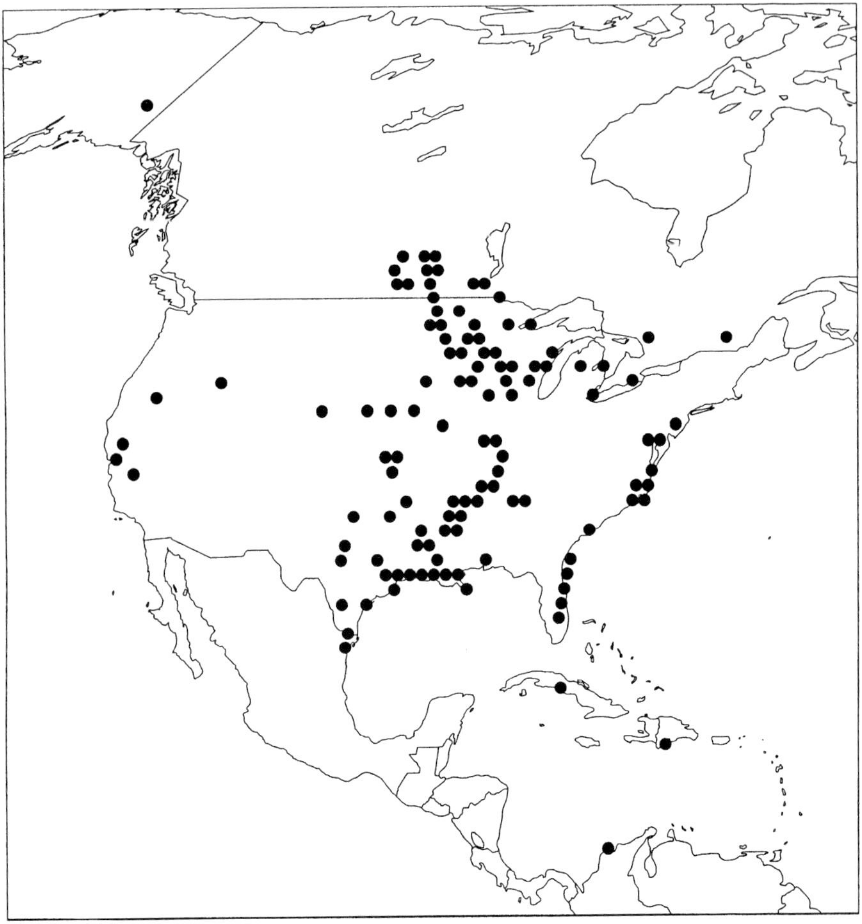

	103 50	103 40	103 30	103 20	103 10	103 00	102 50	102 40	102 30	102 20	102 10	102 00	101 50	101 40	101 30
51 50	●		●	●		●		●							
51 40		●	●	●	●			●	●	●	●				
51 30				●		●				●	●				
51 20		●					●		●	●					
51 10	●		●	●	●	●	●	●	●	●					
51 00						●		●	●	●	●	●			●
50 50						●							●	●	
50 40										●					●

AMERICAN WIGEON

Recoveries, banded outside area, 13 (SK 4, MB 1, AL 1, CA 1, CO 1, MO 2, NE 1, OR 1, SD 1).

Langenburg: Uncommon, 1990 (TH,BL).

DUCK MOUNTAIN: SMNH 1926. SMNH 1951. SNHS 1956. Seen in 1976 (JAS). Fairly common on Bear Head and Batka lakes and several smaller water bodies. Female with brood of nine near Selwood Campground, 12 Aug 1981 (Hatch and Weidl). SNHS 1989.

Point counts: Rank #37, 4 individuals at 1 stop (ARS).

AMERICAN BLACK DUCK *Anas rubripes*

Rare transient. No breeding records.

Apart from an individual shot by C.H. Maddaford at Crescent marsh in the 1920s, and another shot by J.A.M. Patrick near Yorkton on 29 Nov 1923 (Lahrman 1958), the next confirmed records came from duck banding operations, when single adult males were banded at Leech Lake on 21 Jul (Priestly 1945c) and 17 Oct 1945 (JHW).

The first sighting at GSL was one with Mallards on 26 Sep 1954 (WA). A male Black Duck was in company with and apparently mated to a female Mallard, on 19 Apr 1968 at GSL (WA, Hatch 1968c). One was seen near Hyas in May 1982 and another at Key Indian Reserve north of Kamsack on 12 Aug 1985 (CP). A male Mallard/Black Duck hybrid was present at GSL and studied carefully from 25 Apr to 29 May 1990 (WA). Over the years, 53 were banded at the Rousay lakes, 43 at Leech Lake, and others at Yorkton and Otthon, a remarkably restricted local distribution, involving only 4 of 108 10-minute blocks.

Banded 219 (205 adult males, 2 immature males, 12 adult females). Also Black Duck/Mallard hybrid, 66.

Recoveries, banded in area, 39 [17.8%] (SK 3, MB 1, NS 1, ON 3, AL 1, AR 3, IL 8, IN 3, LA 1, MI 1, MO 2, MS 2, NE 1, NJ 1, OH 3, TN 2, VA 1, WI 2). There was a 100% recovery rate from the small sample of two adult females banded at the Rousay lakes in 1945 (Priestly 1945c): one banded 10 Aug was shot 3 Nov 1945 at Wallaceburg, Ontario; the second was in moult and totally flightless when banded 22 Aug, and was shot in November 1945 near Bath, Illinois (CSH).

Recoveries, Black Duck/Mallard hybrid, banded in area, 9 [13.6%] (SK 2, IL 2, NC 1, MS 1, TN 2, VA 1).

Recoveries, banded outside area, 9 (ON 3, IL 2, MA 1, MO 1, MI 1, TN 1; these demonstrate that Black Ducks have been wandering west and often north from their regular breeding and/or wintering grounds in eastern North America since at least 1939. The band of an adult female from the Seney NWR in the Upper Peninsula of Michigan was found the next spring, May 1939, near Hyas; an immature male banded at Havana, Illinois, on 24 Nov 1940 was caught by a dog near Rhein on 4 Aug 1941; a male banded on Cape Cod in Massachusetts on 16 Jan 1940 was shot at Fenwood on 20 Sep 1943).

Recoveries, Black Duck/Mallard hybrid, banded outside area, 1 (ON).

DUCK MOUNTAIN: SMNH 1951.

MALLARD *Anas platyrhynchos*

Abundant summer resident. Breeds. Most common duck.

Frequent on the Assiniboine in September 1881 (Macoun in Seton 1891). J.A.M. Patrick found one nesting in a crow nest in an aspen near Yorkton about 1924 (McKim 1926), and J.B. Paulsen found a Mallard sitting on 10 eggs, 6 feet above ground in a former crow nest (Lawrence 1927). One egg collected north of Saltcoats (NJ,PMA).

Our commonest duck, but numbers are decreasing, especially in

Ducktrap, Upper Rousay Lake, August 1974

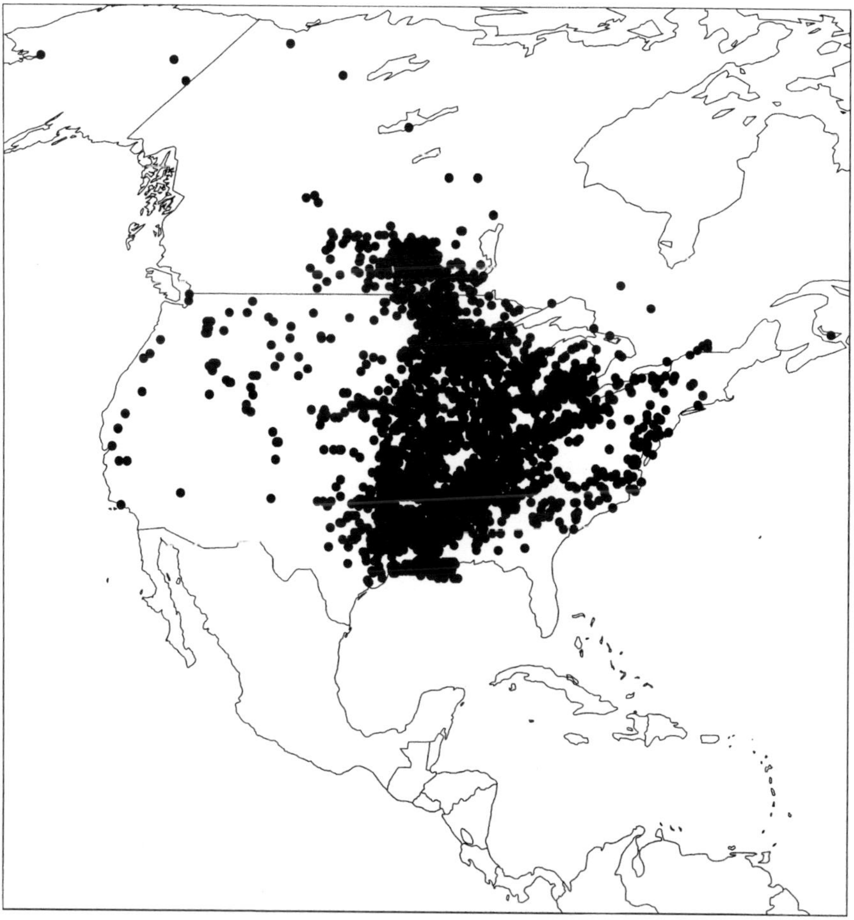

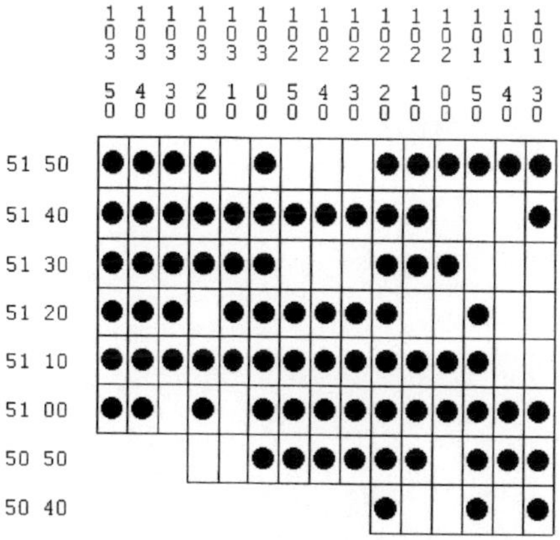
103 50
103 40
103 30
103 20
103 10
103 00
102 50
102 40
102 30
102 20
102 10
102 00
101 50
101 40
101 30
51 50
51 40
51 30
51 20
51 10
51 00
50 50
50 40

MALLARD

comparison with the 1920s. During 1973, counts revealed for Upper Rousay, Lower Rousay, York, Big Leech, Leech marsh, Little Leech, Little Bone, Crescent Lake, and Crescent marsh, respectively, 3.0, 2.3, 3.1, 5.3, 13.6, 5.8, 18.2, 5.2, and 6.3 pairs per mile of shoreline (AS).

Between 1990 and 1993, west of Willowbrook, 83 female Mallards were captured on their nests late in incubation; 34 received transmitters attached with an external harness and 49 had transmitters surgically implanted into their abdominal cavity. Return rates in the following years indicated that the females with internal transmitters survived better (Dzus and Clark 1996). Brood survival was higher for ducklings that hatched earlier in the breeding season, perhaps related in part to a gradual decrease in water levels as the season progressed (Dzus and Clark 1998).

ESA(46): 22 Mar 1945 at Saltcoats (RB) and 23 Mar 1945 at Sheho (WN) and 2000 at GSL (WA), Av 7 Apr. **LFD**(46): 28 Nov 1953, Av 6 Nov. Two hundred still present at GSL on 15 Nov 1970 (WA, Houston 1971a).

Breeding: Early nest, 6 eggs, 27 Apr, had 10 eggs on 1 May, but predated by 4 May 1958; late nest, 9 eggs, 12 Jul 1979 (WA). Early hatching, 22 May 1946 (CSH).

Of 15 nests found in 1946, five were beneath shrubs, willow or rose; three on grassland (all within 150 yards of water); and six in wet locations with aquatic grasses and rushes. They contained 6 to 11 eggs, average 8.1; 83 hatched, 16 were predated, 2 infertile, 20 unknown outcome (Houston 1946b). Another Mallard was incubating 8 eggs in a crow nest in a Black Poplar at Ravine Reserve, 10 Jun 1947 (CSH).

In 1973, six nests found between 28 May and 13 Jun, and one brood of eight three-week-old ducklings on 6 Jul (AS). In 1983, a successful nest in a spruce at Veregin Cemetery on 30 Jun left behind 2 unhatched eggs

A Black-billed Magpie nest 3 miles south of Yorkton was used successfully by a Mallard in both 1946 and 1947. The downy young were observed jumping from the nest on 23 May 1946, the mother quacking on the ground below the nest. The light downy young fell rather slowly, bouncing off several small branches during their descent. Immediately on hitting the ground, each jumped up and ran into the underbrush, apparently none the worse for its experience (Shaw 1946b). The next year Mallard use of the same nest was observed by Lawrence Walkinshaw of Michigan (Walkinshaw 1948).

containing fully developed ducklings; this nest was 4 feet below a Merlin nest with two young! (AS).

PNRS: at Melville, 4, 5 and 9 eggs (GAA,VS); at Saltcoats, 4, 4, 8, 9, 9, 11, 12, 13 eggs (BH); at Stornoway, 2, 3, 3, 4, 4, 5, 5, 5, 6, 6, 6, 6, 7, 7, 7, 7, 7, 7, 7, 7, 7, 8, 8, 8, 9, 9, 9, 9, 9, 9, 9, 9, 9, 10, 10, 10, 10, 10, 10, 10, 10, 10, 11, 11, 11, 12, 14 (SZ,RD); near Kamsack, 4, 7 and 9 eggs (JHJ); at Rhein, 8 eggs (EK); at Canora, 11 eggs on 8 May 1959; at Arran, 8 eggs (LM). Undoubtedly Mallards nest in each of the 108 10-minute blocks of latitude and longitude within the area; United States Game Management Agents banded flightless young in 80 of these blocks.

BBS: overall mean, 32.6. Theodore, mean 32.4 (1970s, 39.7; 1980s, 31.8; 1990s, 31.7; n=30).

CBC: Rank #27. 27 counts, 138 birds. Highest: 24, Yorkton 1999.

Banded 109,948. (One of the first-ever band recoveries in Saskatchewan, apart from Jack Miner's pioneer work at Kingsville, Ontario, occurred a year before licensed Saskatchewan banders began work; the bird had been banded at Browning, Illinois, on 11 Mar 1922 by none other than the future director of North American banding operations, Frederick C. Lincoln, and was killed by a dog at Buchanan on 22 Apr 1922).

Recoveries, banded in area, 17,916 [16.4%](Map; SK 2,450, AB 86, BC 2, MB 290, NF 1, NS 1, NWT 6, ON 105, PEI 3, QU 8, AL 135, AK 4, AR 3,927, AZ 1, CA 14, CO 71, CT 2, DE 6, GA 23, IA 422, ID

A Mallard nesting in Ed Wiley's farmyard near Saltcoats laid 9 eggs in 1953 in a bunch of willows on a little island. When the eggs were half incubated, frequent heavy rains began to engulf the nest with water. The female raised her nest by putting twigs, sticks and leaves under the eggs, the wet eggs resting on sticks instead of being surrounded by down in the former nest cup. After the nine ducklings had hatched, Ed put the nest contents in a pail to assess what had been added: a half gallon of leaves and more than 900 sticks (Wiley 1955). Another interesting nest site was under the manger of an unused barn on the Frick farm, 6 miles north and 2 miles east of Langenburg. The female Mallard entered through an unclosed main door, barely wide enough to allow her entry and egress. On 26 May 1957, I photographed her from a distance of 2 feet, then placed a band on her leg and returned her to her nest (Houston 1957f).

27, IL 1,106, IN 185, KS 365, KY 185, LA 2,064, MA 1, MD 23, MI 194, MN 508, MO 840, MS 1,203, MT 32, NC 42, ND 530, NE 408, NJ 5, NM 5, NV 1, NY 24, OH 96, OK 497, OR 13, PA 31, SC 34, SD 363, TN 651, TX 542, UT 8, VA 41, WA 18, WI 247, WV 2, WY 12, state not given 50).

Recoveries, banded outside area, 1,916 (SK 272, AB 37, MB 266, ON 42, NWT 17, QU 3, AL 24, AR 128, CA 2, CO 39, DE 3, IA 19, ID 6, IL 249, IN 10, KS 44, KY 21, LA 75, MA 1, MD 13, MI 24, MN 58, MO 68, MS 46, MT 15, NC 2, ND 69, NE 33, NM 4, NY 13, OH 7, OK 63, OR 8, PA 18, RI 1, SC 13, SD 75, TN 61, TX 10, VA 3, VT 2, WI 42, WV 2, WY 8).

Langenburg: Second commonest dabbling duck, 1990 (TH,BL).

DUCK MOUNTAIN: SMNH 1926. Broods of four and six young just southwest of Madge Lake, 26 Jun 1948 (CSH). SMNH 1951. SNHS 1956. Seen in 1974, 1976, 1977; broods of four and seven in 1975 (JAS). Common (Hatch and Weidl 1981). SNHS 1989.

MOTTLED DUCK *Anas fulvigula*

This is an example of Earl Baysinger's rule: If a band recovery seems too good to be true, it probably is not true (Baysinger was head of the bird banding laboratory). **Recovery, banded outside area,** 1 (SC). Captured at Marsh Island National Wildlife Refuge in Louisiana, a Mottled Duck was transported to South Carolina and released there by Thomas H. Strange of the South Carolina Fish and Wildlife, 13 Jul 1982. It was then reported as "band only" from Foam Lake, 3 Oct 1989. When Strange phoned the band finder, Edwin Karapita, at Foam Lake, he learned that Karapita had been riding horseback on dry ground near the edge of the lake. He spotted a shiny bird band on a string, and reported it. The best guess is that an American hunter, up to hunt at Foam Lake, carried an unreported band around his neck or in his pocket and inadvertently dropped it. This record cannot be used to add another species to the list of Saskatchewan birds! Mr. Karapita confirmed the gist of this story in a telephone conversation on 2 May 2003 (CSH).

Point counts: 1 individual at 1 stop (ARS).

BLUE-WINGED TEAL *Anas discors*

Common summer resident, our second commonest duck. Breeds.

Rare on the Assiniboine in September 1881 (Macoun in Seton 1891). Three teal eggs collected north of Saltcoats (NJ,PMA).

ESA(46): 19 Apr 1976, Av 29 Apr. **LFD**(29): 26 Oct 1958, Av 5 Oct.

During 1973, at Upper Rousay, Lower Rousay, York, Big Leech, Leech marsh, Little Leech, Little Bone, and Crescent marsh, respectively, there were 3.2, 2.9, 2.3, 4.4, 15.4, 7.4, 4.8, and 7.8 pairs per mile of shoreline (AS).

On 1 Jun 1996 a Northern Harrier was flushed off a partly-eaten female Blue-winged Teal, near her nest with 10 predated eggs, in dense grass along a fenceline (WA).

Breeding: Early nest, 9 eggs, 19 May. Early hatching, 18 Jun 1946. Nests contained 7 to 14 eggs, average 9.6 eggs; 42 hatched, 66 predated, 27 unknown outcome. Eight nests were in open grassland, six under low shrubs such as Wolf Willow, and four were in wet locations of aquatic grasses and bulrushes. The average nest had a cup of 5.7 inches diameter and a depth of 2.5 inches. Late nesting, a female with brood no more than two days old, 2 Aug 1946 (CSH). In 1973, 25 nests found between 24 May and 8 Jul (AS).

PNRS: four nests at Saltcoats in 1958 with 6, 9, 10, 12 eggs (BH); seven nests at Stornoway, 1959-63, with 5, 5, 9, 10, 10, 10, and 11 eggs (SZ,RD); three at Melville in 1960 with 2, 5, 9 eggs (VS). Flightless young banded in 64 of the 108 10-minute blocks.

Also breeds at Wadena, Sheho, Bredenbury, Melville (ARS, *Atlas*).

BBS: overall mean, 7.1. Theodore, mean 6.0; decreasing (1970s, 12.4; 1980s, 3.9; 1990s, 2.7; n=30).

CBC: 1 count, 2 birds. Yorkton 1988.

> On 1 Aug 1958, Maurice G. Street and I, with help from his daughters and my oldest son, about to celebrate his sixth birthday, caught and banded 10 young teal nearly able to fly. There were four recoveries, all shot, from this one brood: near McGregor, Minnesota, on 4 Oct 1958; La Gorgona, Valle, Colombia, on 23 Nov 1958; Tastus, Venezuela, in late 1962; Lake Ariguanabo, Bauta, Cuba, on 7 Dec 1962 (Houston 1963b).

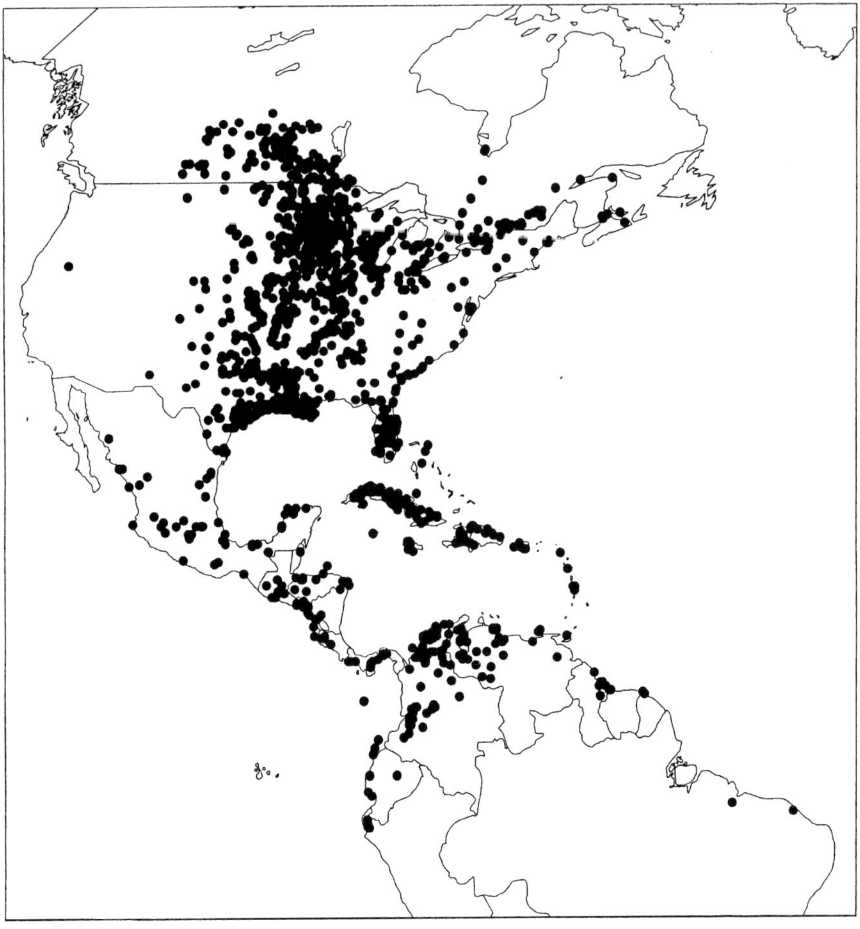

BLUE-WINGED TEAL

	103 50	103 40	103 30	103 20	103 10	103 00	102 50	102 40	102 30	102 20	102 10	102 00	101 50	101 40	101 30
51 50	●	●	●	●		●		●	●			●	●		●
51 40	●	●	●	●		●	●	●	●	●					
51 30	●		●		●	●				●	●	●		●	
51 20	●	●			●	●	●		●	●	●				
51 10		●	●	●	●	●	●	●	●	●	●	●	●	●	
51 00				●				●	●	●	●	●	●	●	
50 50						●	●	●	●	●			●	●	●
50 40															

Banded 105,871.

The Blue-winged Teal is the greatest traveller of all the waterfowl in the Yorkton area, as the following list demonstrates.

Recoveries, banded in area, 3,652 [3.5%] (Map; SK 508, AB 28, MB 56, NB 2, NS 1, ON 32, PEI 2, QU 19, AL 22, AR 24, CA 1, CO 6, FL 100, GA 9, IA 80, IL 57, IN 15, KS 59, KY 6, LA 895, MD 4, ME 4, MI 18, MN 262, MO 49, MS 11, MT 6, NC 7, ND 53, NE 23, NH 1, NM 2, NY 8, OH 16, OK 25, OR 1, PA 2, SC 18, SD 67, TN 8, TX 312, VA 2, VT 1, WI 83, state not given 13, Antigua 1, Bahamas 3, Belize 2, Brazil 2, Cayman Is 1, Colombia 129, Costa Rica 7, Cuba 136, Dominican Republic 13, Ecuador 9, El Salvador 6, French Guiana 14, Guadeloupe 1, Guatemala 18, Haiti 18, Honduras 33, Jamaica 5, Martinique 3, Mexico 227, Nicaragua 12, Panama 16, Peru 4, Puerto Rico 7, Trinidad 1, Venezuela 66).

Recoveries, banded outside area, 233 (SK 92, AB 11, MB 33, NWT 3, ON 14, FL 1, IA 11, IL 1, IN 1, KS 1, LA 8, MD 1, MI 2, MN 7, MO 6, MS 1, ND 16, NE 2, NM 2, NY 1, OK 1, PA 1, SD 12, TN 2, TX 1, WI 1, WY 1).

Langenburg: Commonest dabbling duck, a pair on almost every body of water, no matter how minute, 1990 (TH,BL).

DUCK MOUNTAIN: SMNH 1926. SMNH 1951. SNHS 1956. Seen in 1976, 1977 (JAS). Common, on Madge, Bear Head, Green, and Kipp lakes, and smaller water bodies; nest with 1 egg, 3 Jun; nest with 11 eggs, 2 Jun; adult with three young, 10 Jul 1981 (Hatch and Weidl). SNHS 1989.

CINNAMON TEAL *Anas cyanoptera*

Rare summer resident, not seen every year. No breeding records.

One adult male sighted at Crescent marsh by Frank Baines in spring in the 1920s, and another on slough at Melville, 25 May 1941 (LTMcK). One at Crescent marsh, May 1968 (A. Spelay). Male with female teal at Crescent marsh, 13 Jun 1973 (AS). A male at Stenen on 1 Jul 1988 (CP). A pair near the York Lake Golf Course on 16 May 1992 (AS).

Banded 3, all adult males: banded at Leech Lake, 16 Jun 1944, by J.H. Wilson, shot at Huntington Beach, California, on 9 Dec 1944; banded north of Foam Lake by J.F. Voelzer in late Aug 1993; banded by P. Thorpe west of Wadena in late Aug 2001.

Recoveries, banded in area, 1 [33.3%] (CA 1).

NORTHERN SHOVELER *Anas clypeata*

Common summer resident of shallow waters, but less common in northern part of YDM area. Breeds.

ESA(46): 7 Apr 1963, Av 22 Apr. **LFD**(23): 10 Nov 1983, Av 8 Oct.

Breeding: Early nest, 8 eggs, 30 May 1959; latest nest, 10 eggs, 11 Jul 1956 (WA).

In 1973, three nests found, 31 May-13 Jun; first brood, six week-old ducklings, 6 Jul (AS).

PNRS: 8 eggs, 30 May 1960 (WA). Flightless young banded in 42 of the 108 10-minute blocks.

Also breeds at Kelliher, Sheho, Willowbrook (ARS, *Atlas*).

BBS: overall mean, 2.2. Theodore, 16 years; highest, 14, 1976, mean 1.3.

Banded 676.

Recoveries, banded in area, 41 [6.1%] (SK 6, MB 2, AR 1, CA 2, IL 1, LA 2, MN 5, MO 1, ND 3, NE 2, SC 1, SD 2, TN 1, TX 7, UT 1, Mexico 4).

Recoveries, banded outside area, 5 (SK 4, IL 1).

Langenburg: Third commonest dabbling duck, especially on smaller potholes, 1990 (TH,BL).

DUCK MOUNTAIN: On Highway 57 between park and Kamsack (TH). Uncommon, one pair on Bear Head Lake, 1 Jun 1981 (Hatch and Weidl).

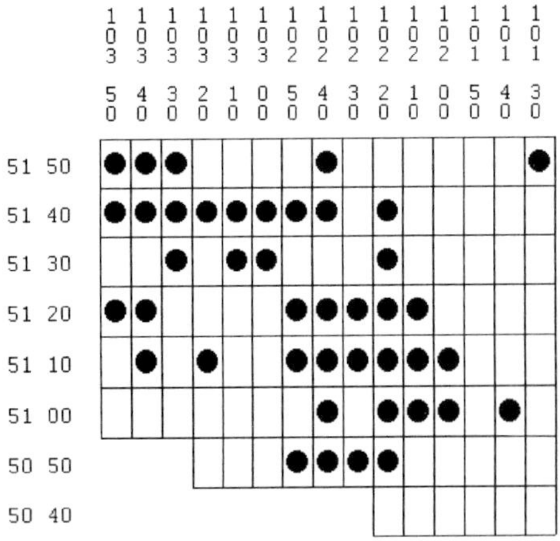

NORTHERN SHOVELER

NORTHERN PINTAIL *Anas acuta*

Common summer resident, but decreasing. Breeds.

Was very common through the 1920s, common through the drought of the 1930s, but numbers are now decreasing, especially noticeable during the dry years of 1988 and 1989. Now disappears at GSL by early June each year (WA). During 1973, there were 4.4 pairs per mile of shoreline at Big Leech Lake and 6.5 at Leech marsh (AS).

ESA(45): 24 Mar 1945 (CSH) and 29 Mar 1986 (WA), Av 9 Apr. **LFD**(13): 2 Nov 1958, Av 9 Oct.

Breeding: Early nest, 10 eggs, at Saltcoats on 4 May 1958 (BH) and Yorkton on 7 May 1946 (CSH); early downy young in nest on 27 May. Eleven nests with 5 to 10 eggs, average 7. Two nests were deserted and three not revisited. Of 53 remaining eggs, 28 hatched, 9 were predated, and 16 had an unknown outcome. Nests were located between 10 and 400 yards from water, three in open stubble, five on prairie, one in a poplar "bluff," one in a willow clump, and one on top of a strawstack. The nest cup averaged 6.3 inches in diameter with a depth of 3 inches (Houston 1946b). Nest, 9 eggs, 22 Jun 1947 (CSH). In 1973, early hatching, 27 May; late incubation, 24 Jun (AS).

Broods of three and five, three weeks old, Crescent marsh, 6 Jul 1973 (AS). Late nest, 7 eggs, 20 Jul 1954. In late Jul 1954, on a small pothole, a Northern Pintail female had a brood of seven pintails and two Northern Shoveler young, all of the same size (WA).

PNRS: two more nests at Saltcoats in 1958 (BH); west of Canora, 9 eggs, 9 May 1959 (WC); 9 eggs, south of Churchbridge on 13 May 1960

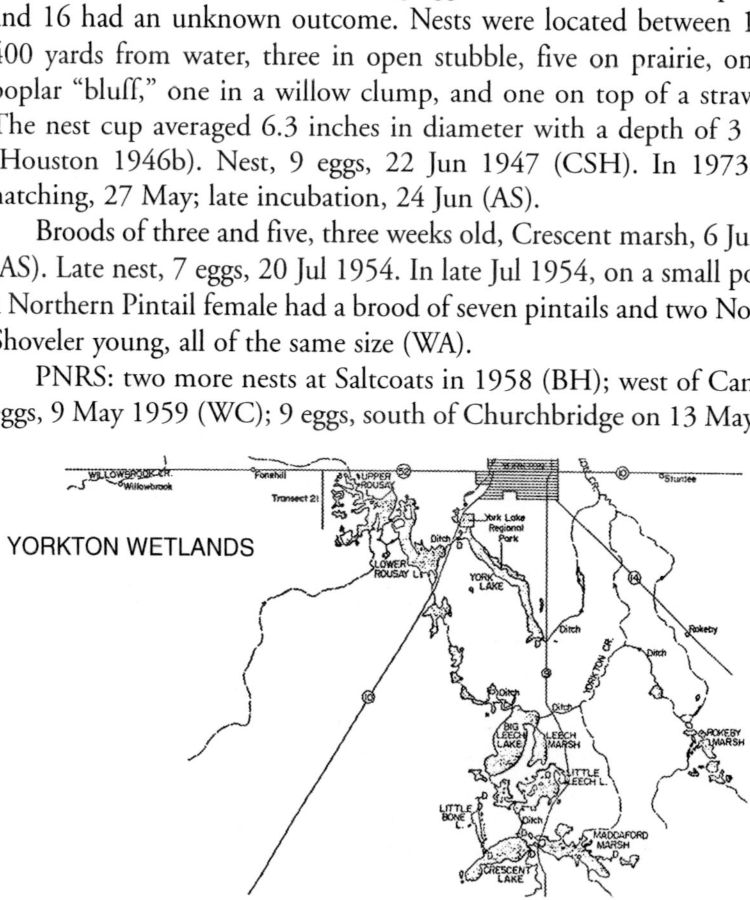

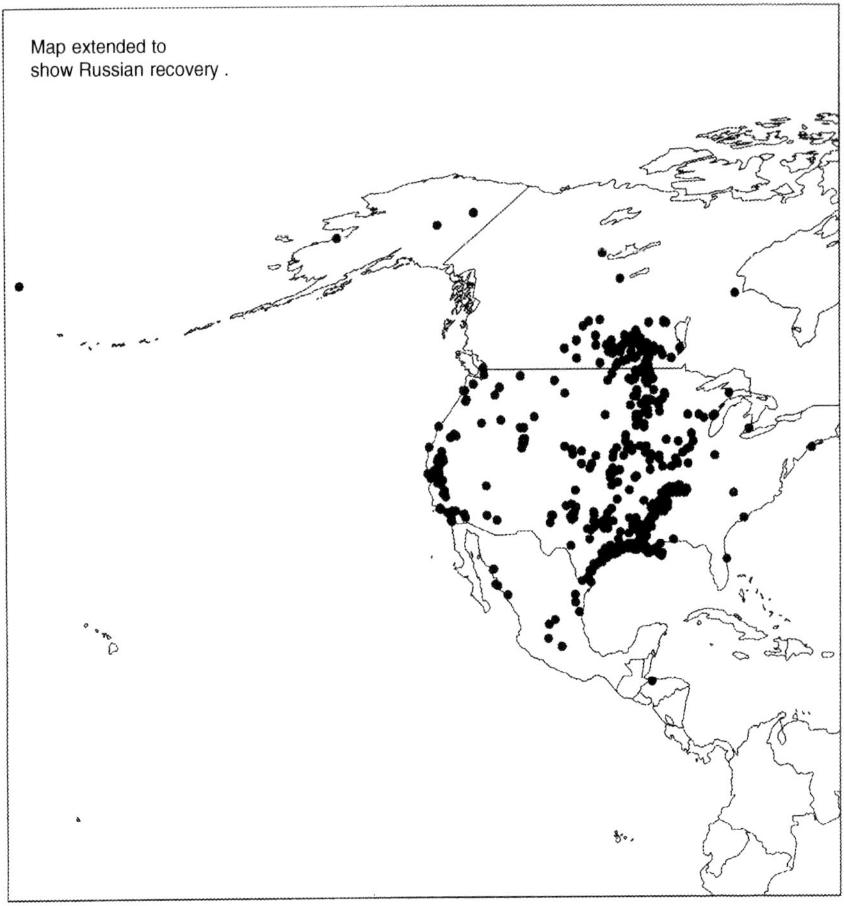

NORTHERN PINTAIL

	103 50	103 40	103 30	103 20	103 10	103 00	102 50	102 40	102 30	102 20	102 10	102 00	101 50	101 40	101 30
51 50	●	●	●	●		●									
51 40	●	●	●	●	●	●			●	●					
51 30	●		●			●			●	●					
51 20			●			●	●		●	●			●		
51 10			●	●	●	●	●	●	●	●	●	●			
51 00								●	●	●	●	●		●	●
50 50							●	●	●	●			●		●
50 40										●					

(John Lane); five nests south of Melville in 1960 (VS); at Stornoway, 1959-1963, when pintails were still common, 11 nests with 5 to 9 eggs were found with seven nests in 1960 alone (SZ); 7 eggs, Stornoway on 7 May 1962 (RD); seven nests at GSL (WA). Flightless young banded in 48 of the 108 10-minute blocks.

Also breeds at Kelliher, Sheho (ARS, *Atlas*).

BBS: overall mean, 2.9. Theodore, 17 years, including 10 of first 12 years; highest, 17, 1970, mean 2.0.

Banded 12,097. An adult female, banded 3 Sep 1945, was found 5 May 1946 at St Michael, Alaska. There is a tendency for pintails to be displaced north and west into Siberia during drought years such as 1959 and 1968 (Henny 1973).

Recoveries, banded in area, 883 [7.3%] (Map; SK 125, AB 6, BC 1, MB 21, AK 3, AL 1, AR 44, AZ 2, CA 124, CO 5, FL 2, IA 3, ID 4, IL 8, IN 1, KS 17, KY 1, LA 174, MD 1, MI 2, MN 7, MO 15, MS 10, MT 4, NC 1, ND 28, NE 11, NM 4, NV 1, NY 1, OK 6, OR 3, SC 1, SD 27, TN 7, TX 137, UT 13, WA 8, WI 5, WY 1, Honduras 1, Mexico 41, Russia 1).

Recoveries, banded outside area, 130 (SK 23, AB 12, MB 10, NWT 5, AK 2, AL 1, CA 23, CO 4, FL 2, ID 1, KS 4, LA 19, MN 3, MO 3, MT 1, ND 4, NM 1, OK 5, OR 1, TX 3, UT 3).

Langenburg: Fairly common, 1990 (TH,BL).

DUCK MOUNTAIN: SMNH 1926. Along Highway 57 between park gate and Kamsack (TH). Uncommon, one pair on Madge Lake on 4 Jun; a nest with 5 eggs near Midway subdivision, 18 Jun 1981 (Hatch and Weidl).

GREEN-WINGED TEAL *Anas crecca*

Fairly common summer resident. Breeds.

In immense flocks on the Assiniboine in September 1881 (Macoun in Seton 1891).

In 1973 there were 4.7 pairs per mile of shoreline at Leech marsh (AS).

ESA(45): 2 Apr 1988, Av 19 Apr. **LFD**(26): 31 Oct 1954, Av 10 Oct. Largest reported group was 500 south of Yorkton on 6 Oct 1993 (DH).

Breeding: Early nest, 4 eggs, 13 May 1946, 50 yards from a transient shallow slough and 2 miles from permanent water, had 1 egg laid each day until 19 May, when the nest was destroyed by a Franklin's Ground

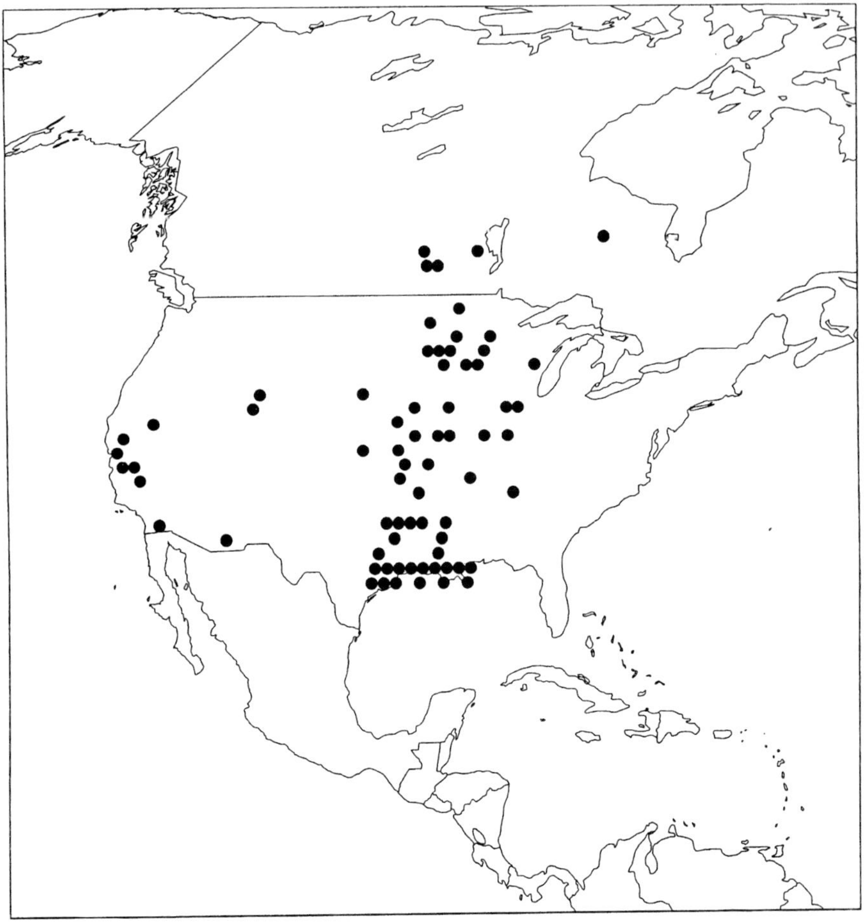

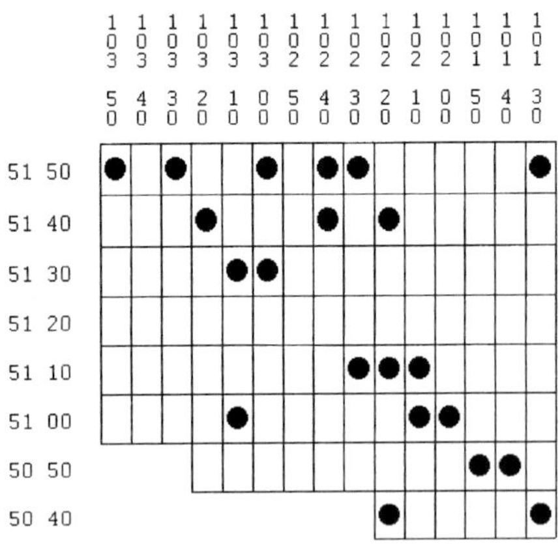

GREEN-WINGED TEAL

Squirrel. Broods of seven and two young, two to three weeks old, 23 Jul 1946 (Houston 1946b). In 1973, one brood of seven three-week-old ducklings on Crescent marsh, 12 Aug, was the only confirmed breeding (AS). Female with eight young, 11 Aug 1985 (WA). Flightless young banded in 21 of the 108 10-minute blocks.

Also breeds at Ituna and Wadena (ARS, *Atlas*).

BBS: overall mean, 0.8. Theodore, 10 years, including 7 of first 12 years; highest, 6, 1977, mean 0.6.

CBC: 1 count, 1 bird. Yorkton 1988.

Banded 2,408.

Recoveries, banded in area, 116 [4.8%] (Map; SK 5, MB 1, ON 1, AL 1, AR 3, AZ 1, CA 13, IA 1, IL 4, KS 7, LA 32, MN 6, MO 2, MS 1, ND 2, NE 3, OK 4, SD 6, TX 17, UT 3, WI 1, Mexico 2).

Recoveries, banded outside area, 4 (SK 2, ND 1, NV 1).

Langenburg: Two pairs, 26 Apr 1976 (DH). Least common dabbling duck, but three males together at one location, 1990 (TH,BL).

DUCK MOUNTAIN: SMNH 1951. Seen in 1976 (JAS). Fairly common, regular on Bear Head Lake and on sloughs beside Highway 57; one of latter had six adults and one duckling on 11 Aug 1981 (Hatch and Weidl).

CANVASBACK *Aythya valisineria*

Common summer resident on larger sloughs and lakes, with some decline in population since the 1940s. Breeds.

It was rare in the Rothbury district, 10 miles south of Wroxton, in the 1890s (Rooke 1947). One of the first reported instances of parasitism of one duck on another was the nest on 20 May 1897 at Crescent Lake containing 9 Canvasback eggs and 7 Redhead eggs (F. Baines in MacFarlane 1908). Seven and 8 eggs, Crescent Lake, 26 May 1898; 7 and 8 eggs, Saltcoats marsh, 2 Jun 1902, W. Raine, WFVZ.

During 1973, at Upper Rousay, Lower Rousay, York, Big Leech, and Little Bone lakes, respectively, there were 3.3, 2.3, 4.7, 4.1, and 4.8 pairs per mile of shoreline.

ESA(47): 9 Apr 2000, Av 22 Apr. **LFD**(25): 27 Nov 1949 (CSH) and 1999 (WA), Av 11 Oct.

Breeding: Three early nests, with 9, 10 and 11 eggs, 24 May 1946. The number of eggs varied from 8 to 17, with an average of 12.7 Canvasback and 0.1 Redhead eggs. Early hatching, nest with 15 eggs, 1

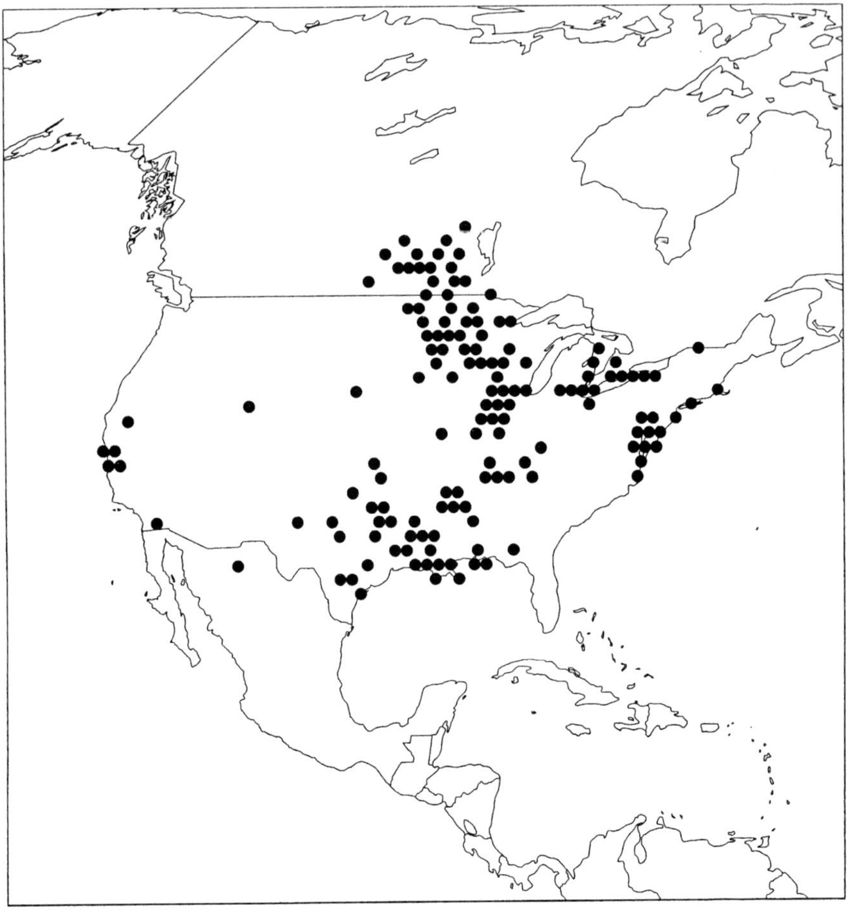

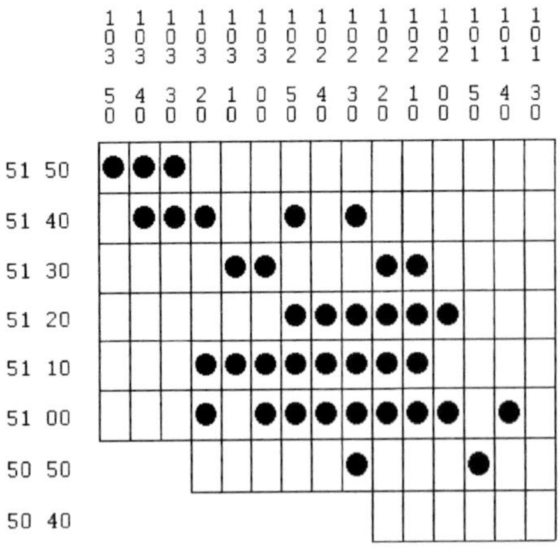
103 50
103 40
103 30
103 20
103 10
103 00
102 50
102 40
102 30
102 20
102 10
102 00
101 50
101 40
101 30
51 50
51 40
51 30
51 20
51 10
51 00
50 50
50 40

CANVASBACK

Jun 1946, indicates laying began about 24 Apr. The latest incubation was observed on 1 Jul. The average nest was made of bulrush, lined with grass, in only 4 inches of water (maximum 24 inches), built 8 inches above water level, with an outside diameter of 20 inches and inside diameter of 8 inches (Houston 1946b). In 1973, first nest, 12 eggs, 23 May, with 11 other nests through 21 Jun; first broods, six and five ducklings a few days old, 13 Jun on Crescent marsh (AS). Broods of seven and six at Horseshoe Lake, 11 Jul 1975 (DH). William Niven saw many more broods in 1976 than for some years.

PNRS: at Saltcoats, 8 and 9 eggs, 18 May 1959 (BH); at Melville, 15 and 7 eggs, 24 and 31 May 1960 (VS); at Stornoway, 4, 6, 7, 8, and 14 eggs and seven young 1960-63 (SZ,RD). Flightless young banded in 37 of the 108 10-minute blocks.

Also breeds at Wadena, Sheho, Springside, Rhein, Goodeve, Willowbrook (ARS, *Atlas*).

BBS: overall mean, 2.7. Theodore, 11 years; highest, 21, 1974.

Banded 2,164.

Recoveries, banded in area, 296 [13.7%] (Map; SK 43, MB 17, ON 13, QU 1, AL 3, AR 4, CA 11, DE 2, FL 1, GA 1, IA 11, IL 15, KY 4, LA 23, MA 1, MD 14, MI 9, MN 33, MO 2, MS 1, NC 3, ND 11, NE 1, NJ 3, NM 1, NY 7, OH 4, OK 4, PA 2, SD 5, TN 11, TX 22, UT 1, VA 6, WI 1, Mexico 5).

Recoveries, banded outside area, 33 (MB 4, DE 1, IL 3, IN 1, LA 4, MO 10, NC 1, NY 5, OK 1, OR 1, VA 1, WI 1).

Langenburg: Common but local on larger marshes; 25 on lake 6 miles northwest of Churchbridge, 1990 (TH,BL).

DUCK MOUNTAIN: Brood with eight young, slough 5 miles southwest of Madge Lake, 26 Jun 1948 (CSH). SMNH 1951. Seen in 1974 and 1976 (JAS). Common, 150 on Bear Head Lake, 10 Jul; female with three young, 22 Jul 1981 (Hatch and Weidl). SNHS 1989.

REDHEAD *Aythya americana*

Common summer resident on larger sloughs and lakes. Breeds.

During 1973, counts at Leech and Crescent marshes, respectively, totalled 10.7 and 7.6 pairs per mile of shoreline (AS).

ESA(42): 15 Apr 1963, 1987, and 1991, Av 27 Apr. **LFD**(13): 30 Oct 1965, Av 2 Oct.

Breeding: In 1946: female flushed from recently completed nest

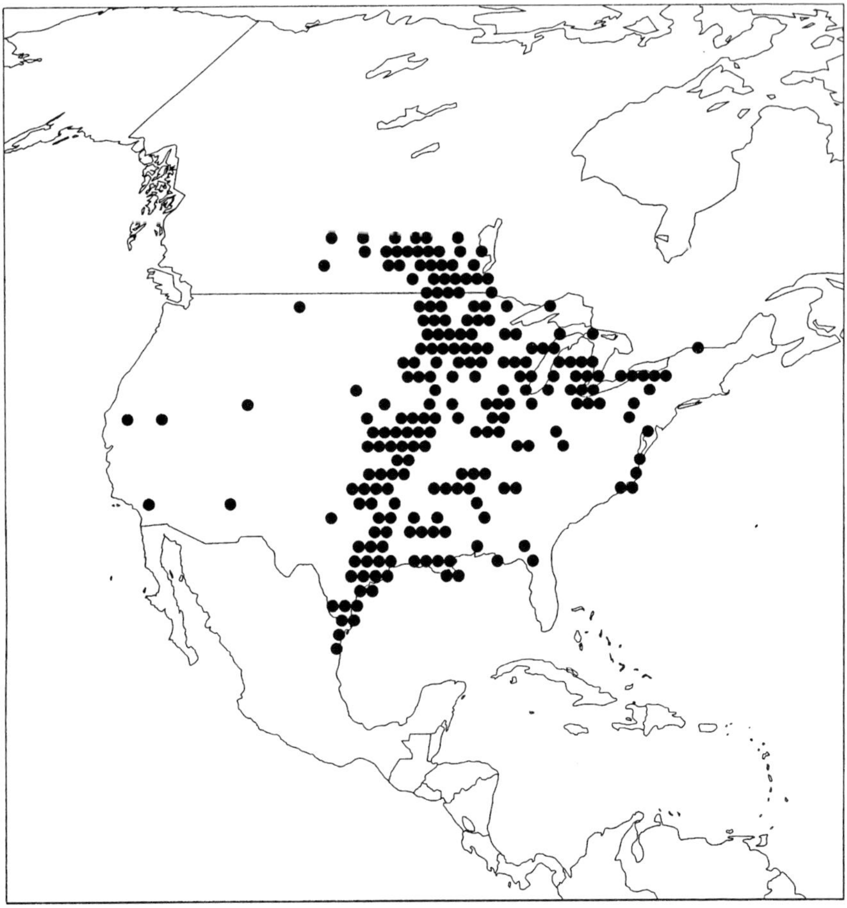

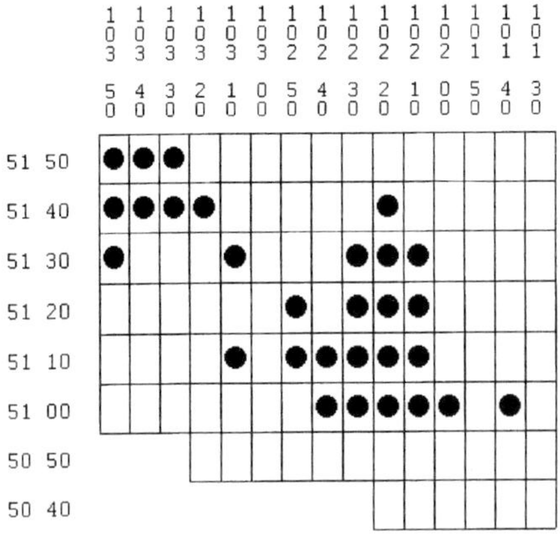
103 50
103 40
103 30
103 20
103 10
103 00
102 50
102 40
102 30
102 20
102 10
102 00
101 50
101 40
101 30
51 50
51 40
51 30
51 20
51 10
51 00
50 50
50 40

REDHEAD

without eggs, 26 May; nest, 11 eggs, 8 Jun; first hatch, nine downy ducklings, 16 Jun; late incubation, three females on 11 Jul. Of 16 nests, 11 had complete clutches with 5 to 15 eggs, and the latter also had 1 Canvasback egg. The average nest was in 10 inches of water, made of bulrushes and grasses, with an outside diameter of 16.9 inches, inside diameter of 7.5 inches, height above water of 8 inches and depth of cup 3.8 inches (Houston 1946b). Early hatching, four young in nest with 3 unhatched eggs on 12 Jun 1960; other nests that month had 8, 9 and 13 eggs (WA).

In 1973, five nests found, the first two on 31 May and the last on 7 Jul; first four broods of eight, two, six, and four young, on Crescent marsh on 6 Jul (AS).

PNRS: at Stornoway, 9 eggs on 25 May 1962 (RD). Flightless young banded in 29 of the 108 10-minute blocks.

Also breeds at Wadena, Rhein (ARS, *Atlas*).

BBS: overall mean, 1.8. Theodore, 9 years; highest, 40, 2000, mean 1.9.

Banded 6,869.

Recoveries, banded in area, 646 [9.4%] (Map; SK 90, AB 3, MB 39, ON 22, QU 1, AL 4, AR 7, AZ 1, CA 2, CO 1, FL 3, IA 10, IL 17, IN 3, KS 22, KY 3, LA 25, MD 17, MI 37, MN 52, MO 6, MS 4, MT 1, NC 8, ND 21, NE 5, NV 1, NY 12, OH 5, OK 21, SD 19, TN 4, TX 117, UT 2, VA 2, WI 15, state not given 3, Mexico 41).

Recoveries, banded outside area, 29 (SK 9, AB 1, MB 2, MD 4, MI 1, MN 1, MT 2, NY 7, TX 1, UT 1).

Langenburg: Locally common on larger sloughs, less common than Canvasback, 1990 (TH,BL).

DUCK MOUNTAIN: SNHS 1956. Common, 500 on Bear Head Lake, 17 Jun 1981 (Hatch and Weidl).

RING-NECKED DUCK *Aythya collaris*

Fairly common summer resident. Breeds.

In 1973, a loose flock of 31 post-breeding or non-breeding males was seen at Upper Rousay Lake on 4 Jul and 16 at Crescent marsh on 6 Jul (AS). A pair and a male at Ravine Reserve, 31 May 1986 (DH).

ESA(42): 8 Apr 1984, Av 27 Apr. **LFD**(8): 7 Nov 1998, Av 11 Oct.

Breeding: Young seen, but no nest found yet. Three broods on 3 Aug 1951. Seven broods in 1973: early brood of five five-day-old young at pothole east of Little Bone Lake on 18 Jul, one on Upper Rousay Lake on 19 Jul, and five near Crescent Lake on 19, 21, and 25 Jul (AS, Houston 1973c); the last three broods also contained Redhead ducklings.

Seventeen flightless young have been banded: eight at the Rousay lakes by R. Brevig, 18 Jul 1956; three at Insinger by R. Brevig on 28 Jul 1956; six south of Kuroki by F.C. Kniffin on 7 Aug 1960.

BBS: overall mean, 0.7. Theodore, 6 years; highest, 12, 1994, mean 0.6.

Banded 109.

Recoveries, banded in area, 14 [12.8%] (SK 1, AR 2, GA 1, KY 1, LA 2, MN 2, NC 1, SC 1, TN 1, TX 1, VA 1).

Recoveries, banded outside area, 1 (AL).

Langenburg: Three drakes seen, including two on the lake 6 miles northwest of Churchbridge, 1990 (TH,BL).

DUCK MOUNTAIN: SMNH 1926: New to museum collection, mounted male, #1873; female, #1874; study skin males, #1889 and #1890. Common here but scarce elsewhere in province; "they seem to prefer the numerous sloughs or small lakes, especially those...where water lilies grew." (Mitchell 1927). Common, SMNH 1951: nest collected, #11768. Seen in 1976, 1977 (JAS). Fairly common, on several small water bodies; a destroyed nest on a beaver dam, 15 Jun 1981 (Hatch and Weidl). SNHS 1989.

GREATER SCAUP *Aythya marila*

Uncommon transient.

Two supposed and presumably misidentified nests: Ten eggs. Size 2.50 x 1.70. Nest in the grass; a depression lined with down and dried grasses, Saltcoats marshes, 15 June 1901. (W. Raine in Reed 1904:76). [The egg width is appropriately larger than Raine's egg of the Lesser Scaup collected five days earlier, yet the identification remains in question-CSH]. Seven eggs, Crescent Lake, 10 Jun 1901, W. Raine, AMNH.

A male was studied, swimming and flying, at GSL, 27 Apr 1980 (RK, Gollop 1980a). A male seen near Preeceville, 24 Jun 1991 (MIC,MIH,CSH). Two females were sighted at GSL on 28 Sep 1997 (RK, Bain and Holder 1997d).

LFD: One on 11 Nov 1994 (WA).

Banded 2: an immature male, flying well, banded by Leonard Walker and Larry Owens at Upper Rousay Lake in 1974 on 7 Aug, and an adult male a week later.

Recoveries, banded outside area, 1 (NY, banded 27 Apr 1958, shot south of Pelly, 1965 hunting season).

LESSER SCAUP *Aythya affinis*

Very common summer resident on larger sloughs and lakes. Breeds.

Nest, 10 eggs, size 2.55 x 1.55, on grass and down at the edge of a lagoon, Saltcoats marshes, 10 Jun 1901 (W. Raine in Reed 1904:76). Thirteen eggs, Crescent Lake, 20 May 1915, "for H. James," WFVZ.

Large flocks stop over during migration. Until the 1920s, rafted in countless thousands on GSL. Seventy years later, on 26 Sep 1997 there were 5,000 on the same lake (WA, Bain and Holder 1997d), and there were 2,000 from 1 to 10 Oct 2002 (WA, Bain 2002b). During 1973, at Upper Rousay, York, Big Leech, Little Leech, and Crescent lakes and Crescent marsh, respectively, there were 3.3, 4.7, 3.4, 6.6, 11.5 and 6.1 pairs per mile of shoreline (AS).

ESA(48): 8 Apr 1991, Av 21 Apr. **LFD**(44): 18 Nov 1975, Av 1 Nov.

Breeding: In 1946: early nest, 11 eggs, 8 Jun; early hatching, 23 Jun; another female still incubating, 26 Jul. Fifteen nests varied between 7 and 13 eggs, average 10.8. Of 96 eggs of known outcome, 75 hatched, 19 were predated, and 2 were infertile. One nest with 10 scaup eggs also had 2 Ruddy Duck eggs. Most of the 21 nests were made of grasses, in an average of 3.2 inches of water, but one was built on a muskrat house and three nests were on dry land, between 3 and 24 yards from water (Houston 1946b).

In 1973, six nests found between 10 and 26 Jun, with first three broods on 17 Jul (AS). Two broods of 10 on Horseshoe Lake, 11 Jul 1975 (DH). Flightless young banded in 19 of the 114 10- minute blocks.

Also breeds at Wadena (ARS, *Atlas*).

BBS: overall mean, 11.1. Theodore, mean 14.0; increasing (1970s, 9.9; 1980s, 10.9; 1990s, 25.4; n=29).

Banded 611.

Recoveries, banded in area, 53 [8.7%] (SK 12, MB 2, AL 1, AR 1, FL 2, IL 2, LA 6, MN 11, MO 1, MS 1, OK 4, SD 1, TN 1, TX 5, UT 1, WI 1, Honduras 1).

Recoveries, banded outside area, 19 (Map; AB 3, AK 1, IL 1, LA 7, MD 1, MI 3, MN 1, OK 1, VA 1). An adult male banded on the Porcupine River near Shuman House in extreme northeastern Alaska on 7 Aug 1960 was shot near Finnie during Sep 1963.

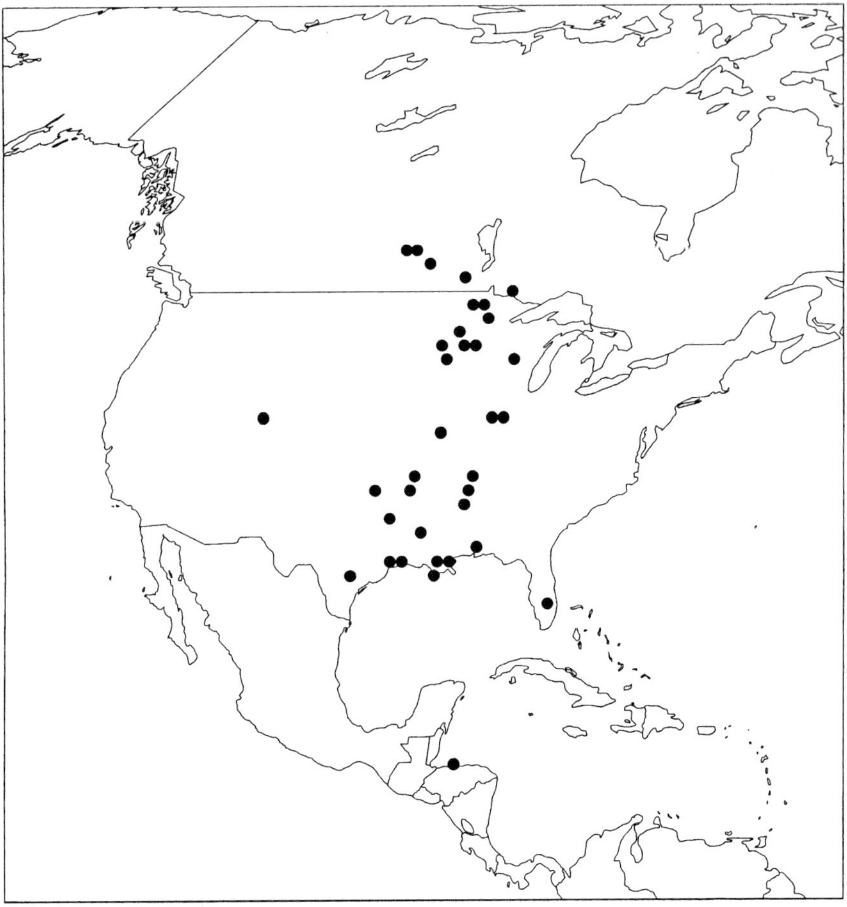

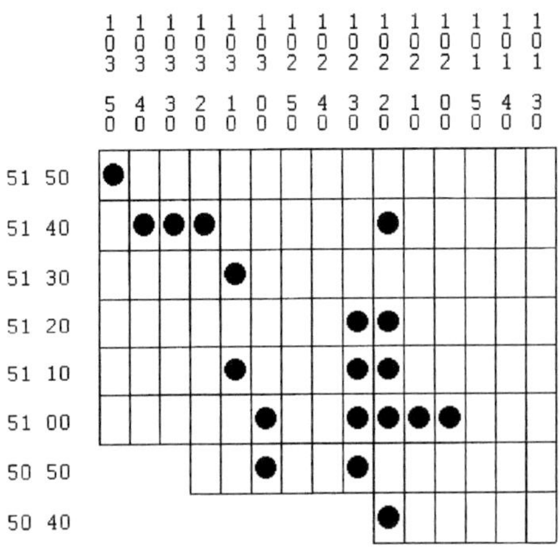

LESSER SCAUP

Langenburg: Local and uncommon on larger sloughs, 1990 (TH,BL).

DUCK MOUNTAIN: SMNH 1926. SMNH 1951. SNHS 1956. Seen in 1976, 1977 (JAS). Common, nests with 2, 4, 5, 6, 9, 8, 10, 11 eggs 1981 (Hatch and Weidl). SNHS 1989.

SURF SCOTER *Melanitta perspicillata*

Vagrant.

Good view obtained at 40 yards, by Ronald and Donald Hooper at York Lake, 26 Oct 1953 (Hooper and Hooper 1953; Gammell et al. 1954a).

WHITE-WINGED SCOTER *Melanitta fusca*

Formerly an uncommon summer resident on some of the deeper bodies of water, now rare. Last known breeding in 1981.

From the 1940s through the 1970s it was regularly seen at York, Crescent, and Anderson lakes, several lakes north of Sheho, and Duck Mountain. Decreased in numbers at Sheho since the 1920s; last sighting of four resident pairs was on 8 May 1976; the last migration date was 30 May 1983 (WN). This drop in numbers is part of a general decline, with virtual extirpation from North Dakota and almost complete elimination from the lower third of Saskatchewan (Kehoe and Alisauskis, unpubl.). The 1974 survey of Foam Lake marsh revealed only one pair present, 28 May and 16 June (AS). Twenty-eight adults counted at Horseshoe Lake, 9 and 10 Jul 1975; some of these may have been breeding (DH).

ESA(22): 8 May 1939 (WN) and 9 May 1954 (WA), Av 25 May. **LFD**(12): 27 Oct 1982, Av 8 Oct.

Breeding: Early brood of four young, 21 Jul 1963; late brood with 10 downy young, 18 Aug 1963, remaining until mid-October (WA). After water levels were restored at York Lake in 1942, three pairs were present in 1944, 10 pairs in 1945 and over 30 pairs in 1946. A nest with 10 eggs, on the ground at the base of a clump of willows, on 3 Jul 1946, was almost exactly half a mile from the nearest water (York Lake), a most unusual site for this slow-moving, heavy, clumsy duck which prefers to nest on islands, fairly close to water.

Broods seen on Salt Lake in the early years and larger sloughs north of Sheho as late as 1975 and 1976 (WN). In 1973, a brood of seven on

Crescent marsh on 17 Jul (AS).

In 1960, F.C. Kniffin banded 15 flightless young, eight of them near Saltcoats in early August, and seven (with one adult female) north of Sheho in mid-August.

BBS: overall mean, 0.06. Theodore, 1 year, 2 in 1979.

Banded 16, all by FC Kniffin.

DUCK MOUNTAIN: SMNH 1926. SMNH 1951. SNHS 1956. Uncommon; three pairs on Green Lake; a gang brood of 21 ducklings on 12 Aug 1981 (Weidl 1983). Six on 29 May 1986 (JFR et al.). At Green Lake, 12 on 22 May 1987, and eight each on 21 May 1988 and 3 Jul 1990 (TH).

BLACK SCOTER *Melanitta nigra*

Vagrant.

DUCK MOUNTAIN: One sighted 8 Aug 1990 (RK, Koes and Taylor 1991a).

LONG-TAILED DUCK *Clangula hyemalis*

Vagrant.

A pair seen at Ketchen on 4 May 1986 (Dave Weiman). An immature male at Little Quill Lake, 21 Jun 1999 (B. Korol, D. Sawatsky, Bain and Shanahan 1999d).

BUFFLEHEAD *Bucephala albeola*

Common transient, fairly common summer resident, common at Duck Mountain. Breeds.

Breeding in Duck Mountains, 1884 (Seton 1891).

The legendary "rain of ducks," otherwise known as the "Bufflehead crash" of between 400 and 500 birds near Foam Lake, in heavy fog between 9:30 p.m. and 2 a.m. on the night of 4-5 Nov 1940 was reported by H.S. Swallow (1941): "Some had the unpleasant experience of receiving a Butterball duck right on the head" and discussed in greater detail by Finley (2000 and 2001). Loose flock of 200 at GSL on 28 Sep 1998 (WA, Bain and Shanahan 1998c).

ESA(46): 15 Apr 1987, Av 25 Apr. **LFD**(45): 19 Nov 1983, Av 24 Oct.

Breeding: Buffleheads elsewhere nest in flicker holes; although such a natural nest has not yet been identified in our area, it could be the source of a brood of seven on Upper Rousay Lake, 19 Jul 1973 (AS); a brood of four at Horseshoe Lake, 12 Jul 1975 (DH); and broods of three, near Jedburgh, and of five and six near Willowbrook, all on 13 Jul 2002 (AS).

Nine nest boxes were erected by William Anaka in 1987 and 1989 at Horseshoe Lake and at a 50-acre slough between it and GSL. On 3 Jun 1990 a female Bufflehead was on 7 eggs plus a Common Goldeneye egg. On 4 and 11 June 1991, a female Bufflehead flushed from a box containing 1 egg plus 9 goldeneye eggs; a second female was on 5 eggs plus 10 Common Goldeneye eggs; and a third Bufflehead nest had 12 eggs, abandoned. A female with 11 downy young, 7 Jul 1991 (WA, Koes and Taylor 1991c). [Such double-laying is not parasitism in the usual sense, but occurs when there is a shortage of nest cavities. One duck "continues laying, even if another bird lays there during the same period....Either bird may finish laying and start incubating before the other bird is ready to do so" (Erskine 1990).]

Flightless young have been banded north of Foam Lake (three on 27 Jul 1957, W.E. Price); east of Rhein (nine on 17 Jul 1972, W.E. Sanders); and singles at Sturgis about 3 Aug 1956 (R. Brevig) and north of Mikado on 12 Jul 1959 (W.E. Price).

BBS: overall mean, 0.7. Theodore, 6 years; highest, 24, 1996.

Banded 82.

Recoveries, banded in area, 12 [14.6%] (ON 2, MN 4, ND 1, NE 1, NY 1, TX 2, Mexico 1).

Recoveries, banded outside area, 1 (NY).

Langenburg: Three of the largest sloughs had two, three and four birds, respectively, 1990 (TH,BL).

DUCK MOUNTAIN: SMNH 1926. SMNH 1951. SNHS 1956. Seen in 1974, 1976, 1977 (JAS). Fairly common on Madge, Kipp, and Bear Head lakes and small water bodies; female with one young, 21 Jul; six large ducklings, 11 Aug 1981 (Hatch and Weidl). SNHS 1989.

COMMON GOLDENEYE *Bucephala clangula*

Uncommon summer resident at Horseshoe and Anderson lakes, common at Duck Mountain. Breeds.

Ten males were in eclipse plumage, 21 Aug 1991 (BL, RK, Bain and Holder 1997d).

On 13 Jun 1989, Joyce and I joined Stuart and Mary Houston and Marc and Joe Bechard to band a Great Horned Owl nest nearby. We decided to check the duck box, so five of us went into the aspen bluff to the box. Marc shinnied up 8 feet to the box and I held a fish net over the nest hole. The female duck popped out of the hole and into the net. Then the fun began. The duck wormed its way out of a hole in the net and ran along the ground. Four of us, on our knees, tried in turn to catch her but she outran us and took wing. We were disappointed until Joyce called out "Were you looking for a duck?" The duck had flown directly at her; reactivating skills from playing softball at school, she simply reached out and gently caught it (WA).

ESA(45): 14 Mar 1981 at Sheho (WN) and 23 Mar 1997 at GSL (WA), Av 14 Apr.

Breeding: A pair nested in a blind chimney at the James Harris farm, 4 miles south of Yorkton (and half a mile from York Lake), from 1924 through 1927. Nested in barrel modified as a goldeneye nest box at Anderson Lake, Saltcoats, where one female was banded on her nest and 8 eggs on 1 Jun 1958; she also had 8 eggs on 20 May 1959 (BH,CSH).

Brood of four at Horseshoe Lake, 12 Jul 1975 (DH). Later, pairs regularly used nest boxes at Horseshoe Lake; 13 eggs on 29 May 1988, 12 of which hatched; two dump nests of 17 and 20 eggs, 29 May 1988; 8 and 14 eggs, 4 Jun 1989; 9 eggs, 7 Jun 1989 (WA). An adult female was caught at her nest and banded on 13 Jun 1989 (WA,CSH). Female with at least 12 downy young, 15 Jun 1991. Three times mixed clutches of goldeneye and Bufflehead eggs have been encountered in a nest box (WA, Koes and Taylor 1991c).

F.C. Kniffin banded nine flightless young at Leech Lake about 23 Jul 1961 and P. Thorpe banded one flightless young west of Wadena on about 23 Aug 1995.

BBS: overall mean, 0.06. Theodore, 1 year; 2, 1995.

CBC: 1 count, 2 birds. Yorkton 1996.

Banded 21.

Recoveries, banded in area, 4 [19.0%] (SK 1, BC 1, MN 1, NE 1).

Langenburg: A single male on a major slough, 1990 (TH,BL).

DUCK MOUNTAIN: SMNH 1926: The common duck on Madge Lake; study skins of young, #1887, #1888. Broods of six and eight young, 27 Jun 1948 (CSH). SNHS 1956. Seen in 1974, broods with 15 in 1976 and six in 1977 (JAS). Uncommon, birds in June and July in non-breeding plumage; up to 200 on Madge Lake, 5 Nov 1981 (Hatch and Weidl). SNHS 1989.

Aspen count totals: OA 2, YA 1 (TH).

BARROW'S GOLDENEYE *Bucephala islandica*

SPECIAL CONCERN, COSEWIC, NOVEMBER 2000

Vagrant.

An immature at Little Quill Lake, 21 Jun 1999, was a most unusual find (B. Korol, D. Sawatsky in Bain and Shanahan 1999d).

HOODED MERGANSER *Lophodytes cucullatus*

Sporadic transient through spring, summer, and fall. No breeding records.

An adult male shot at Crescent Lake, mid-October 1944, by John Smith. One shot at Upper Rousay Lake, 18 Sep 1951, by R.M. Baldwin. An adult female seen north of York Lake on 24 Jun 1952 (Houston 1959b). In 1958, a male and a female were shot by Bert Moore at Leech Lake, 20 Oct, and a female was shot nearby by C.J. Houston at Crescent marsh on 25 Oct. In 1960, one at GSL on 6 Nov (Krause 1961). In 1973, an adult male at Little Bone Lake, 29 May, and three in eclipse plumage on 9 and 18 Jul (AS). One male and two females at Foam Lake marsh on 28 May 1974 (AS). Sighted on Assiniboine River near Tadmore on 3 May 2003 (DF,DGJ,SAJ).

One female banded at Leech Lake by J.H. Wilson, 12 Jul 1945 (Priestly 1945c). Immature males were banded at Leech Lake on 28 Aug 1984 by A. Novara, and two at the Rousay lakes, 18 Aug 1996 by J.F. Voelzer.

ESA(3): 20 Apr 1995. **LFD**(14): 18 Nov 1983, 11 present (WA,JA, Harris 1984), Av 29 Oct.

Banded 4.

Recoveries, banded in area 1 [25.0%] (AR 1 [banded by A.N. Novara at Leech Lake on 26 Aug 1984, shot near Humphrey, AR, on 27 Dec 1989]).

DUCK MOUNTAIN: One at Little Boggy Creek near ski hill, 18 Jun 1988, and one male at small boggy pond on road to Batka Lake, 20 May 1989 (TH).

COMMON MERGANSER *Mergus merganser*

Uncommon transient in spring and fall. Uncommon summer resident at Duck Mountain. Breeds.

A group of about 35 on CNR dam at Melville, 29 Apr 1945, in company with over 1,000 Lesser Scaups, 250 Canvasbacks, 25 Redheads and 25 Buffleheads, the largest aggregation of diving ducks he had ever seen (LTMcK).

ESA(38): 1 Apr 2000, Av 19 Apr. **LFD**(14): 16 Nov 1983, Av 1 Nov.

Summer: one or two males, each seen 10 days through 20 Jul 1973, south of Yorkton (AS). One seen at SNHS summer meet, GSL, 11-12 Jun 1977 (RLS).

Breeding: Flightless young at Horseshoe Lake, 19 Jul 1991 (JFR, Lawrence Beckie, Koes and Taylor 1991c), and Crystal Lake (13 young) in August 2003 (Peter and Fran Buryk).

CBC: 1 count, 1 bird. Yorkton 1997.

Banded 1 (J.H. Wilson at Leech Lake, July 1944).

DUCK MOUNTAIN: SMNH 1926. SMNH 1951. SNHS 1956. Brood of 15 in 1976 (JAS). Uncommon, two females on island, 16 Jul, and one female at Little Boggy Creek, 17 Jul 1981 (Hatch and Weidl). One juvenile seen 25 Aug 1989 on secluded part of Madge Lake (TH).

RED-BREASTED MERGANSER *Mergus serrator*

Rare transient.

An adult male was shot by Olie Sherwin on the Little Whitesand River near Yorkton, about 1933; the mounted specimen was identified later by CSH and IMP.

Pair at York Lake, 25 Apr 1954 (CSH). In 1973, sightings of two males and two females on 1 May and two males on 23 May (AS). One male along ice edge at Newburn Lake, north of Sheho, on 20 Apr 1976 (DH).

ESA(18): 17 Apr 1993, Av 28 Apr. **LSD**: 16 Jun 1995. Two fall dates: 24 Oct 1987 and 29 Oct 1990.

CBC: 1 count, 1 bird. Yorkton 1998.

Banded 2. F.C. Kniffin banded two adult males, both in 1960: at Saltcoats on 18 Jul, and east of Wishart on 3 Aug.

DUCK MOUNTAIN: SMNH 1926. SMNH 1951, "a typical species." SNHS 1956. Seen in 1974, 1975 (JAS). SNHS 1989.

RUDDY DUCK *Oxyura jamaicensis*

Common summer resident on larger sloughs and lakes. Breeds.

Nest, 8 eggs, size 2.40 x 1.75, nest made of aquatic grasses, lined with down. Built in a tuft of rushes in Saltcoats marsh, 6 June 1901 (W. Raine in Reed 1904:82). Nesting, 13 June 1901 (W. Raine in Macoun and Macoun 1909:119). Met with more regularly between Yorkton and Saskatoon (George Atkinson in Macoun and Macoun 1909:119).

A regular breeder at Sheho between 1920s and 1940s, until none was seen in dry years of 1988 and 1989 (WN). Decreased at GSL between 1950 and 1989 (WA). Fifty adults on Horseshoe Lake on 7 Jul 1975 (DH). Two large but loose groups of more than 20 near Crescent Lake, 20 Jun 1989 (CSH,MIH)

ESA(36): 22 Apr 1984 and 1987, Av 9 May. **LFD**(26): 17 Nov 1983, Av 13 Oct.

Breeding: In 1946: early nest, 6 eggs, 19 Jun 1946; early hatching, with a single three-day-old duckling, 26 Jun, but a nest with fresh eggs was found on 18 Jul. Of 15 nests, with 5 to 12 eggs (average 7.6 eggs), six were in bulrushes, seven in a mixture of bulrush and aquatic grasses, and one in aquatic grass alone. Of 61 eggs with known outcome, 8 hatched, 30 were deserted, 19 were predated, 3 rolled out of the nest, and 1 was infertile. The eggs have a recognizable white, chalky surface and are incredibly large for the size of the duck. The average nest was in 9.8 inches of water, made of rushes and grasses, with an outside diameter of 11.5 inches, inside diameter of 7.3 inches, depth of cup of 2.9 inches, and height above water 4.6 inches (Houston 1946b).

Adult pair with 12 downy young, 30 Jul 1954 (WA). In 1973, two broods, each with six five-day-old young, Crescent marsh, 17 Jul (AS). A late brood of downy (group Ib in size) ducklings at York Lake on 23 Aug 1995 (DH). Flightless young banded in only 8 of the 108 10-minute blocks: East of Otthon, the Rousay lakes, Willowbrook, Rhein, Amsterdam, south of Kuroki, Kelliher, and Kylemore.

Also breeds at Kelliher (ARS, *Atlas*).

BBS: overall mean, 3.4. Theodore, 17 years; highest, 50, 1991, mean 3.4.

Banded 70.

Recoveries, banded in area, 4 [5.7%] (SK 1, CA 1, MN 1, Mexico 1).

Langenburg: Uncommon, but one group of six and one of 12, 1990 (TH,BL).

DUCK MOUNTAIN: Highway 57 (TH). Uncommon, pair on small water body on 2 Jun, a few on Green Lake, 10 Jul 1981 (Hatch and Weidl). SNHS 1989.

OSPREY *Pandion haliaetus*

Rare transient and rare summer resident. Breeding near Preeceville since 1999.

Early dates: two records at Sheho, single birds flying over on 10 May 1949 and 2 Jun 1974 (WN). At GSL, seen appreciably more often during migration in the 1980s than in the 1960s. On 5 May 2002, one soared over a large beaver pond and perched on a snag east of Parkerview (AS).

One seen at Preeceville on 19 Jul 1988 by D.F. Hooper and L. Baker. Fall migration dates include 3 Sep 1990 and 10 and 13 Sep 1991 (WA).

ESA(12): 9 Apr 1992, Av 30 Apr. **LFD**(11): 25 Sep 1988, Av 8 Sep.

Breeding: In 1999 and 2000, there were three active nests immediately south of 52°: at Loch Lomond; on the Assiniboine River near Preeceville; and near Sturgis (Pollock 1999) [as well as another three nests immediately north of 52° at Hinchcliffe, Nelson Lake, and Lady Lake]. Active nest on Lillian River between Preeceville and Sturgis, 2002 (Russ and Vicki Peet). Nest built near shore at Crystal Lake in 2002, occupied for first time in 2003, with one young banded by Brenton Terry on 8 Aug (Fran and Peter Buryk).

Langenburg: One flying north, 6 miles south of MacNutt, 26 Apr 1976 (DH).

DUCK MOUNTAIN: SMNH 1951. Rare; an adult with fish in its talons was harrassed by Bonaparte's Gulls, 17 Jun 1981; it may have come from a nest across the border in Manitoba (Hatch and Weidl, Weidl 1983).

BALD EAGLE *Haliaeetus leucocephalus*

Uncommon transient. Rare summer resident. Breeds irregularly.

Individuals were seen half a dozen times at Crescent Lake in the 1910s and 1920s, when fish were plentiful. One eagle brought a fish up onto a telephone pole near the deserted house of Frank Baines' father, and proceeded to devour the fish. John A. Gunn saw one flying south over GSL in late October and on 20 Nov 1945, but had seen them during spring migration two or three times over the previous 58 years.

Uncommon to rare transient at Yorkton and Sheho. Regular migrant at Ketchen, including an immature on 4 Mar 2000, and a mature adult on 18 Mar 2001 (DF). A late spring migrant seen 15 miles southwest of Willowbrook on 25 May 1980 (WH).

The first fall date at Sheho was 31 Oct 1945; another stayed from 19 Oct-1 Nov 1980 (WN). One adult at Horseshoe Lake on 7 Oct 1975 (DH). At GSL, seen appreciably more often during migration in the 1980s than in the 1960s. In 1992, on 9 Nov, five flew south over York Lake; the next day another five were seen at Willowbrook, and the day after that, one flew over the Rousay lakes (WH); one to three were present daily at GSL until the lake froze over (WA). Four adults and 10 young at Crescent Lake on 6 Oct 1993 (DH).

ESA(20): 18 Mar 1985, Av 30 Mar. **LFD**(27): 28 Nov 1988, Av 11 Nov.

Breeding: Conservation Officer Brad Tokaruk observed three large young at the top of an aspen on the south shore of Crescent Lake on 28 Aug 1995; they did not fledge until September. A large nest about 5 feet in diameter and over 3 feet deep was built 35 feet above ground in a 45-foot aspen on the north side of Crescent Lake; it contained two large young on 28 Jun 2001. Pasture manager Cliff Chadney said this was the fifth year of use of that nest; the adults raised only one young in 2000. Two dozen dead suckers along the shore indicated a fish winterkill; Tokaruk observed the eagles feeding on geese and ducks rather than their usual diet of fish (AS).

Brad Tokaruk and Al McCutcheon saw a pair of adults near a large and presumed active nest on the west side of Cutarm Reservoir, which contains walleye and suckers, on 4 Aug 1999 (AS).

Summer: fully plumaged adults at GSL: 16 Jul 1994 and 21 July 1998.

CBC: Rank #45. 5 counts, 11 birds. Highest: 4, Kamsack 2001.

DUCK MOUNTAIN: SMNH 1926. SMNH 1951. SMNH 1956. One adult seen on Roy Island, 15 Jun, and one seen 16 Jul at northwest corner of Madge Lake in 1981 (Hatch and Weidl, Weidl 1983). Subadult over Madge Lake on 21 Jun 1988 (TH). Active nest on shore of northwest bay, 10 Jun 1989 (CSH,MJB); adults were visible that summer with spotting scope; adult fishing at Green Lake, 24 Aug 1989 (TH).

NORTHERN HARRIER *Circus cyaneus*

Common summer resident, but numbers declining. Breeds.

Exceedingly abundant on the marshy prairies from Livingston southward along the Assiniboine, September 1881 (Macoun in Seton 1891). Nest, 5 eggs, photographed at Crescent Lake, 5 June 1901 (W. Raine in Macoun and Macoun 1909:244). Two eggs collected north of Saltcoats (NJ,PMA).

Common throughout area in 1930s and 1940s, and in years when mice were plentiful, as in 1960. On 17 Jul 1946, Farley Mowat and I watched a harrier drop down and seize a young Willet, nearly ready to fly (CSH). At Yorkton, 0.4 nests per 10 km^2 (Sargeant et al. 1993). Numbers at GSL decreased appreciably between 1950 and 1989 (WA).

ESA(47): 18 Mar 1945 (CSH) and 21 Mar 1963, 1977, 1988 (WA), Av 4 Apr. **LFD**(38): 27 Nov 1953, Av 22 Oct.

Breeding: Early nest, 3 eggs, 9 May 1959 at Saltcoats; other nests that year had 5 and 4 eggs (BH). In 1956, 4 eggs on 23 May and 5 eggs on 3 Jun; four young fledged (BH).

Nest, 2 eggs, at Upper Rousay Lake on 29 May 1946 was on a platform of bulrushes, 20 inches in diameter, in 18 inches of water. Three or four small twigs were used in the construction above water, and the nest was lined with grass; 1 egg was infertile but the surviving young was banded on Jun 20. A second nest, at Lower Rousay Lake on June 3, was 18 inches in height, built in 2 inches of water but was on dry land when the three young hatched; they were banded Jun 24. Within 50 yards of this nest were eight duck nests and a coot nest, and within 100 yards, another nine waterfowl nests, the heaviest density encountered in that year's waterfowl nesting study (CSH; Shaw 1946a; Belcher 1955). A later harrier nest at Lower Rousay Lake in 1946 contained one young on 20 Jun, three on 24 Jun and five young when they were banded on 11 Jul.

Nest, five young and 1 egg, oldest two large enough to hold a band, Saltcoats, 13 Jun 1959 (CSH). In 1960, a vole year, there were two successful nests at Invermay, two at Kamsack, and one each at Orcadia, Rhein, and Fenwood (CSH). In 1973, nest with 2 eggs near York Lake and nest with 4 eggs near Upper Rousay Lake on 27 Jun, but both nests failed (AS).

At GSL, Anaka banded 22 young in eight nests: 29 Jun 1955, three; 8 Jul 1956, four; 30 Jun 1957, two; 20 Jul 1958, three; 15 Jul 1959, three; 19 Jul 1959, one; 18 Jul 1961, three; 21 Jun 1964, three.

PNRS: at Stornoway in the vole year 1960, 2, 3, 3, 4, and 5 eggs, and a nest with 3 eggs and three young (SZ).

BBS: overall mean, 1.6. Theodore, mean 1.5 (1970s, 1.5; 1980s, 1.4; 1990s, 1.8; n=24).

CBC: 1 count, 1 bird. Melville 1986.

Banded 53 (WA 22 [in eight nests, one nest per year except two nests in 1959], CSH 31) + pre-1955, 33 (CSH 13, Wotherspoon 20).

Recoveries, banded in area, 12 [14.0%] (FL 1, GA 3, LA 2, MO 1, OK 1, TX 4). From 20 nestlings banded at Hyas by the Wotherspoons, there were a remarkable eight recoveries: from two nests of three and two young banded in 1933, all five were shot in the United States: Beaumont and Edna, TX, and New Orleans, LA, from the first nest; Archie, MO, and Wynnewood, OK, from the second nest. The Beaumont recovery warranted mention in Lincoln (Lincoln 1939; Houston 1945). Houston's birds were shot the same year as banded at Martin, GA, and Monroe, LA (Houston 1968a), while Anaka's were both shot in Georgia, at Blakely the same year and near Woodbury the following year.

Langenburg: Only a single male sighted, 1990 (TH,BL).

DUCK MOUNTAIN: SMNH 1926. SMNH 1951. SNHS 1956. Rare, one seen near Little Boggy Creek in June and one along Highway 57 just outside the park on 15 Jul 1981 (Hatch and Weidl).

SHARP-SHINNED HAWK *Accipiter striatus*

Fairly common transient. Probable 1901 breeding record.

Saw a number at Cote's Reserve and shot one at Livingston, 11 September 1881—Macoun in Seton 1891. Three eggs, 35 feet in aspen at Crescent Lake, 3 Jun 1901, W. Raine, Prov. Museum of British Columbia.

In 1973, one seen 12 May at Little Bone Lake (AS).

ESA(30): 27 Mar 1984 (WN), 31 Mar 2003 in Ketchen (DF), and 3 Apr 1997 (WA), Av 18 Apr. **LFD**(26): 14 Oct 1995, Av 29 Sep.

DUCK MOUNTAIN: SMNH 1926. SMNH 1951. SNHS 1956. Seen in 1977 (JAS). Uncommon, one chased by robins at Little Boggy Creek on 16 Jun and one near Bear Head Lake on 15 Jun 1981 (Hatch and Weidl). One soaring above Little Boggy Creek valley, 8 Jun 1989 (TH).

Aspen count total: OA 1 (TH).

COOPER'S HAWK *Accipiter cooperii*

Fairly common summer resident. Breeds.

Secretive and rather inconspicuous in aspen woods. If one finds one of these birds in summer, the nest is likely nearby. The second commonest nesting falconiform at GSL.

A minor flight on 19 Apr 1959 (WA, Krause 1959c). In 1973, one sighting only, at York Lake on 1 May (AS). Seen near Pelly, 24 Jun 1991 (MIC,MIH,CSH).

ESA(45): 31 Mar 1976, Av 19 Apr. **LFD**(30): 23 Oct 1979, Av 21 Sep.

Breeding: Early nest, 4 eggs, 22 May 1960, had two young on 17 Jul (WA). A nest with 3 eggs was found at Leech Lake on 28 May 1945 (JHW) and visited on 8 Jun 1945 by WHC. Another nest 22 feet above the ground at Ravine Reserve had 3 eggs on 10 Jun 1947; two young were banded on 11 Jul (CSH). Maurice G. Street found two nests northwest of Sheho on 26 Jul 1948 (Street 1949). Nest with 2 eggs at York Lake, 30 May 1953, two young banded 12 Jul; feathers from domestic poultry were present at the foot of the tree (Houston 1953b).

In 1956, a nest with 3 eggs at Saltcoats and another with 4 eggs at Bredenbury (BH). In 1958, within 2 miles of Saltcoats, Bill Horseman found three nests, including one 15 feet above ground in a poplar crotch with 1 egg on 28 May, destroyed by crows on 6 Jun. A very late renesting attempt in the same nest had 3 eggs on 17 Jul, and a single young still in the nest on 2, 15 and 21 Aug 1958 (BH,CSH).

Stanley Zazelenchuk had one nest in 1959, two in 1960, and one nest in 1961, all near Stornoway, then a nest with 5 eggs, 30 feet in an aspen near Pelly on 7 Jun 1964. Two nests each year, 1959 and 1960, at Invermay, one nest with 3 eggs and one with 1 egg; one nest was used both years (Ferrie 1960). Four nests within 2 miles of Saltcoats in 1959: 5 eggs on 19 May, 4 eggs on 26 May, 4 eggs on 10 Jun, and 3 eggs on 26

> The above nests through 1958 were reported by Houston (1958), together with nine other nests from elsewhere in Saskatchewan. These nests proved the Cooper's to be one of the commonest hawks, in striking contrast to the fact that no Saskatchewan specimen had been collected through 1924 (Mitchell 1924) and that the AOU Check-List (1957) failed to mention Saskatchewan as being within its breeding range!

Jun (BH). At GSL, nestlings banded yearly, 1955-61, and 1963, 1964, and 1967, a total of 38 young in 13 successful nests, all between 5 and 25 Jul. Most fledge in mid- to late July; only one nest failed (WA).

Nest at Sheho, 1971 (WN, Houston 1971c). Three were banded on 15 Jul 1992 south of Kelliher in a nest found by Ian Lochtie. Three active nests: 7 miles west of Willowbrook, 35 feet in a spruce and only 100 feet from main door of farmhouse, 27 Jun 1987; 18 feet in an aspen west of Goodeve on 14 Apr 1998; in an aspen 13 miles east of Melville, 23 Apr 1998 (AS).

PNRS: 1, 2, 4, 4, and 5 eggs at Stornoway, 1959-63 (SZ).

Also breeds at Duff (ARS, *Atlas*).

BBS: overall mean, 0.2. Theodore, 10 years; only in 1998 were 2 counted, mean 0.3.

CBC: 1 count, 1 bird. Kamsack 1993.

Banded 47 (WA 38, CSH 9) + pre-1955, 5 (CSH, including one that entered a duck trap on dry ground on 30 Aug 1945).

Recoveries, banded in area, 1 [1.9%] (MN 1, found dead nearly four years later, southwest of Sauk Center).

Langenburg: One adult in a coulee leading into Assiniboine valley, 1990 (TH,BL).

DUCK MOUNTAIN: SMNH 1951. Uncommon, one at Little Boggy Creek on 18 Jun and one on Pelly Point Nature Trail on 24 Jun 1981 (Hatch and Weidl). One at Ministik Beach, 23 Aug 1989 (TH).

NORTHERN GOSHAWK *Accipiter gentilis*

Uncommon but unpredictable fall and winter transient. One recent breeding record.

Nest with 2 eggs found at Crescent Lake on 24 May 1890 by Fred Baines. The eggs measured 2.40 x 1.68 and 2.33 x 1.65 inches (Raine 1892). Four eggs sets in WFVZ collected by the Baines brothers from Crescent Lake were critically examined by Lloyd Kiff and CSH; eggs in two sets were unequivocally too large for Cooper's Hawk eggs: 3 eggs, 25 feet in scrub oak on 29 May 1893; 2 eggs in cottonwood, 21 May 1899. Sets of 3 and 2 eggs in poplars on 4 and 7 Jun 1901, W. Raine, WFVZ.

Common during winters of 1942-43 and 1943-44 (CSH). Large movement in 1972, 12 seen during fall and nine during winter (WA). East of Parkerview, territorial adults noted on 9 Apr 1998 and May 1999, but no nest was located (AS).

EFA(36): 30 Aug 1992, a fresh carcass predated by a Great Horned Owl, Av 19 Oct. **LSD**(23): 12 May 1983, Av 6 Apr.

Breeding: With settlement, the goshawk retreated northward, only to reappear as a nesting species in the 1990s. In 1992 a nest was found by Bev McLaren, 1 mile north of Highway 15, 11 miles west of Bredenbury, with identification confirmed by good video footage obtained when the young were out of the nest and learning to fly on 11 and 12 Jul (Houston 1998).

CBC: Rank #32. 22 counts, 30 birds. Highest: 5, Yorkton 1943. One reported on Yorkton CBC, 26 Dec 1954 (Gammell et al. 1955).

Banded 1 (WA), an adult was caught by one toe in a mink trap and banded at GSL in January 1957.

Recoveries, banded outside area, 4 (AB 2, MN 2). An adult female banded east of Clyde, AB, during the nesting season by Ray Cromie on 15 Jun 1996, was found on skeletal remains north of Marchwell on 6 May 2002.

DUCK MOUNTAIN: SMNH 1951. SNHS 1956. Rare; a female flew across the road near Bear Head Lake on 1 Jun 1981 (Hatch and Weidl). An immature in Little Boggy Creek valley, 18 Jun 1988 (TH). One flew over the road to the ski hill, 14-20 Jun 1993, distant from any survey stop (TH).

[RED–SHOULDERED HAWK *Buteo lineatus*

HYPOTHETICAL

Vagrant.

One possible sighting along Highway 35, 5 miles north of Bankend, on 14 Sep 1990 (Hooper 1993).]

BROAD-WINGED HAWK *Buteo platypterus*

Uncommon summer resident. Breeds.

Breeds in dense aspen forest, especially near lakes. Present on territory near study plots at GSL on 1 Jun 1969 (Flack 1976). In 1973, one sighted in migration on 1 May (AS). A pair on territory with one melanistic adult at the time of the SNHS summer meet, GSL, 11-12 Jun 1977 (WA,FAS). A territorial pair present east of Ketchen at Loch Lomond in 2001 and 2002 (DF,DGJ,SAJ).

ESA(38): 25 Apr 1985, 1988, and 1989, Av 7 May. **LFD**(24): 24 Sep 1980, Av 5 Sep.

Breeding: Early nest had 2 eggs on 22 May 1956, 4 eggs on 29 May, eggs hatching 22 Jun, three young banded 24 Jul (WA). Nest, 2 Jun 1957, but only one nestling was banded. Other nests fledged two young in 1959, three in 1960, two in 1962 (Nero 1962c), two in 1963, and three in 1970. In 1971, the first year after Anaka gave up his banding permit, there were two nests (WA, Houston 1971c). East of Parkerview in 1997, adults were observed building a nest 25 feet above ground in an aspen on 23 May; three young fledged in July (AS).

BBS: Theodore, 3 years (one each in 1977, 1981, 1985), mean 0.1.

Banded, 13 (in six nests, 1957-1970, WA).

DUCK MOUNTAIN: Seen in 1976, 1977 (JAS). Common and regular through aspen stands, the most frequently encountered hawk in the park in summer; two adults with an immature on Red Squirrel Nature Trail on 16 Jul 1981 (Hatch and Weidl). Seen regularly along road to Batka Lake; one on Green Lake road, 21 May 1988; pair seen on Pelly Point Trail, 17 Aug 1990 (TH). Incubating in crotch of White Birch south of golf course on 19 Jun 1994 (ARS).

Point counts: 1 individual at 1 stop (ARS).

Aspen count totals: YA 1, MA 1.

SWAINSON'S HAWK *Buteo swainsoni*

Fairly common summer resident, decreasing. Breeds.

Whitesand River, 4 May 1884 (R. Miller Christy). Set of 2 eggs taken at Crescent Lake, 29 May 1890, "are very handsome. The ground colour is clear bluish-white; one of the eggs has heavy blotches of rich chestnut brown at the butt end, and the other has the smaller end capped with the same heavy colouring. These 2 eggs measure 2.25 x 1.73, and 2.15 x 1.63 inches" (Raine 1892:187). Three eggs, Crescent Lake, 24 May 1898, USNM. Between 5 and 13 June 1901 at Crescent Lake, Raine examined "close upon 30 nests" (Raine in Macoun and Macoun 1909:258; this included 3 eggs on 7 Jun 1901 (ROMZ # 8005). Three Raine sets from Crescent Lake, on 2, 5, and 8 Jun 1901, with 4, 3, and 3 eggs, in WFVZ are mislabelled as Rough-legged Hawk eggs, as is a Baines set from 9 May 1898.

This species has decreased noticeably since the 1930s, and particularly since the 1970s, as the area of pasture has decreased. Paradoxically, its numbers have decreased at Sheho even as more bush was cleared, but coincident with a drop in the numbers of its chief prey, the

Richardson's Ground Squirrel, in the early 1940s (Niven 1973). It showed a further decline in the late 1970s, becoming "scarce" by 1979 and last seen locally in 1987 (WN). Not yet noted nesting in cropland in this area, as it does farther west. A tour of both Rousay lakes and York and Leech lakes on 13 Jun 1989 resulted in three sightings (and seven Redtail sightings; CSH,MJB). One seen east of Sturgis, 24 Jun 1991 (MIC,MIH,CSH).

ESA(25): 7 Apr 1952 and 10 Apr 1953 at Saltcoats (RB), 12 Apr 1950 at Sheho (WN), and 15 Apr 1976 at GSL (WA), Av 6 May. **LFD**(19): 25 Sep 1983, Av 5 Sep.

Breeding: Early nest, 1 egg on 22 May 1958 had three young on 13 Jul; no breeding records at GSL since 1958 (WA). Late nest, still with young, Leech Lake, 5 Aug 1944 (CSH).

Nest with two downy young, Lower Rousay Lake, 20 Jul 1945; there was an active House Sparrow nest at the base of this nest. Nest, 3 fresh eggs, Lower Rousay Lake, 25 May 1946, had three downy young on 6 Jul. Nest, two young, 10 Jul 1953. In 1973, several territorial pairs seen but the only nest found failed by 5 Jul (AS).

Oology: two eggs near Canora, 1965-66 (H. Cymbalisty, SMNH).

BBS: overall mean, 0.5. Theodore, 6 years, including 2 in 2000, mean 0.2.

Banded 4 (WA) + pre-1955, 12 (CSH).

Recoveries, banded in area, 3 [18.8%] (AL 1, WI 1, TX 1). One banded at Lower Rousay Lake on 6 Jul 1946 was shot at Selma, AL, in October, the first record of the Swainson's Hawk for that state. On this basis, the species was included in *Birds of Alabama* (Imhof 1962). Since then, it has become appreciated that there is a small wintering population, perhaps 300 Swainson's Hawks, in Florida that fails to take the usual 7,200-mile route to Argentina. The second was banded at Bredenbury, 10 Jul 1953, and caught in a trap at Whitehall, WI, in October that year (Houston 1968b). Anaka's recovery was retrieved the next 19 Apr about 15 miles south of San Antonio, TX.

Langenburg: One presumed territorial pair 1 mile north of Langenburg, 26 Apr 1976 (DH). Two singles seen separately along southern edge of area, 1990 (TH,BL).

DUCK MOUNTAIN: SMNH 1951. Rare in open areas, one seen along North Boundary Trail on 11 Jul 1981 (Hatch and Weidl).

RED-TAILED HAWK *Buteo jamaicensis*

Common summer resident. Breeds.

Shot one fine specimen at Livingston, 13 Sept 1881 (Macoun in Seton 1891). "An abundant species in the Duck Mountain wooded region, observed breeding....June 18 shot a Red-tailed Hawk, female, length 20, extension 48...this species seems to be the complement of Swainson's Hawk...adhering to the forest regions. Each species is adapted to a set of conditions slightly different from those of its nearest congeners" (Seton 1891). One egg collected north of Saltcoats (NJ,PMA).

Increased at Sheho and around the lakes south of Yorkton since the 1930s, in part replacing the decreasing Swainson's Hawk. At Yorkton, 1.5 pairs per 10 km^2 (Sargeant et al. 1993).

On 10 and 11 Apr 1954, mixed, loose flocks of at least five hawk species flew over GSL; one flock contained 58 hawks and another, 47. Red-tailed Hawks predominated, with Rough-legs second commonest. In five days, 600 to 800 hawks passed over (Gunn 1954b, Anaka 1954).

On 10 Apr 1974, 24 Redtails and one Roughleg passed over Yorkton between 9:30 and 10:30 a.m. (Shaw 1949). In 1974, migration peaked from 18 through 21 Apr, with up to 20 seen daily (WA, Houston and Shadick 1974).

ESA(47): 19 Mar 1945 at Saltcoats (RB), 20 Mar 1945 at Sheho (WN), and 21 Mar 1981 at GSL, Av 6 Apr. **LFD**(37): 3 Nov 1981, Av 10 Oct.

Breeding: Early nest, 1 egg on 20 Apr 1958, was frozen on 24 Apr and deserted (WA). Another early nest had three fledglings, two-thirds grown, on 8 Jun 1946 (CSH).

A nest at Leech Lake contained four Richardson's Ground Squirrels and one Sharp-tailed Grouse as food for the young ready to fledge on 14 Jul 1945; there was an active House Sparrow nest in the base of the hawk nest (CSH). Near Saltcoats, eight banded in five nests in 1958 and 12 in seven nests in Jun 1959 (BH,CSH); others were banded at Leross, Kelliher, Willowbrook, Otthon, Rhein, and Gorlitz (CSH).

At GSL, nine young were banded in six nests in 1959 (WA, Krause 1959d), while another 35 nests were located in other years, including a nest with 2 eggs on 29 May and two young on 20 Jun 1960 (WA).

In 1969 and 1970, Lorne Scott banded young in successful nests at West Bend, Duff, and Melville. In 1973, 11 nests found, with nine young raised in five climbable nests; some young were flying by 4 Jul. One pair had one normally coloured bird mated with an extremely light (Krider's) individual (AS).

Oology: nest, 2 eggs, near Canora in 1965-66 (H. Cymbalisty, SMNH).

PNRS: Stornoway, three nests with 1 egg, seven with 2, two with 3 eggs, 1959-63 (SZ); 1 egg near Buchanan, 6 May 1959 (WC); two young near Kamsack, 20 Jul 1961 (RD); three young at Mikado, 22 May 1966 (EK).

Also breeds at Wadena, Preeceville, and Togo (ARS, *Atlas*).

BBS: overall mean, 4.3. Theodore, mean 4.5; increasing (1970s, 2.9; 1980s, 5.2; 1990s, 5.4; n=31).

CBC: 1 count, 1 bird. GSL 1981.

Winter: A single bird at GSL on 26 Dec 1981 (WA,JA, Gollop 1982).

Banded 153 (CSH 85, WA 49, RLS 19) + pre-1955, 14 (Blakeley 3, Wickencamp 2, Wotherspoon 2, CSH 7).

Recoveries, banded in area, 12 [7.2%] (SK 2, MB 1, AL 1, AR 1, LA 2, MS 2, TX 3). Blakeley's two and Mrs Emma Wickencamp's one were shot in Texas (Houston 1967a).

Recoveries, banded outside area, 3 (SK 2, MO 1).

Langenburg: Four breeding pairs in coulees, two leading into Assiniboine, and two into Cutarm, plus one pale *kriderii* subspecies, 1990 (TH,BL).

DUCK MOUNTAIN: SMNH 1951. SNHS 1956. Seen in 1975 (JAS). Fairly common, soaring overhead at Bear Head, Kipp, and Batka lakes and along Highway 57, 1981 (Hatch and Weidl). SNHS 1989.

Aspen count totals: OA 2.

FERRUGINOUS HAWK *Buteo regalis*

SPECIAL CONCERN, COSEWIC, APRIL 1995

A summer resident in open grasslands maintained by prairie fires until after 1900. A rare breeder until the 1930s. Now a rare wanderer.

Four fresh eggs collected by Frank Baines at Crescent Lake, 7 May 1898, are #1357 in the Richter collection, University of Wisconsin-Green Bay (T.C. Erdman). Walter Raine found a dozen nests in the Saltcoats-Crescent Lake area between 5 and 13 June 1901 (Raine in Macoun and Macoun 1909:265). Regular and common from west of Yorkton to Edmonton in 1906 (Atkinson in Macoun and Macoun 1909:264). It was rare in the Rothbury district, 10 miles south of Wroxton, in the 1890s, but a few more were seen in the 1920s (Rooke 1947).

One sighted at Hyas in the spring of 1979 (CP). An adult was seen in potential grassland habitat near Shellmouth, Manitoba, on 11 May 1989 (Ken De Smet, *fide* R. Wapple 1999). Seen flying north of Saltcoats, 9 Apr 2003 (JRJ).

Two fall dates: 13 Sep 1942 (IMP) and one perched on a post in Garry PFRA Pasture north of Homefield on 8 Aug 1992 (AS).

ESA: at Upper Rousay Lake: 24 Apr 1962 (CSH,MIH), 14 May 1943 (CSH), 13 May 1992 (AS).

Breeding: Nested at the Louis Kirk farm, 8 miles north of Neudorf, until the 1930s, but disappeared as a nesting species by 1940.

ROUGH-LEGGED HAWK *Buteo lagopus*

Fairly common transient in early spring and late fall.

ESA(19): 16 Mar 1982, Av 7 Apr. **LSD**(6): 27 May 1980 south of Yorkton (DH) and 2 May 1991 at GSL (WA). **EFA**(26): 28 Sep 1998, Av 19 Oct. **LFD**(18): 20 Dec 1981, Av 12 Nov.

The supposed Rough-legged Hawk nest at Crescent Lake, on 3 May 1890, was evidently that of a Red-tailed Hawk (Bechard and Houston 1984).

CBC: 2 counts, 2 birds. GSL 1981 and Preeceville 1998.

Winter: 20 Dec 1981; 11 and 19 Dec 1987 (WA).

DUCK MOUNTAIN: Rare spring and fall transient; one seen at Little Boggy Creek, 10 Sep 1981 (Hatch and Weidl).

GOLDEN EAGLE *Aquila chrysaetos*

Rare wanderer.

Frank Baines shot three specimens in the 1910s, including one with a 7-ft. 2-in. wing span. L.T. McKim reported that two had been shot at Melville in the 1930s. J.R. Foreman noted three flying over the west edge of Yorkton on 6 Nov 1938, and one on 6 Apr 1943, frightened by the approach of an airplane.

Spring: at Saltcoats, sightings of one on 1 Mar 1949, 29 Mar 1943, and 14 Apr 1947 (RB); at GSL, 19 Apr 1968, 24 Apr 1994, and 1 May 1981 (WA); at Crescent marsh, William McInnes watched one catch a duck about 3 May 1973 (AS).

EFA(7): 5 Oct 1977, Av 22 Oct. **LSD:** 1 May 1981 (WA).

Fall: at GSL, 5 Nov 1966 (WA, Hatch 1967a); at Sheho, 19 Nov 1974 (WN).

CBC: Rank #48. 7 counts, 10 birds. Highest: 3, Kamsack 1993. One during CBC at GSL, 31 Dec 1953 (WA).

Winter: Recorded at Saltcoats, 2 Jan 1948 and 6 Dec 1948 (RB), and at Parkerview on 1 Dec 1988 and 4 Dec 2002 (AS).

Recoveries, banded outside area, 1 (SK, banded as a nestling by Doug Whitfield near Askwatam Lake on the Churchill River on 9 Jul 1968, caught in trap near Norquay on 12 Nov 1970). A nestling had a radio transmitter applied in North Dakota; the eagle was electrocuted south of Melville on 26 Jun 1995, but no leg band was reported (AS).

DUCK MOUNTAIN: The Kamsack-Madge Lake CBC had one in each of 1959, 1994, 1996, and 1997, two in 1995, and three in 1993.

AMERICAN KESTREL *Falco sparverius*

Fairly common transient and summer resident. Breeds.

Frank Baines found a nest in a flicker hole in a hollow tree at Maple Island, Crescent marsh, about the turn of the century. More common in the 1930s, decreased by the 1940s.

Summer resident in the early years at Sheho, using flicker holes in dead trees, but now seen only during migration (Niven 1973). At GSL, common in migration, but rare summer resident with only 10 summer sightings, 1982-2000 (WA). Common in migration, rare summer resident, with 10 sightings at GSL between 1982 and 2000.

Heavy spring migration noted on 20 Apr 1939. In 1973, brisk migration on 26 Apr, then recorded regularly until 16 May and one on 11 Jun (AS).

ESA(42): 14 Mar 1985 at Foam Lake (WN), 26 Mar 1946 at Yorkton (CSH), and 1 Apr 1981 at GSL (WA), Av 19 Apr. **LFD**(34): 14 Oct 1990 (WA) and 31 Oct 1945 (WN), Av 20 Sep.

Breeding: Pair nested in a hollow tree at the Ravine Reserve in 1942. Young fledged from a cramped space in the soffit of a pumphouse near the Melville Golf Course in 1983. After a nest box was placed in a poplar at Melville in 1983, four or five young were raised annually for about 10 years until the tree was cut down. A female flushed from a nest with 4 eggs in a previous magpie nest 27 feet up a 39-foot spruce at Landestrew Emmanuel Church, 10 miles north of Langenburg, on 22 Jun 1988. Nest, five young, at Wadena on 30 Jun 1992 (CSH).

Also breeds at Calder (ARS, *Atlas*).

BBS: Theodore, 8 years; highest, 3, 1994, mean 0.3.

Banded 29 (CSH 28, Anweiler/Fox 1) + pre-1955, 1 (Briggs).

Recoveries, banded in area, 1 [3.3%] (TX 1; banded 19 Apr 1981, found dead 25 Mar 1984 at Robstown, TX, CSH).

Langenburg: Two singles sighted, 1990 (TH,BL).

DUCK MOUNTAIN: SMNH 1926. SMNH 1951. Seen in 1975 (JAS). Fairly common along Highway 57 near west park boundary and core area, 25 Jun and 14 Jul 1981 (Hatch and Weidl). One on 13-15 Jun 1990 (MB).

MERLIN *Falco columbarius*

Fairly common summer resident. Breeds.

F.c. richardsonii shot at Fort Pelly 13 September 1881 (Macoun in Seton 1891, Seton 1909a). Three nests collected by W. Raine at Yorkton, 2, 5, and 5 June 1901 (specimens in Chicago Museum of Natural History, in Houston and Schmidt 1981). Two sets from Crescent Lake: 6 eggs collected by Raine on 3 Jun 1901, and 5 eggs collected by Edward Arnold from a former crow nest, are in the Museum of Vertebrate Zoology, Berkeley, CA (Houston and Schmidt 1981). Two sets of 4 eggs from Crescent Lake, 7 Jun 1893, 20 feet up in an aspen, and in a crow nest in a poplar, found by Raine on 4 Jun 1901, are in WFVZ. [Eggs collected by W. Hopkins in the Beaver Hills, 20 May 1901, were misidentified—L.F. Kiff and CSH].

Almost totally extirpated in the early 1960s, when dieldrin was heavily applied by farmers (Houston and Hodson 1997). Seen as a transient at Sheho until the 1950s, then none through 1973 (WN). First

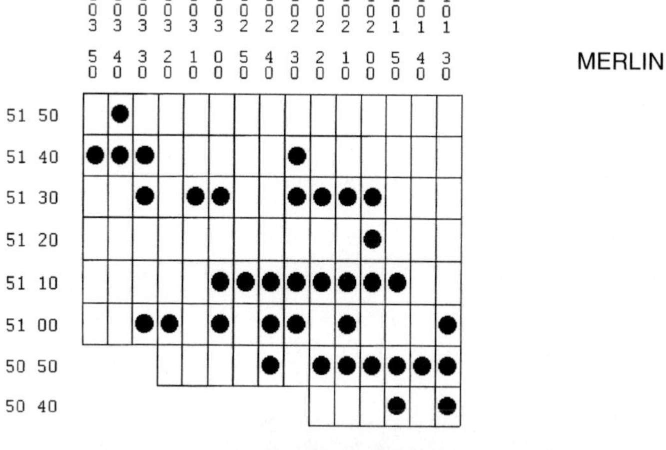

reappearance noted in 1975, then increased since 1980 (WA).

ESA(29): 24 Feb 1976, Av 25 Mar. **LFD**(15): 27 Nov 1994, Av 14 Oct.

Breeding: Beginning in the 1970s, as crows moved into the two cities of Yorkton and Melville, as well as Saltcoats and other towns, and as spruce trees reached maturity, Merlins began nesting in crow nests in conifers in cities, in the cemeteries of a number of smaller towns and villages, and in spruce trees in both active and abandoned farmyards. Two pairs used a former magpie nest in a Green Ash. The earliest returns of territorial Merlins to a nest site were 2 Mar 1990 and 1992 and 7 Mar 1986 (AS). A small number of territorial pairs and some of their nests were outside the YDM area; including them, there were territorial Merlins at 298 sites: 92 in occupied farmyards, 65 in unoccupied farmyards, 60 in rural cemeteries, 45 in urban cemeteries and 36 in cities and town parks or residential areas. Time permitted only 159 nests to be located, of which 54 were in magpie nests and 105 in previous crow nests (AS). Nests were found in the Yorkton cemetery in three consecutive years, with five young banded 28 Jun 1980 (CSH), and at the Old Victoria school grounds on Fourth Avenue.

Nesting in spruce trees in Good Spirit PP, probably each year since 1980. Early nest with four newly fledged young, 8 Jul 1985 (WA). Banded within 37 10-minute blocks. Much the commonest Merlin prey item throughout the YDM area, even for urban-nesting birds, is the Horned Lark (AS).

BBS: overall mean, 0.3. Theodore, 12 years, all since 1981; 2 in 1984 and 1996, mean 0.4.

CBC: Rank #44. 11 counts, 12 birds. Highest: 2, Yorkton 1998. The first winter record in the area was in Yorkton on 26 Dec 1951, the second was on 24 and 26 Dec 1958 (Krause 1959b), and the next on the Saltcoats CBC on 26 Dec 1964. Seen outside of count day at GSL, 29 Dec 1958.

The new propensity for a small proportion of Merlins to winter so far north appears to be related to the maturing of Mountain Ash and Cotoneaster, their berries allowing increasing numbers of Bohemian Waxwings to winter. It took time for northern wintering to become well known: as late as 1969 some birders considered that even wintering in Colorado and Kansas was "farther north than usual" (Williams 1969).

Banded 423 (CSH 10, Schmidt 413).

Recoveries, banded in area, 8 [1.9%] (SK 6, MB 1, SD 1). A young

Merlin, banded at Homefield, east of Ituna, on 23 Jun 1985, was recovered as a prey item in a Great Horned Owl nest near Semans on 8 Jun 1986 (AS, Lyon 1987). One banded near Hubbard on 21 Jun 1989 was dead on a highway north of Yankton, SD, on 20 Jan 1990; one banded at Melville was found dead at Forrest, MB, the following spring (AS).

Recoveries, banded outside area, 1 (SK, banded near Crooked Lake by Adam Schmidt on 19 Jun 1986, electrocuted at Tuffnell on 30 Mar 1987).

Langenburg: One male sighted, 1990 (TH,BL).

DUCK MOUNTAIN: Seen in 1976, 1977 (JAS). Rare, only recorded sighting east of Batka Lake, 9 Sep 1981 (Hatch and Weidl). Seen regularly at Ministik Beach from June 1988 through 1991 (TH). SNHS 1989.

GYRFALCON *Falco rusticolus*

Irregular winter transient.

Almost 20 sightings along Cutarm Creek south of Saltcoats, throughout the winter of 1963-64, including: 5 Nov; dark phase on 18 and 25 Nov; light phase on 6 Dec; dark phase on 26 Dec (BH). Brown-phase birds were seen west of Hamton on 24 Feb 1988 and 4 miles southwest of Melville on 26 Mar 1991 (AS). A white-phase bird was seen 5 miles north of Canora on 14 Feb 1996 (AS). An immature brown morph was sighted 15 Feb 1999 near Yorkton (WA,JA, Bain and Shanahan 1999b).

EFA: 5 Nov 1963 (BH). **LSD:** 20 Mar 1959 (a white-phase bird seen 5 miles south and 1 mile east of Kamsack by Jacob H. Jmaeff; Jmaeff 1959).

CBC: 2 counts, 2 birds. Bangor 1984, 1996.

PEREGRINE FALCON *Falco peregrinus*

SPECIAL CONCERN, COSEWIC, APRIL 1992 (*TUNDRIUS* SUBSPECIES) THREATENED, COSEWIC, MAY 2000 (*ANATUM* SUBSPECIES)

Uncommon transient, seen most often in late summer and fall.

One shot at Cote's Reserve on the Assiniboine, 22 September 1880. I first watched it pursue and capture a "prairie chicken" [=Sharp-tailed Grouse] (Macoun in Seton 1891). One at Little Boggy Creek near Cote's Reserve, October 1883 (Seton 1891).

Seen at Upper Rousay Lake on 21 Sep 1942 and 18 Jul 1944, and at Leech Lake on 5 Aug 1944 (CSH). One at Yorkton on 26 May 1945 (IMP). William Niven had only a single record at Sheho, one on 2 Jun 1945.

Seen at Leech Lake on 24 May 1973; five transients sighted: 2 Oct 1983 near Melville, 10 Oct 1985 south of Veregin, 28 Apr 1993 north of Langenburg, 6 May 1995 near Parkerview, 4 May 1996 near Veregin, and 11 May 1996 near Melville (AS).

On 13 May 1984, a Peregrine Falcon harassed female Great Horned Owls as they flushed from their nests: the first south of Kelliher and the second near Crescent marsh, west of Saltcoats (Houston 1985). One found fatally injured, 6 miles south of Okla, 7 May 1989 (Alvin Reisner, *fide* Donald F. Hooper). One at Little Quill Lake on 10 Aug 1989 (ARS). One killed and fed on an American Avocet on the southwest shore of GSL, 5 Sep 1991 (WA). Two chasing shorebirds at Little Quill Lake on 30 May 1998 (DH). On 19 May 2002, at a slough southwest of Preeceville, a female Peregrine killed a wigeon; she and the smaller male Peregrine were eating the duck, whose legs were still kicking, when we approached (DF,DGJ,SAJ).

ESA: one on 1 Apr 1983 at Norquay (CP), and 29 Apr 1956 and 14 May 1983 at GSL (WA). **LFD**: 21 Oct 1942 at GSL (JAG). Other fall dates at GSL: 5 Sep 1991, 27 Sep 1959 and 1999.

Recoveries, banded outside area, 1 (FL, banded 14 Oct 1975 in the Dry Tortugas, west of Key West, found injured 18 May 1978 near Arran)

PRAIRIE FALCON *Falco mexicanus*

Occasional visitor.

One individual 8 miles west of Springside on 17 Jul 1946 (F.M. Mowat). One sighted at Hyas in May 1977 (CP). One at Little Quill Lake on 30 Jul 1988 (ARS). Three records: 7 Sep 1983 near Leech Lake, 29 Jul 1992 south of Melville, and 8 Aug 1992 north of Homefield (AS). One flying over Rokeby marsh, 25 Jul 1994 (DH).

CHUKAR *Alectoris chukar*

Introduction attempts have been unsuccessful. Bred only briefly. Extirpated.

After a release at Melville in 1940, 60 wintered there. The next spring, scattered reports came from localities up to 30 miles apart, including 12 near the Old Folks' Home at Melville and one in a downtown back lane in Melville on 21 Jun 1942. The last report was of one pair seen by John Kosa near Brewer in 1945.

Fifty were released by the Yorkton Fish and Game League in 1941, from which there were only two reports: one near Tonkin that year, and one 3 miles west of Yorkton in the spring of 1947. There have been no sightings since.

Breeding: A female with young at Fonehill in 1944.

GRAY PARTRIDGE *Perdix perdix*

Introduced. Fairly common summer resident, especially around cultivated fields and farmyards. Breeds.

First introduced into Alberta in 1908, it appeared at Yorkton and Sheho about 1928, and quickly established itself, increasing rapidly until there were two coveys around almost every farmyard (WN). For a few years it became about as common as the native Sharp-tailed Grouse. An albino was seen a mile from Ituna on 28 Mar 1933 (Fred Caswell, LTMcK, Lawrence 1933a). A decrease became evident in 1968, when only one covey was seen in the 5 miles between the Niven farm and the village of Sheho.

Breeding: The nests are very difficult to find. One nest, 10 eggs, 8 Jun 1946 (CSH). A female had a brood of eight downy chicks, only about three days old, 6 miles south of Melville, on the unusually late date of 21 Sep 1962 (Schmidt 1964). In 1973, roadkill of a one-third-grown bird on 2 Aug (AS). One adult with 12 very small young, 22 Jun 1952; adult with two young, 24 Aug 1989 (WA). Brood near Little Quill Lake on 26 Jul 1989 (ARS).

Also breeds at Sheho, Kamsack, Togo, Jasmin, Yorkton (ARS, *Atlas*).

BBS: overall mean, 0.3. Theodore, 4 years; 2 in 1976, 1990, and 1997, mean 0.2.

CBC: Rank #13. 76 counts, 1,139 birds. Highest: 133, Yorkton 1941. Yorkton's 18 individuals in the 1952 CBC tied for first place continent-wide.

Langenburg: small flock near Churchbridge, 1990 (TH,BL).

DUCK MOUNTAIN: Nine in 1947, 12 in 1958, in Kamsack CBC.

RING-NECKED PHEASANT *Phasianus colchicus*

Introduced. Uncommon summer resident. Has bred.

Saskatchewan introductions began about 1934. In 1937, after mention of the introduction of this species to Saskatchewan "three or four years ago," comes the first mention of spread to the YDM area: "fair numbers are reported from Canora" (Forsyth 1937). A few individuals reached the Yorkton area in 1940, probably from the Qu'Appelle valley and from releases by various branches of the Fish and Game League. Major propagation efforts by the renamed Saskatchewan Wildlife Federation, with government assistance, took place at the Beaver Creek Pheasant Farm south of Saskatoon from 1945 through 1982 (Leighton et al. 2001). One pair seen 5 miles east of Saltcoats only once, about 1955 (RB). A male, probably a released bird, crossed Highway 8, south of Wroxton, on 28 Nov 1961 (Ninowski 1963). One stray reached GSL 18 Nov 1965 (WA). None recorded anywhere in the YDM area in 1973 (AS).

BBS: Theodore, 1 in 1989, possibly an escapee from a breeder, James Hupka, whose pens were 2 miles distant from where the pheasant was sighted.

Breeding: A few broods have been seen. One egg collected north of Saltcoats, probably around 1945 (NJ,PMA).

CBC: Rank #40. 2 counts, 18 birds. Highest: 17, Yorkton 1990.

DUCK MOUNTAIN: SMNH 1951.

RUFFED GROUSE *Bonasa umbellus*

Fairly common year-round resident. Breeds.

Along Swan River, 1881 (Macoun in Seton 1891). On 19 June 1884, flushed an agitated female, who may have had tiny, downy young, but neither nest nor young was located (Seton 1891). Two eggs collected north of Saltcoats (NJ,PMA). Nest with 12 fresh eggs at the base of a tree, beneath an overhanging branch, near Melville on 10 Jun 1918 (H.C. Grose, SMNH).

Once common in wooded areas, especially at Sheho and north of Yorkton, but decreased as land was cleared, especially between 1942 and

1948 (WN). Two and six in the two mature aspen study plots at GSL, 1969 (Flack 1976). In 1973, rare, drumming heard several times (AS).

Breeding: Early nest, 11 eggs, Otthon, 18 May 1958. Nest, 11 eggs, York Lake, 25 May 1953 (PPa, Houston 1953b), and 8 eggs, Saltcoats, 25 May 1958 (CSH).

Adult with young at Cussed Creek, 26 Jun 1952, and with two chicks at Horseshoe Lake, 12 Jul 1975 (DH). Female with 10 young, 9 Jun 1955, with broods noted every summer at GSL (WA). A female, whose nest was burned on 11 May, kept incubating her hard-boiled eggs through 7 Aug (12 weeks!) at GSL (Anaka 1959; Krause 1959d). Nest, 15 eggs, GSL, 22 May 1969 (JADF).

PNRS: at Kamsack, 10 eggs, 20 May 1958 (JHJ); at Stornoway, 10 eggs, 27 Jun 1959 (SZ), and five young, 22 Jun 1963 (RD); at Mikado, 5 eggs predated, 7 May 1969 (EK).

BBS: overall mean, 0.2. Theodore, seen or heard on 1 of 33 counts, mean 0.03.

CBC: Rank #18. 128 counts, 462 birds. Highest: 20, Kamsack 1994 and 1999. Seen during 19 of the first 22 CBCs at GSL, 11 of 19 at Yorkton and 8 of 11 at Sheho. The three species that best measure CBC effort on foot through aspen bush, in contrast to observing along roadsides from a car, are the Ruffed Grouse and the Downy and Hairy woodpeckers.

Winter: Regular at feeder at the edge of Ketchen, up to 12 at one time (DF).

Banded 1 (CSH, 25 Sep 1958).

Langenburg: Two drumming, 26 Apr 1976 (DH).

DUCK MOUNTAIN: SMNH 1926. SMNH 1951. SNHS 1956. Seen in 1974 and 1977; broods in 1975, 1976 (JAS). Fairly common year-round, heard drumming at different locations in Jun 1981 (Hatch and Weidl). SNHS 1989.

Point counts: Rank #48, 2 individuals at 2 stops (ARS).

Aspen count totals: OA 2, MA 1 (TH).

SPRUCE GROUSE *Falcipennis canadensis*

Rare year-round resident at Duck Mountain. No breeding records.

Very numerous in the poplar woods to the north of Fort Pelly in Aug 1881 (Macoun in Seton 1891).

One very tame bird was photographed in the garden and in a spruce, at GSL on 7 Nov 1954, by W.J. McDonald and Joyce Gunn (1956a). This is an unusual extralimital record.

CBC: Rank #37. 2 counts, 20 birds. Highest: 10, Kamsack 1986 and 1987.

DUCK MOUNTAIN: Rare. Ten seen during Kamsack-Togo CBC, 2 Jan 1988.

[Present in Duck Mountains, Manitoba: a female was collected 11 Aug 1937 by W. Watkins, who encountered young on several occasions, 6-27 Aug (W.E. Godfrey, pers. comm.); a female was accompanied by four small, late-hatched juveniles on 1 Oct 1997 (Koes and Taylor 1998a)].

WILLOW PTARMIGAN *Lagopus lagopus*

Formerly a winter vagrant.

Until the early 1920s, it rarely wandered as far south as the Qu'Appelle valley (Mitchell 1924). One was shot by an Indian south of Invermay in the winter of 1929 and mounted as a specimen by H. Deacon (Lawrence 1931b). In 1930, "J.D.S." reported arrival near Norquay in mid-December and H. Dutton saw one a few miles north of Norquay on 17 Dec (Lawrence 1931a).

In the 1930s and 1940s, occasional individuals reached the north boundary of the area at Norquay. One wintered near Sheho in early 1958 and stayed on at a neighbour of Niven's until 21 May (WN).

SHARP-TAILED GROUSE *Tympanuchus phasianellus*

Common year-round resident, but now in a general decline. Breeds.

Very abundant on the prairies from Livingstone southward in 1881 (Macoun in Seton 1891). Northwards in prairie localities in Cote's Reserve, 1884 (Seton 1891). So common at GSL in 1887 that John A. Gunn never wasted a shot on only one bird; he would line up two and get both with one shot. Two eggs collected north of Saltcoats (NJ,PMA).

In the Rothbury district, south of Wroxton, R.P. Rooke said of the Sharp-tailed Grouse: They "almost invariably roost under the snow after it has fallen to a sufficient depth to allow them to do so, and the weather gets severe. In very severe weather they will also lie under the snow a good part of the day. Many times...I have walked into a patch of deep snow at midday and the ['chickens'] have risen from under the snow on all sides of me. It is not as easy as it would appear for animal enemies to capture them under these circumstances, as naturally it is soft drifts they use for this purpose and the enemy cannot approach them very silently, having to jump or flounder to some extent. The birds also roost a considerable distance, usually 5 to 10 feet from where they enter the snow bank and the enemy has therefore to flounder around scenting them out before locating them, which gives the birds the necessary warning. There is no doubt, of course, that these animals take a certain toll of birds, but to no greater degree than the large owls and hawks would do if they were roosting in trees" (Lawrence 1921).

At Sheho, they showed better adaptation than the Greater Prairie-Chicken, especially in the days when grain was harvested by stooking and threshing. In the dry years of the 1930s, flocks of 100 and more gathered to feed at stacks of wheat and oats containing immature grain, cut for feed, on the Niven farm. At that time there were three dancing grounds within 3 miles of the Niven farm: a half-mile west; 1 mile north; and 3 miles northeast. After the widespread crash in numbers in 1942, they increased again, so that 50 could be seen at a strawstack in early 1949 (Niven 1949). Following a cold spell in late May and early June of 1972, only the latter of Niven's three leks remained active; he saw only two small broods in 1972, with one and five chicks, respectively, and the largest group that year numbered 25 (Niven 1973). Several broods seen in 1975, and three broods of 10 to 14 young in 1976, the best success for years, and a covey of 25 in the winter of 1977-78, and several broods in 1978 and 1979. Another drop in numbers occurred at Sheho in 1980, with a slight temporary increase in 1981 when one group of 18 was seen, and in 1982, when there were only six birds left at the remaining dancing ground; in 1983, no breeding birds were evident; low numbers persisted through the end of Niven's record-keeping in 1989 (WN).

Near Yorkton they suffered a sharp decline in 1942, but had increased by March 1951 when 194 were counted in two days along country roads near Rokeby and Saltcoats (Deighton 1951). Six leks were located near lakes south of Yorkton in 1973 and one lek a mile east of Foam lake marsh on 26 May 1974 (AS). At Saltcoats, a covey in each square mile in the early 1940s, but in fall of 2002, only three were seen in a drive of 100 miles, an estimated drop to less than 10% of former numbers (JRJ).

Near GSL, numbers vary from year to year, but have never returned to the high numbers of the 1950s and 1960s. At Ketchen and Preeceville, numbers are drastically diminished in comparison with the mid-1980s (Norman Johnson), as described a bit farther west at Rose Valley (Anderson 1998).

After 200 years of study, the mechanisms behind the remarkable synchronicity of the 10-year cycle of game abundance (affecting Sharp-tailed Grouse, Ruffed Grouse, Great Horned Owl, and Northern Goshawk, in addition to the Lynx and the Snowshoe Hare) are still poorly understood (Houston 2003b). Though assessment of abundance is difficult in such cyclical species, there is no question that the Sharp-tail numbers have declined dramatically.

Breeding: Early nest at GSL, 10 eggs on 13 May 1988 and 12 eggs on 7 Jun, all hatched; 14 eggs on 11 Jun 1959 and 11 on 24 Jun 1950 (WA). An adult with young at GSL, 23 Jun 1936 (JDS). Nest with 14 eggs found under an old tin stove on the edge of Yorkton, 7 Jun 1940; these eggs hatched the following week (CSH). A pair with at least one partly-grown young, Rokeby marsh, 19 Jul 1994 (DH).

PNRS: 4, 12, and 13 eggs, and 14 young at Stornoway, 1959-61 (SZ).

BBS: overall mean, 2.6. Theodore, 5 years; highest, 6, 1977, mean 0.5.

CBC: Rank #10. 134 counts, 2,258 birds. Highest: 96, Saltcoats 1954. Yorkton totals of 46 in 1949 and 63 in 1956 topped all other counts on the continent. Since then, a marked drop, so that Sharp-tails were recorded only three times in 11 years, 1990-2000 inclusive. At GSL, none sighted from 1996 to 2001, but 12 were counted in 2002 (WA).

Banded pre-1955, 3 (CSH).

Langenburg: Three leks located on 26 Apr 1976, and a fourth on a rural road allowance on 31 May 1976 (DH).

DUCK MOUNTAIN: Kamsack CBC, 21 in 1958, 28 in 1959. Seen Jul-Aug 1970 (RJL).

GREATER PRAIRIE-CHICKEN *Tympanuchus cupido*

Fairly common year-round resident, 1910 to 1930.

Extirpated, COSEWIC, 1994

The "Pinnated Grouse" (this terminology avoids confusion with the Sharp-tailed Grouse, which is popularly called "Prairie Chicken") followed soon after settlement, moving northwesterly from Minnesota and the Dakotas, to invade the Prairie Provinces from east to west (Houston 2002). A nest with 11 eggs was collected by Frank and Fred Baines at Crescent Lake on 24 May 1897 (WFVZ, Houston 2002a), where this species increased to become "just as plentiful as the Sharptail between 1910 and 1920, when it gradually thinned out and disappeared" (Baines 1956). H.S. Swallow reported that a few scattered individuals were present in the Yorkton area when he arrived in the fall of 1902. John A. Gunn, at GSL, and Sheho old-timers all estimated "about 1905" when queried as to their arrival date, but Baines failed to remember the 1897 nest when he gave this estimate. Known as the "Manitoba Prairie Chicken," it arrived in the Rothbury district, 10 miles south of Wroxton, about 1905 (Rooke 1947). A "few dozens" were present near the future townsite of Kamsack in 1907 and 1908; P. Fraser concluded correctly that township 29, range 32 was "probably the northern limit of their range" at this longitude (Fraser 1961). At Ituna, H.M. Rayner first saw them about 1910. One egg collected north of Saltcoats, probably around 1910 (NJ,PMA). Nest with 10 eggs on the ground at the edge of a clump of willows near Melville on 30 May 1920 (H.C. Grose, SMNH). At Sheho, Greater Prairie-Chicken had a dancing ground on the Niven farm until 1925 or 1926, but disappeared entirely from the Sheho area in the 1930s (Niven 1973). Once half the land in a given area was broken, numbers declined (Houston 2002).

As a larger percentage of the land became cultivated and the fields became larger, the birds found it difficult to adapt, as they seemed to thrive on a mixture of some cultivated fields with retention of fairly large areas of wild grassland. They were common in the Theodore district in the 1930s (Brother S. Matthias). One along Highway 9, south of Yorkton, on 5 Jun 1938, was reported to IMP.

At Ituna, H.M. Rayner saw an occasional covey until 1940 and flushed his last covey of eight in the fall of 1943. After they had disappeared from the rest of the district, as agriculture became more intense, they were still present on the wet flats south of Lower Rousay Lake and also west of Saltcoats. One was shot by Robert Cock in the fall of 1942 (Shaw 1942). Five on 17 Jan 1946 were believed to be the very last remaining at GSL (John A. Gunn). Three were seen by W. Dean Lightbody in early November 1946 (Shaw 1947).

Breeding: 1897, as above.

Langenburg: There was a hypothetical sighting of a dozen, which lit briefly in Gilbert Johnson's yard in Marchwell, on 16 Feb 1967 (Johnson 1967, Houston 2002a).

WILD TURKEY *Meleagris gallopavo*

Introduced. Breeds at Duck Mountain.

DUCK MOUNTAIN: Introduced at the Horse Academy in the park between 1986 and 1988, by the Kamsack Wildlife Federation. Two males and one female at round bales near Maynard Whitehawk riding stables, 13 Jun 1990; Linda Griffiths saw up to 20, with a displaying male, on her drive in early morning to the lodge (MB).

Breeding: Several adults, with groups of fuzzy young, wandered around the Kamsack Beach Subdivision acting like shameless barnyard fowl (TH). Female with eight young, 13 May 1989 (FAS, Gollop 1989b); two adults with five young on 9 Jul 1989 (ARS,PC); an adult with two chicks at Kamsack Beach, 19 Aug 1989 (TH).

CBC: Rank #34. 4 counts, 29 birds. Highest: Kamsack 14, 2000.

YELLOW RAIL *Coturnicops noveboracensis*

SPECIAL CONCERN, COSEWIC, NOVEMBER 2001

Rare summer resident. Breeds.

[A few rails in a marsh near Fort Pelly, September 1881, that he "took to be this species"; they would fly and drop in the sedge (Macoun in Seton 1891].

Rarely seen. One heard 9 miles southwest of Sheho, 22 Jun 1976 (WN). Two heard after midnight at Horseshoe Lake, 21 May 1988, at the start of the "Big Day Count" (CP,RK,RE,TH)."Tic tic" call, similar to tapping of two stones together, heard at Rokeby marsh, 19 Jul 1994, and at Leech Lake, 28 Jul 1994 (DH). On the latter occasion Dale was accompanied by Richard Urbanek, who had censused and banded this species at Seney Wildlife Refuge in Michigan. Resident and probably breeding at Barvas marsh in 1996; at least three males calling (JRJ, Kornder and Belcher 1996). One heard at Little Quill Lake by Burke Korol, 20 Jul 1999 (Bain and Shanahan 1999d).

Breeding: Frank Baines found one Yellow Rail nest in Crescent marsh about 1900. In the early 1930s, Charles H. Maddaford ran over two nests with a mower, 9 miles west and 3 miles south of Saltcoats. Both nests were in long grass on damp ground at the edge of the water, each with 6 to 8 eggs. A small yellow rail, definitely not a Sora, flushed from one of the nests and the eggs were quite different from those of the Sora. The identification was checked at the time in Chamberlain's 1905 Canadian revision *of Nuttall's Ornithology* (Houston 1969c).

VIRGINIA RAIL *Rallus limicola*

Uncommon summer resident. Breeds.

Two dark birds, probably juveniles of this species, were noted at the Ravine Reserve on 17 Jul 1942 (IMP). A more convincing identification is that by AS on 10 Jun and 12 Aug 1973. One seen near outlet of Foam Lake marsh on 9 Jul 1974 (AS). Two adults seen at close range, Upper Rousay Lake, on 6 Aug 1977 (CSH,MIH). One heard after midnight at Horseshoe Lake, 21 May 1988 (CP,RK,RE,TH). Seen at Little Quill Lake on 29 Jul 1989 and 18 Jul 1990 (ARS). Resident and probably breeding at Barvas marsh in 1996; at least three males calling (JRJ,SJ, Kornder and Belcher 1996).

Breeding: Two confirmed records: one flightless young rescued from a cat at GSL on 15 Aug 1965 (WA). Five downy young in a dry pothole near GSL on 25 Jul 1967 (WA, Hatch 1967b). Present at Horseshoe Lake (TH).

Also breeds at Wadena (ARS, *Atlas*).

BBS: Theodore, 2 in 1971, mean 0.06.

DUCK MOUNTAIN: SMNH 1951. Uncommon, heard calling at Yellow Lake near west park boundary, 1 Jun 1981 (Hatch and Weidl). One seen at close range in marina area, Madge Lake, 23 Aug 1989 (TH).

SORA *Porzana carolina*

Very common summer resident in marshes. Breeds.

Have taken half a dozen nests in a day on Duck Mountains (Seton 1891). One egg collected north of Saltcoats (NJ,PMA).

As with other rails, it is more often heard than seen. Less common during the drought of the 1980s.

ESA(37): 30 Apr 1957 and 1977, Av 14 May. **LFD**(18): 27 Sep 1952, Av 2 Sep.

At GSL, Soras were absent during the dry years of 1959 (Krause 1959d) and 1961-64, inclusive, then returned in 1965 (WA, Hatch 1965), and decreased drastically again by the dry year of 1989 (WA).

Breeding: Early nest, 7 eggs, 1 Jun 1958, increased to 9 eggs by 7 Jun; the next day, a second nest found with 8 eggs. Nest, 9 eggs, Upper Rousay Lake, 20 Jun 1946. Early hatching, 24 Jun 1946. Another nest had 6 eggs and one newly hatched young on 24 Jun 1956 (CSH). Two tiny young swimming, 13 Jul 1992 (WA).

Nest, 7 eggs, one mile west of Melville, 26 May 1957 (GAA). Largest clutch, 11 eggs, 8 Jun 1958 (WA). In 1973, nest, 4 eggs, 2 Jun and adult with chick on 25 Jul (AS).

PNRS: at Saltcoats, 8 eggs, 28 Jun 1959 (BH); at Stornoway, 6 eggs on 8 Jun 1960 and 4 eggs on 15 Jun 1961 (SZ).

Also breeds at Kamsack, Yorkton (ARS, *Atlas*).

BBS: overall mean, 10.7. Theodore, mean 12.5 (1970s, 10.5; 1980s, 13.1; 1990s, 14.9; n=32).

Banded pre-1955, 4 (CSH 3, Wickencamp 1)

Langenburg: Heard at most small and medium sloughs, 1990 (TH,BL).

DUCK MOUNTAIN: SMNH 1926. Seen in 1974 and 1975; nest with eggs destroyed in 1976 (JAS). Uncommon, three heard on 3 Jun; one nest with 3 predated eggs on an island on 18 Jun 1981 (Hatch and Weidl). Commonly heard in small ponds and marshes, and along edges of lakes, 1988, 1989, and 1990 (TH). SNHS 1989.

PNRS: an empty nest found on 30 Jun 1986 (HWRC).

Point counts: 1 individual at 1 stop (ARS).

AMERICAN COOT *Fulica americana*

Very common summer resident on sloughs and lakes. Breeds.

Eleven eggs, Crescent Lake, 20 May 1897, Baines for Raine, WFVZ. Three eggs collected north of Saltcoats (NJ,PMA).

Stages in large numbers. Spring: flocks of 2,000 at GSL, 7 May 1989 (JA,WA, Gollop 1989a). Foam Lake marsh survey, 23-25 May 1974, counted 646 (AS). Fall: large flock of 3,000 on Horseshoe Lake, 8 and 9 Oct 1967 (WA, Hatch 1968a), and 500 at GSL on 24 Aug 2002 (WA,JA, Bain 2002a).

ESA(48): 15 Apr 1976 and 1977, Av 26 Apr. **LFD**(33): 31 Oct 1954, Av 10 Oct.

Breeding: In 1946: early nest, 2 eggs, 23 May; early hatching, 11 Jun. Of 175 nests found, 140 were in bulrush stands, 3 in cattails, 3 in mixed bulrush and cattail, 1 in willows growing in water, 11 in pure whitetop grass, and 18 in a mixture of bulrush and aquatic grasses. Nests in 8 to 44 inches of water, average 15.3 inches. On approach of an intruder, the incubating female would quietly slip off her nest and dive under water—and several times bumped my rubber boot when I came closer to count the eggs. Nests contain up to 15 eggs, with an average of 9 eggs in 132 complete clutches. Once I found eight consecutive nests with 10 or more

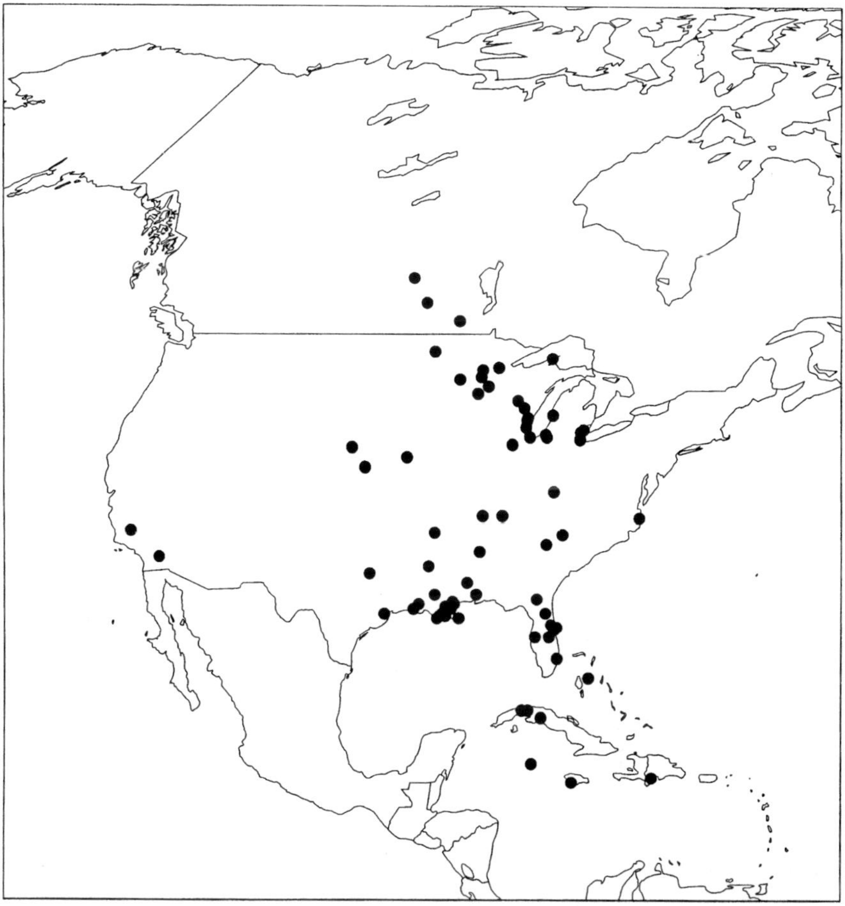

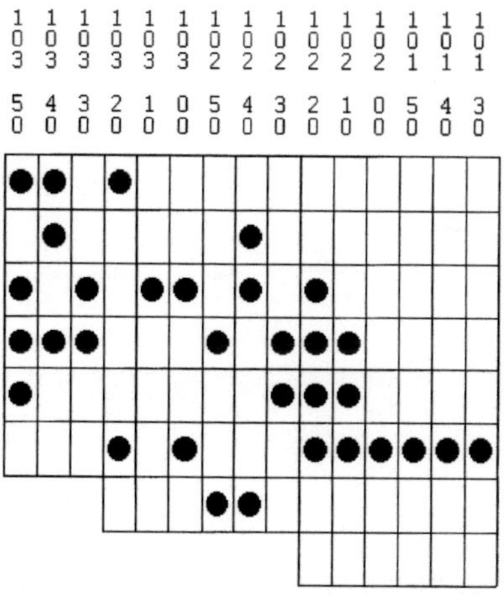

AMERICAN COOT

eggs. Three nests had a dead adult on or beside the nest. Hatched coot eggs do not leave a membrane behind as do duck eggs, only tiny eggshell fragments—so I could not calculate egg success, but eggs hatched from at least 90% of nests. Nest diameters average 12.3 inches outside and 5.9 inches inside, 4.9 inches above water with a cup depth of 2.2 inches (Houston 1946b).

Nest with 14 eggs on 5 Jun 1955 (WA). Nests with three young and one young still in nests in sloughs west of Churchbridge on 21 Jun 1959 (CSH,MIH). In 1973, 49 nests found between 27 May and 10 Jul (AS). In 1975, nests with 9 and 10 eggs at Horseshoe Lake, 3 Jun; 45 young on 11 Jul and 30 on 12 Jul (DH).

PNRS: 1, 1, 3, 7, and 8 eggs at Stornoway, 1959-62, and 4, 5, 8, 8, 9, 10, and 11 eggs near Arran in 1963 (SZ). Flightless young banded in 32 of the 108 10-minute blocks.

Also breeds at Margo, Preeceville, Kelliher, Kamsack (ARS, *Atlas*).

BBS: overall mean, 15.9. Theodore, mean 18.9; peaked in 1980s (1970s, 19.1; 1980s, 29.5; 1990s, 12.0; n=30).

Banded 2,234 including pre-1955, 7 (Patrick 5, Wickencamp 1, Wotherspoon 1)

Recoveries, banded in area, 84 [3.8%] (Map; SK 2, MB 1, ON 1, AL 1, AR 1, CA 2, FL 8, GA 2, IL 3, KS 1, KY 1, LA 20, MI 6, MN 7, MS 2, NC 1, ND 1, NE 2, OH 1, SC 1, TN 3, TX 3, WI 5, Mexico 2, Bahamas 1, Cuba 3, Dominican Republic 1, Jamaica 2).

Recoveries, banded outside area, 8 (SK 1, AB 1, MB 1, CA 2, FL 1, LA 2).

Langenburg: Abundant, 1990 (TH,BL).

DUCK MOUNTAIN: SMNH 1926. SMNH 1951. Fairly common throughout park; 50 on Bear Head Lake and 250 on Ketchamonia Lake, 17 Jun; about 500 staging on Bear Head Lake on 10 Sep 1981 (Hatch and Weidl). One on 13-15 Jun 1990 (MB).

Point counts: Rank #48, 2 individuals at 2 stops (ARS).

SANDHILL CRANE *Grus canadensis*

Uncommon summer resident. Breeds.

Assiniboine near the Whitesand, and at the mouth of the Shell River, September 1881 (Macoun in Seton 1891:492). "I am informed by Mr Ripon [E. Rippon] of Toronto that this bird commonly breeds on the Upper Qu'Appelle" (Seton 1893a).

Breeding: Female Little Brown Crane nest and 2 eggs collected at Crescent marsh, 27 May 1899, by Fred Baines, in the Museum of Comparative Zoology, Harvard University; and 2 eggs collected 18 May 1899, in the Chicago Museum of Natural History (Walkinshaw 1949). Eggs collected near Grayson, 11 April 1914 by Allen Palmer (Walkinshaw 1949). Soon after 1900, Peter Fernie found two nests in a wet meadow southwest of Springside (Houston 1972c), but cranes then deserted the area. One fresh egg was collected from a nest at the edge of the File Hills, west of Birmingham, by H.C. Grose on 30 May 1918 (Houston 1993). Walter Raine found two nests, each with 2 eggs, at Saltcoats marsh on 16 and 20 May 1890 (Raine 1892:167), and another large nest with 2 eggs in Saltcoats marsh, 6 June 1901 (full-page photo in Reed 1904:101). Other Raine sets of 2 eggs each from west of Saltcoats in the Western Foundation of Vertebrate Zoology: F. Baines, 28 May 1898; unnamed Indian, 30 May 1901 [mentioned by Lattin 1901]; 19 May 1903. In 1910 or 1911, there was a nest with one egg on NE 31-23-3W2, and in 1918, two nests on the west side of Rokeby marsh (C.H. Maddaford). Frank and Fred Baines also remembered a nest 1 mile east and over 2 miles south of Yorkton, on 1 July in the 1890s.

J.A. Gunn found a nest about 2 miles south of GSL about 1890; cranes were regular there from 1888 to 1896, but not thereafter (Walkinshaw 1960).

R.P. Rooke knew of springs in a ravine 1.5 miles long and 200 yds. wide about 1 mile west and 3 miles south of the present village of Willowbrook; here in June 1893 there were 10 nests, each with 2 eggs (Houston 1949).

Still a regular transient and an uncommon summer resident. Cranes have increased noticeably since about 1970. In 1973, cranes were seen and heard sparingly on five occasions between 27 May and 9 Jul (AS). Sixty-three observed at Little Quill Lake on 10 Jul 1976 (Dan Nieman, *fide* DH). Sixteen seen near Upper Rousay Lake on 14 Jun 1989 (CSH,MJB). Forty-one adults seen on tour of south Yorkton wetlands, 6 Oct 1993 (DH, Brian Johns). On 1 Jul 1994, 26 at Leech Lake and 3 at Rokeby marsh (DH). A flock of 80 non-breeding individuals summered west of Rokeby marsh from Jun to late Aug 1996; 130 were on the marsh from Apr to Jun 1997 (Burke 2001). A pair in a field south of Margo on 8 Jun 2001 (AS).

ESA(47): 1 Apr 2000, Av 15 Apr. **EFA:** 2 Sep 1942 (CSH) and 33 on 15 Sep 1973 (AS). **LFD**(33): 1,000 over Sheho on 10 Nov 1970 (WN, Houston 1971a) and a few on 18 Oct 1984 at GSL (WA), Av 21 Sep.

Breeding: Nested sparingly near Sheho in the early days; about 1920, a pair had young on the flats 4 miles north of the Niven farm. Fred Bard found a resident pair at Rokeby marsh in Jun 1941 but did not find a nest. In 1942, David Baines found a nest with eggs already destroyed at McNicholl's slough, 3 miles west of Saltcoats (Walkinshaw 1960). William McInnes found five nests on the west side of Rokeby marsh between 1938 and 1940. On the east side of Rokeby marsh, John Domon saw young, flightless cranes on several occasions. E.C. Major of Wroxton came upon an adult and two small young ones, 5 miles south and 3 miles west of Calder about 20 Jun 1945. No further nests until 10 May 1954, when a nest with 1 egg, in 21 inches of water, was found by John Maddaford on NW12-24-3W1 and photographed on 10 May 1954 (Houston 1955; Gammell and Huenecke 1954). In the marsh at Lower Rousay Lake on 2 Aug 1946, a crane egg was found floating in the water (CSH).

A pair with a 10-day-old chick in Rokeby marsh on 8 Jun 1993 (DH, David H. Ellis). A pair with two non-flying young in a field NE8-25-4W2, 1 mile south of York Lake, 26 Jul 1994 (DH). A study by Ann M. Burke revealed a pair of cranes with an empty nest at Crescent marsh in 1995, and in 1996 an apparently non-breeding pair at Crescent Lake, five nests with eggs, and a sixth pair with young in the Rokeby marsh area (Burke 1996). Sixteen nesting attempts, 1995-97, fledged 12 offspring (0.75 fledged chicks per pair, Burke 2001).

BBS: Theodore, 1 in 1997, mean 0.03.

Banded 2 (Johns at Leech Lake, 1994).

Recoveries, banded outside area, 1 (NE, Platte River west of Kearney on 28 Mar 1966, shot west of Wadena on 5 Sep 1966). A crane, captured on its breeding area along the Overflowing River north of Hudson Bay, SK, on 29 Jul 1994, staged in the Yorkton wetland area, 140 miles to the south, from 11 to 24 Sept, before it went on to winter near McFaddin, TX (Hjertaas et al. 2001).

DUCK MOUNTAIN: SMNH 1926.

WHOOPING CRANE *Grus americana*

ENDANGERED, COSEWIC, 2000

Rare spring and fall transient. Rare breeder until about 1900.

Seen along the Whitesand River in May 1884 by R. Miller Christy (Christy 1885).

Nest in the late 1880s, with 2 white eggs, on top of a muskrat house on "Battersby's Lake" or "Scheller's Slough," west of Horseshoe Lake, section 25-28-5W2, or 8 miles west-northwest of the present site of Gorlitz. Fernie told Cliff Shaw in 1955 that the "whoop" call could be heard for 5 miles on a quiet morning. Nest, about 1900, with 1 dirty white egg, on top of a muskrat house in the "Fernie meadow," in a marsh near Cussed Creek, 4 miles west and 1 mile south of the present village of Springside (Houston 1972c).

Nest, a mass of marsh hay, 3 feet in diameter, with 2 eggs, on the prairie near Yorkton, 16 May 1900, found by "Cowbry [=Cowboy] Brown," from the oological collection of Walter Raine (Reed 1904:99 and Allen 1952:60). These eggs are extant in the Thayer Collection, Museum of Comparative Zoology at Harvard University. Nest, 2 eggs, collected "at foot of Beaver Hills" 21 May 1901 by A.Y. Arnold, set in WFVZ. [Was this a misprint for Edward Arnold?] (Houston and Bechard 1982).

A probable nest was reported by E.H.M. Knowles along Cutarm Creek, 3 miles west and 1 mile south of Bredenbury, in 1893 (Hjertaas 1994).

In the 1890s, flocks of up to 20 or 30 flew over Rothbury settlement each spring (R.P. Rooke). John A. Gunn saw the species only once—a pair in a meadow south of GSL, in the spring of about 1890. Between 1883 and 1899, Frank Baines saw small flocks of up to 15 birds nearly every year in migration, and occasionally they rested on the fields near Crescent marsh. About 1895, Jim Nelson shot a Whooping Crane, with sandy splashes on its neck, near Crescent Lake. The bird was skinned by Frank Baines and sent to E.J. Cousins, a Toronto dealer. Robert Rousay shot four in the field on the north edge of Upper Rousay Lake about 1896; they were regular during migration at the time. W.G. Sharpe shot a Whooping Crane which flew low over his pit near Crescent Lake in the fall of 1924. At the time, it was thought that this might have been the last crane to have existed in the wild in North America [William T. Hornaday in *Our Vanishing Wild Life*, published in 1913, had ascribed the honour of seeing the last Whooping Crane to John F. Ferry (actually, R.M. Barnes) at Big Quill Lake in June 1909!].

Yorkton is near the eastern edge of the regular Whooping Crane migration route. In the fall of 1942, John Domon saw a single bird in a flock of Sandhills at Rokeby marsh. One seen along the Whitesand River, 3 miles east and 2 miles south of Canora in June 1945 (students and Pearl J. Achtemichuk, reported in Hjertaas 1994). One seen in spring migration west of Saltcoats, May 1946, by Walter Domon (Walkinshaw

1949). Three were seen by Robert Barnhart near Saltcoats in the 1960s and one on 8 Nov 1986. One flew over, a few miles south of Melville, on 12 Apr [about 1960] (Victor Schmidt).

Three sightings from the YDM area were reported by Stephen (1979): three adults flying near Wadena by Mrs H.D. Meyers, 4 Sep 1977; 10 adults near Margo by V. and C. Lowey, 7 May 1978; nine adults and one immature near Bangor by Mrs J. Hilton on 21 Sep 1978.

Four flew over the Anaka yard at GSL on 10 April 1995.

The following confirmed records since 1957 have been provided by Brian W. Johns:

Spring migration: 4 miles south and 2 miles east of Elfros, 22-28 Apr 1988; GSL, three adults, 2 May 1995; east of Jasmin, one adult, 2 May 1998; Hubbard, one adult, 3 May 2000.

Summer: One adult summered at the southeast corner of Little Quill Lake, 30 May through 31 Aug 1981 (Johns 1987).

Fall migration: Mozart, three adults, 4-5 Oct 1957; Paswegin, one, 12 Oct 1957 only; Wadena, one adult, 6 Oct 1958 only; Elfros Community Pasture, one adult, 6-10 Oct 1958; 3 miles west and 12 miles north of Margo, two, 3-10 Oct 1992; Foam Lake, one adult, 24 Aug 1995 only; 2.5 miles northwest of Runnymede, three adults, 22 Oct 1995 only; 5 miles west of Preeceville, two, 14 Sep 2000 only.

Cutarm valley, 6 miles south of Churchbridge, six adults, 8-13 Oct 1999 (Jim Trowell, JRJ, SJ, Bain and Shanahan 1999f).

The Rousay-Leech-Crescent lakes-Rokeby marsh area and perhaps Horseshoe Lake were studied as possible sites for future reintroduction of a wild Whooping Crane population. However, the wetland area was small, many power lines were present nearby, the invertebrate population was low, use of agricultural chemicals was high, the wetland complexes were prone to drought, the area was too close to Yorkton with its potential for human interference, and it was within the migratory path of the Wood Buffalo population, which would have allowed potential for contact between the two populations (Burke 1996 and 2001; Lyon et al. 1995).

ESA:10 Apr 1995. **LSD:** 3 May 2000. **EFA:** 24 Aug 1995. **LFD:** 22 Oct 1995.

Breeding: historic, as above.

Recoveries, banded outside area, 1 (Wood Buffalo Park, NWT; banded as a flightless young on 8 August 1985, read by telescope near Leslie on 22 Apr 1988).

BLACK-BELLIED PLOVER Pluvialis squatarola

Uncommon transient.

Small flocks of 25 to 50 seen daily at Leech Lake from 19 May through 1 Jun 1945 (WHC). In 1973, single individuals seen on 17 and 24 May (AS). Three at Foam Lake marsh on 25 May 1974 (AS). Over 200 loafing on rocks and reefs along south shore of Little Quill Lake, 27 May 1994 (DH). Observed beside slough southwest of Preeceville on 4 May 2001 (DF) and 1 Jun 2002 (DF,DGJ,SAJ).

ESA(8): 24 Apr 1939 at Sheho (WN) and 10 May 1991 at GSL (WA), Av 22 May. **LSD**(11): 11 Jun 1981, Av 1 Jun. **EFA**(20): 2 Jul 1955 (Anaka 1955), Av 2 Sep. **LFD**(17): 9 Nov 63, Av 13 Oct. Stayed from 18 to 26 Oct at GSL (WA, Krause 1961).

DUCK MOUNTAIN: Rare transient, one at Bear Head Lake, 10 Sep 1981 (Hatch and Weidl).

AMERICAN GOLDEN-PLOVER *Pluvialis dominica*

Uncommon transient.

A flock seen at Fort Pelly, 18 September 1881 (Macoun in Seton 1891).

Seventy-five seen at slough within Melville town limits in May 1944 (LTMcK). Twelve at GSL on 26 May 1966 (WA, Hatch 1966). Five flocks, totalling 127 individuals, seen in pasture east of Lower Rousay Lake on 27 May 1973 (AS). A dozen near Hyas on 21 May 1985 (CP). At slough southwest of Preeceville on 4 May 2001 (DF).

This species generally follows a mid-continental northern path in spring and an offshore nonstop Atlantic route in fall (Johnson and Connors 1996). Hence, it is not surprising that there were no fall sightings in the Yorkton area--until 17 were seen at Horseshoe Lake on 17 Sep 1961 (Anaka 1961a; Nero 1962a).

ESA(8): 13 May 1978, Av 22 May. **LSD**: 26 May 1966, 27 May 1989 (WA), and 29 May 1945 (WN). **EFA**: 1 Aug 1964. **LFD**(4): 1 Oct 1972 at Sheho (WN, Houston 1973a) and 27 Sep 1964 (WA).

SEMIPALMATED PLOVER *Charadrius semipalmatus*

Uncommon transient.

At Sheho, 21 May 1939 and 1942 (WN). Three on 12 May and 20 on 17 May 1973 at Leech Lake (AS). Two at Foam Lake marsh on 23 May and two on 29 May 1974 (AS). Seen throughout Aug 1964 at GSL

(WA, Lister 1965).

ESA(6): 10 May 1959, Av May 20. **LSD**: 23 May 1989, 26 May 1957 (WA), and 9 Jun 1947 (WN). **EFA**(12): 17 Jul 1946 (Farley Mowat, CSH) and 1 Aug 1964 (WA). **LFD**(6): 24 Sep 2000, Av 4 Sep.

Banded 55 (Dickson at Little Quill Lake).

PIPING PLOVER *Charadrius melodus*

ENDANGERED, COSEWIC, MAY 2001

Formerly a rare summer resident. Bred at GSL until 1962.

Six spring dates: One at York Lake, 29 May 1939 (IMP); 30 Apr 1969, 17 May 1953 and 1968 at GSL (WA), and 3 May 1942 at Sheho (WN); one on south shore of Little Quill Lake, 27 May 1994 (DH). One adult on a small island in Stonewall Lake south of Invermay on 8 Jun 2001 (AS).

Two fall dates: 27 Sep 1953 at GSL (WA) and 14 Aug 1989 at Little Quill Lake (ARS).

Breeding: A nesting pair on the south shore of GSL, 23 Jun 1936 (JDS). Pair with four downy young on the east side of GSL, 23 Jun 1952 (CSH). Two flightless young seen and one banded by Anaka on a mudflat in a cultivated field 2 miles southeast of GSL, when high water levels in the lake forced them off the lakeshore beach in July 1955 (Anaka 1955). "Disappeared as a breeder in the 1950s" (Renaud et al. 1979). Territorial pair present at GSL, 10 Jun 1962 (Dorothy E. Wade, Nero 1962c). By the time of the International Piping Plover Census in 1996, a survey of 10 potential nest sites (including GSL) in the Yorkton area failed to detect a single breeding pair (Skeel, Duncan and Wiltse 1997, map p. 160). Smith (1996) maps a pre-1966 breeding record for Norquay.

Banded 3 (WA 1, Dickson 2).

KILLDEER *Charadrius vociferus*

Very common summer resident in vicinity of grassy uplands and sloughs throughout the area. Breeds.

Two eggs collected north of Saltcoats (NJ,PMA).

At Sheho, numbers increased with more intensive farming, since the Killdeer requires little or no grass cover for nesting (Niven 1973).

ESA(48): 21 Mar 1945 at Saltcoats (RB) and 27 Mar 1981 at GSL (WA), Av 8 Apr. **LFD**(40): 29 Oct 1961 (WA, Nero 1962a), Av 30 Sep.

Breeding: Early nest, 4 eggs, 4 May 1963 (WA). Between 27 May and 10 Jul 1958, 14 nests located but only four successful (WA). Late nest, 2 eggs, hatched 29 Jul 1950 (Darcy Wershler and Wayne Bjorgan).

In 1973, nest, 4 eggs, 12 Jun, and nest with four newly hatched young, 7 Jun (AS). Two nests each had 4 eggs at Little Quill Lake on 30 May 1998 (DH).

PNRS: 32 WA and two JG nest records from GSL; at Melville, seven nests in 1960 and two in 1961, all with 4 eggs (VS); at Stornoway, 14 nests, 1959-63, with 2, 2, 3, 3, 4, 4, 4, 4, 4, 4, 4, and 4 eggs, and two with three newly hatched young (SZ,RD); at Saltcoats, seven nests in 1958 and five in 1959, with 2, 3, 3, 4, 4, 4, 4, 4, 4, 4, and 4 eggs and one with three newly hatched young (BH); at Amsterdam, 14 May 1959 (WC); near Togo, 14 Jun 1964 (LM); 4 eggs at Ebel's marsh, 17 miles east of Foam Lake, on 10 Jun 1981 (Doug Graham); three young with one cowbird young at Kuroki Lake on 12 Jun 1981 [a coincidental association, one suspects] (Doug Graham), and three flightless young near Homefield, 27 Jun 1992 (Brenda Dale and Dan Sawatsky).

Also breeds at Sheho, Canora, Kamsack, Jasmin (ARS, *Atlas*).

Killdeers appeared to have had more success when they nested on bare summerfallow fields, even though some nests were destroyed by cultivating machinery, than they do now with the better cover provided by chemical summerfallow (WA).

BBS: overall mean, 12.9. Theodore, mean 13.2; decreasing (1970s, 21.3; 1980s, 13.2; 1990s, 6.6; n=33).

Banded 21 (CSH 8, 13 Dickson) + pre-1955, 23 (CSH 10, Blakeley 6, Wickencamp 7).

Langenburg: Very common at margins of open potholes, 1990 (TH,BL).

DUCK MOUNTAIN: SMNH 1926. SMNH 1951. SNHS 1956. Seen in 1974, 1975, 1976, 1977 (JAS). Common throughout the park; adults with two young on 4 Jun and 17 Jul 1981 (Hatch and Weidl). SNHS 1989.

Point counts: Rank #48, 2 individuals at 2 stops (ARS).

AMERICAN AVOCET *Recurvirostra americana*

Uncommon summer resident since the 1940s. Breeds.

A new arrival in the area, it first appeared at Yorkton about 1940. At Sheho, first seen on 2 Jun 1945, when a flock of 20 remained on a small alkali slough all summer (Priestly 1945c); seen there in 12 subsequent years.

At Crescent marsh, Dean Maddaford saw none between 1957 and 1973, when AS found them regular but uncommon at Crescent marsh and Leech Lake. Ten at Foam Lake marsh on 12-15 Jun 1974 (AS). Over 1,100 non-breeding birds seen in aerial survey of Little Quill Lake on 4 Jul 1988 (Michael Gollop, Gollop 1988b) and over 200 on a ground survey on 12 Jul 1988 (DH). Twenty present in DU's Foam Lake marsh complex, 14 and 17 Jun 1989 (CSH,MIH,WA,FAS).

ESA(22): 25 Apr 1999, Av 9 May. Unusual fall presence at GSL, from 2 Sep to 6 Oct 1991, with **LFD**: 6 Oct 1991.

Breeding: Eight pairs nested on a gravelly island in Leech Lake in company with Ring-billed Gulls, June 1945 (WHC). Three nests, with 4 eggs each, 17 May 1964, in slough at Dave Baines' farm south of Crescent marsh (CSH,MIH). One breeding record at GSL, 4 eggs, 12 Jun 1977 (WA). On nests on island at dam on west side of Melville, 22 Jun 1989 (CSH,MIH). Nest, 4 eggs, in a wet field on 15 Jun 1990, 1 mile northwest of Bangor (John Dycer, AS). Two nests each had 4 eggs at Little Quill Lake on 30 May 1998 (DH).

PNRS: Nest, 4 eggs, 10 Jun 1981 at Ebel's marsh east of Foam Lake (Doug Graham).

Also breeds at Sheho (ARS, *Atlas*).

BBS: overall mean, 0.1. Theodore, 1 in 1996, mean 0.03.

Langenburg: 1, 1 Jun 1976 (DH). One pair seen, 1990 (TH,BL).

DUCK MOUNTAIN: Rare transient, one pair on shore of Watt Island on 24 Jun 1981 (Hatch and Weidl).

GREATER YELLOWLEGS *Tringa melanoleuca*

Fairly common transient. No breeding records.

Very abundant on the Assiniboine, 1881 (Macoun in Seton 1891). In 1973, recorded from 26 Apr through 2 May; an adult, acting as though on territory, was seen at Upper Rousay Lake on 21 Jun (AS). In a given year, regarding both species of yellowlegs, the spring migrants have been gone for only a week or so before the first fall migrants return. With variations of a week or so between years, the latest spring departure date on record approaches (Greater) or overlaps (Lesser) with the earliest fall arrival date, but, of course, in different years.

ESA(46): 7 Apr 1943 (IMP) and 1976 (WA), Av 20 Apr. **LSD**(22): 10 June 1992 (WA), west of Churchbridge, two on 21 Jun, and one near Yorkton golf course, 22 Jun 1989 (CSH,MIH), Av 15 May. **EFA**(36):

seven on 28 Jun 1973 (AS), and 13 Jul 1989 (WA), Av 21 Jul. **LFD**(44): 5 Nov 1990, Av 19 Oct.

BBS: overall mean, 0.06. Theodore, 1 in 1998, mean 0.03.

Banded 5 (Dickson).

DUCK MOUNTAIN: Seen in 1975, 1977 (JAS). Uncommon, one each at Watt Island on 18 Jun and Bear Head Lake on 25 Jun, and two near Pelly Bay, 15 Jul; during migration, one each on 12 Aug and 10 Sep 1981 (Hatch and Weidl). Four at sewage lagoon, 4 Jul 1990 (TH).

LESSER YELLOWLEGS *Tringa flavipes*

Common transient. No breeding records.

Abundant on the Assiniboine, 1881 (Macoun in Seton 1891).

In a given year there are only a few days near the first of July when one is not likely to see or hear this species. Fifty, probably late transients, on Foam Lake marsh on 17 Jun 1989 (WA,FAS). Hundreds at Little Quill Lake on 12 Jul 1988 (DH). One to six individuals present at GSL throughout the summer of 1989, but no evidence of breeding (WA).

Present near Preeceville, 24 Jun 1991 (MIC,MIH,CSH).

ESA(32): 8 Apr 1946 and 17 Apr 1949 at Saltcoats (RB), 12 Apr 1976 at Sheho (WN) and 20 Apr 1977 at GSL (WA), Av 4 May. Three observations between 2 and 8 May 1973 (AS). **LSD**(22): 23 Jun 1963, Av 20 May. **EFA:** west of Churchbridge, 21 Jun 1989 (CSH,MIH) and 22 Jun 1989 (WA), Av 21 Jul. **LFD**(29): 15 Oct 1961, Av 16 Sep.

BBS: overall mean, 0.09. Theodore, 2 in 2000, mean 0.06.

Banded 365 (Dickson) + pre-1955, 6 (CSH).

Recoveries, banded in area, 1 [0.3%] (state not given).

DUCK MOUNTAIN: SMNH 1951. Seen in 1975 (JAS). Uncommon, one each at Bear Head Lake on 1 Jun and Green Lake on 3 Jun; in migration, three at Batka Lake, 9 Sep, and Bear Head Lake, 10 Sep 1981 (Hatch and Weidl). At sewage lagoon: four on 24 Aug 1989 and 17 on 4 Jul 1990 (TH).

SOLITARY SANDPIPER *Tringa solitaria*

Fairly common transient. Uncommon summer resident at Duck Mountain. No breeding records.

One or two seen between 2 and 11 Aug 1973 (AS). Three along Whitesand River north of Yorkton, 11 Aug 1988 (DH).

ESA(15): 22 Apr 1946 (WN) and 30 Apr 1962 (WA), Av 8 May. **LSD**(9): 23 May 1967, Av 14 May. **EFA**(24): 26 Jun 1988 (WA) and 7 on 28 Jun 1973 (AS), Av 2 Aug. **LFD**(16): 14 Sep 1986, Av 27 Aug.

Banded 2 (Dickson)+ pre-1955, 2 (Blakeley).

DUCK MOUNTAIN: SMNH 1926: study skin male, #1897. SNHS 1956. Rare summer resident, one seen at southeast corner of the park on 19 Jun 1981 (Hatch and Weidl). SNHS 1989. At sewage lagoon, six on 24 Aug 1989 and two on 4 Jul 1990 (TH). One calling and scolding at the marl bog, Jun 1993 (Herriot 1995). One, upper Assiniboine River, 7 Sep 2002 (TH).

WILLET *Catoptrophorus semipalmatus*

Common summer resident in vicinity of sloughs and lakes. Breeds.

Three eggs, Crescent Lake, 8 Jun 1903, Baines for Raine, WFVZ. Four eggs, Crescent Lake, 6 Jun 1903, F. Baines, AMNH.

Nine during Foam Lake marsh survey on 24-25 Jun 1974 (AS). Eighteen at Leech Lake, 21 Jul 1994 (DH).

ESA(44): 23 Apr 1943 and 1945 at Sheho (WN) and 27 Apr 1953 and 1964 at GSL (WA), Av 4 May. **LFD**(22): 12 Sep 1940 (CSH) and 6 Sep 1990 (WA), Av 14 Aug.

Breeding: Early nest, 3 eggs, 17 Jun 1955, and 4 eggs, 17 Jun 1961, two of which hatched on 3 Jul (WA). An immature, not yet able to fly, was banded at Upper Rousay Lake on 11 Jul 1946 (CSH).

PNRS: at Melville, nest, 3 eggs, 31 May 1960 (VS).

BBS: overall mean, 1.0. Theodore, 15 years; highest, 5, 1994, mean 0.8.

Banded 37 (Dickson)+ pre-1955, 6 (CSH).

Langenburg: One sighted, 1990 (TH,BL).

DUCK MOUNTAIN: Rare, one seen near Ranger Bay on 4 Jun 1981 (Hatch and Weidl).

SPOTTED SANDPIPER *Actitis macularia*

Fairly common summer resident. Breeds.

Two eggs collected north of Saltcoats (NJ,PMA).

Seen chiefly along gravelly shores of lakes and creeks, including the Little Whitesand River and Cussed Creek. Pair on shore of GSL, 23 Jun 1936 (JDS). Niven's first-ever sighting at Sheho was on 20 May 1951 (WN). Pair on shore of Stonewall Lake south of Invermay on 8 Jun 2001 (AS).

ESA(44): 2 May 1977, Av 12 May. **LFD**(35): 27 Sep 1983, Av 10 Sep.

Breeding: Early nest, 4 eggs, 31 May 1957 (WA).

Three immatures, unable to fly, were seen at Birch Island Lake on 28 Jul 1954 (Thompson 1954). An immature, four-fifths grown, was banded at Crescent Lake on 13 Aug 1958 (CSH).

PNRS: at Melville, nest with 2 eggs on 10 Jun 1960 and another with 4 eggs on 9 Jul 1960 (VS); at GSL, 2 eggs on 11 Jun 1977 (WCH).

Also breeds at Wadena, Buchanan, Kamsack (ARS, *Atlas*).

BBS: overall mean, 0.5. Theodore, 10 years; highest, 4, 1972, mean 0.5.

Banded 4 (CSH 1, Dickson 3)+ pre-1955, 2 (Wickencamp).

Langenburg: Two pairs seen along Cutarm Creek, 1990 (TH,BL).

DUCK MOUNTAIN: SMNH 1926. SMNH 1951. SNHS 1956. Seen in 1975, 1976, 1977 (JAS). Common summer resident along shores throughout the park; nest, 4 eggs, Watt Island, 18 Jun 1981 (Hatch and Weidl). Four downy young running with adult at Boomerang Lake, 3 miles northeast of Runnymede, 21 Jun 1988 (TH). SNHS 1989.

PNRS: Nest, 4 eggs, 17 Jun 1979 and one flightless young on 20 Jun 1986 (HWRC).

UPLAND SANDPIPER *Bartramia longicauda*

Now an uncommon summer resident on pastures, formerly common. Breeds.

When Fred Langstaff settled in the Wallace area near Rhein in 1892, the "Bartramian Sandpiper" was one of the most plentiful and tastiest birds on the grassy uplands, even at some distance from water. As land was broken, Upland Sandpipers decreased, although they remained common and widely distributed until about 1910 (Frank Baines, Fred Langstaff). Reed (1904:122) features a full-page photograph of a nest and 4 eggs taken by Walter Raine at Crescent Lake on 9 June 1901. Common east of Saltcoats until the late 1920s (RB). Two eggs collected north of Saltcoats (NJ,PMA).

At Sheho, Upland Sandpipers disappeared almost entirely about 1927, then reappeared in smaller numbers about 1939, became scarce again after 1960 (Niven 1973), and disappeared once more for about 10 years. One was seen in 1975, a few in 1977, one on 11 May 1978, one in 1979, one each on 7 Jun 1984 and 9 May 1985, and several on 13 May 1986 (WN).

Near Yorkton, they disappeared almost entirely during the 1930s and reappeared about 1940 in small numbers on grassy flats around Leech, Crescent, and both Rousay lakes. Two were seen on the pastures beside Upper Rousay Lake, 14 Jun 1989, and three near Crescent Lake, 20 Jun 1989 (CSH,MIH).

Four adults near Horseshoe Lake on 9 Jul 1975 (DH). Six near Rokeby marsh on 23 May 1982 (DH). Two sighted on sandy prairie in a community pasture, 3 miles north of Preeceville, on 8 Jul 1985 (D.F. Hooper, L. Baker).

ESA(38): 27 Apr 1940 (CSH) and 2 May 1987 (WA), Av 13 May. **LFD**(14): 24 Aug 1954 and 1955, Av 25 Jul.

Breeding: Early nest, 4 eggs, 31 May 1991 (WA) and 5 Jun 1955 at York Lake (CSH). Nest, 3 eggs, Upper Rousay Lake, 8 Jun 1946, had 4 eggs the following day. Immature captured and banded at Lower Rousay Lake on 11 Jul 1944, and immature, several days old, seen on 27 Jun 1947 (CSH).

At GSL, adult with one or more young on 15 Jul 1987. In 1991, a nest had 4 eggs on 31 May and 18 Jun and was empty on 26 Jun (WA).

BBS: overall mean, 0.7. Theodore, 15 years; highest, 7, 1994, mean 1.0.

Banded pre-1955, 1 (CSH).

WHIMBREL *Numenius phaeopus*

Vagrant.

One noted at Melville on 20 May 1945 (LTMcK, Priestly 1945b). One sighted along Highway 9, several miles north of Yorkton, by Marilyn Pollock, about 21 May 1985 (Pollock 1986c).

LONG-BILLED CURLEW *Numenius americanus*

Vagrant.

One seen at York Lake on 22 May 1938 (IMP), and one at Little Quill Lake on 2 Aug 1989 (ARS).

HUDSONIAN GODWIT *Limosa haemastica*

Rare transient.

One seen at slough on southwest edge of Yorkton, 14 May 1939 (Priestly 1945b). One at Leech Lake, 25 May 1945 (JHW). Three on 20 and 26 May 1956 at GSL (WA). Two at Crescent marsh, 17 May 1973 (AS). At Foam Lake marsh, a flock of 135 on 17 Jul 1973 grew to 394 the next day, with 70 Marbled Godwits (Harris 1974). Four at Hyas, 21 May 1984 (CP). At Little Quill Lake, 1,800 on 10 Aug 1989 (PC, Koes and Taylor 1990a).

Banded 110 (Dickson).

MARBLED GODWIT *Limosa fedoa*

Fairly common summer resident in vicinity of sloughs and lakes. Breeds.

Reed (1904:118) has a full-page photograph by Walter Raine of nest with 4 eggs at Crescent Lake, 6 June 1901; 2 eggs, 2 June [1902?], ROMZ #8796].

At Sheho, godwits decreased with heavy overgrazing of pastures in the 1940s and 1950s (Niven 1973). A godwit attacked a crow passing by; the godwit stabbed the crow in the neck, breaking it, and the crow fell dead to the ground (Baines 1942). Seven at Foam Lake marsh on 25 June 1974 (AS).

ESA(42): 21 Apr 1943 at Saltcoats and Sheho (RB,WN), and 23 Apr 1981 (WA), Av 2 May. **LFD**(9): 15 Aug 1991, Av 28 Jul. South of Yorkton, three groups totalling 21 individuals gathered on 13 Aug 1973 (AS).

Breeding: Early nests, 4 eggs, 22 May 1963 and 31 May 1961 (WA); 5 eggs, 24 May 1973 (AS).

Nest at Upper Rousay Lake, 1 egg, 25 May 1946 (CSH). Nest near Salt Lake north of Sheho in 1979; eggs hatched successfully (WN).

PNRS: nest with 1 egg near Amsterdam on 1 Jun 1959 had 4 eggs on 4 Jun (WC).

Also breeds at Yorkton (ARS, *Atlas*).

BBS: overall mean, 0.6. Theodore, 18 years; highest, 4, 1975, mean 0.9.

Banded 29 (WA 2, Dickson 27) + pre-1955, 2 (JHW 1, CSH 1).

Recoveries, banded in area, 2 [6.5%] (SK 1, CA 1 [banded at Leech Lake 3 Aug 1944 by JHW, picked up dead at Oxnard, north of Los Angeles, CA, in December 1944]).

RUDDY TURNSTONE *Arenaria interpres*

Uncommon transient.

One at CNR Dam, Melville, 15 May 1940 (LTMcK). Several at York Lake, 26 May 1946. One at Salt Lake near Sheho on 22 May 1959 (WN). Two at York Lake, 28 May 1971 (AS). Six at Foam lake marsh on 28 May 1974 (AS).

At GSL, three on 10 Jun 1962 (Dorothy E. Wade, Nero 1962c), six on 27 May 1989 (RK,CP,RE,TH), and one on 18 May 1990 (WA). Two at Little Quill Lake, 14 Aug 1989 (ARS) and again on 30 May 1998 (DH).

ESA(8): 18 May 1990, Av 25 May. **LSD**(5): 11 Jun 1982. One fall date: 17 Aug 1994.

DUCK MOUNTAIN: Recorded by R.J. Long, August 1970.

[**SURFBIRD** *Aphriza virgata* HYPOTHETICAL

Vagrant.

One sighted by Josh Bilyk at Little Quill Lake, 19 Aug 1998, the first-ever for Saskatchewan (Bain and Shanahan 1999e).]

RED KNOT *Calidris canutus*

Rare transient.

Four sightings, all at Little Quill Lake: two on 14 Aug 1989 and two on 2 Aug 1990 (ARS); one along south shore on 27 May 1994, and eight on 30 May 1998 (DH).

Banded 2 (Dickson, Little Quill Lake, late May 1990).

SANDERLING *Calidris alba*

Uncommon transient.

Seen occasionally in migration at Melville during the 1930s and early 1940s (LTMcK).

At York Lake on 6 Jun 1945 (CSH), also 12 on 31 May 1986 (DH). Eight at Little Quill Lake on 30 May 1998 (DH). A flock of about 100 was chased by an adult male Merlin at Usinneskaw Lake, 3 miles south and 1 mile west of Margo, on 4 Jun 2001 (AS).

At York Lake, 4 Oct 1944 (IMP); a flock of over 100 was seen at GSL, 6 Sep 2000 (WA,JA, Bain and Shanahan 2000c).

ESA(13): 9 May 1992, Av 22 May. **LSD**(13): 22 Jun 1990, Av 6 Jun. **EFA**(14): 4 Aug 2000, Av 30 Aug. **LFD**(12): 29 Oct 1989, Av 27 Sep.

DUCK MOUNTAIN: SMNH 1926. Recorded by R.J. Long, Aug 1970. One, 14 Jun 1990 (MB).

SEMIPALMATED SANDPIPER *Calidris pusilla*

Common transient.

At Little Quill Lake on 16 Aug 1989, Paul Chytyk counted 3,000 (Koes and Taylor 1990a), and on 23 May 1990, 10,800 Semipalmateds among 91,500 shorebirds (Koes and Taylor 1990b). Bill lengths of this species banded at Little Quill indicate that spring flocks contain birds destined for both the western Arctic and central Arctic. Those in fall are mainly the shorter-billed birds from the western Arctic, which return south through the prairies, whereas those from the central Arctic tend to return south along the Atlantic coast. Many winter in French Guiana (Gratto-Trevor and Dickson 1994). At Little Quill Lake on 30 May 1998, 1,500 present (DH).

ESA(5): 15 May 1940 (IMP) and 18 May 1998 (WA). **LSD**: 5 Jun 1996 (WA) and 7 Jun 1942 (CSH). **EFA**: 17 Jul 1973 (AS), but only one fall date at GSL: 13 Aug 1998 (WA).

Banded 3,449 (Dickson) + pre-1955, 4 (CSH).

Recoveries, banded in area, 9 [0.3%] (SC 1, Cuba 1 [banded 19 Aug, mistnetted 13 Sep, roughly 2,300 miles in 25 days], Guyana 7 [six degrees north of the equator, four died, three released alive]).

Recoveries, banded outside area, 1 (CT, mistnetted again at Little Quill Lake).

DUCK MOUNTAIN: Uncommon transient; eight at Bear Head Lake on 10 Sep 1981 (Hatch and Weidl).

WESTERN SANDPIPER *Calidris mauri*

Rare transient.

Banded 2 (Dickson at Little Quill Lake, 4 Jun 1989 and 18 Aug 1991; Smith 1996).

LEAST SANDPIPER *Calidris minutilla*

Common transient.

Leasts overfly GSL when lake water levels are high.

ESA(6): 10 May 1959 (WA) and 12 May 1973 (AS). **LSD**: 23 May 1956 and 1989, 28 May 1954 (WA), and 7 Jun 1942 (IMP). **EFA**: 7 Aug 1999 (WA) and five on 13 Aug 1973 (AS). **LFD**(5): 11 Sep 1991.

Banded 452 (Dickson) ı pre-1955, 5 (Blakeley 1, CSH 4).

DUCK MOUNTAIN: Luterbach, July 1973.

WHITE-RUMPED SANDPIPER *Calidris fusicollis*

Uncommon spring transient, except occasionally abundant at Little Quill Lake.

Duck Mountain, June 1884 (Seton 1891).

One at Ravine Reserve, 1 Jun 1941 (IMP). Large flock at Melville, 23 May 1943 (LTMcK). Fair numbers at Leech Lake in late May 1945 (WHC). Seen only twice at GSL, 31 May 1956 and 1964 (WA). Four on marsh near Yorkton, 24 May 1982 (DH).

At Little Quill Lake on 23 May 1990, Paul Chytyk counted an astounding 15,300 White-rumpeds among 91,500 shorebirds (Koes and Taylor 1990b).

No fall dates.

DUCK MOUNTAIN: Rare transient, flock of 17 at Bear Head Lake on 1 Jun 1981 (Hatch and Weidl).

BAIRD'S SANDPIPER *Calidris bairdii*

Uncommon transient.

A largest-ever (for YDM area) flock of 1,500 at Little Quill lake on 30 May 1998 (DH).

ESA: 19 May 1942 (IMP) and 23 May 1943 (LTMcK). Seen four times between 20 and 31 May 1956 (WA). **EFA**: 7 Aug 1999 and 16 Aug 1991. **LFD**: 7 Oct 2000.

Banded 6 (Dickson).

DUCK MOUNTAIN: Seen 1977 (JAS).

PECTORAL SANDPIPER *Calidris melanotos*

Common transient.

ESA(9): 4 May 2001 at GSL (WA), Av 14 May. **LSD**: 29 May 1956 (WA). **EFA**: 17 Jul 1938 (IMP) and 21 Jul 1955 (WA). **LFD**(25): 25 Oct 1964, Av 7 Oct.

Banded 99 (Dickson) + pre-1955, 1 (Blakeley).

DUCK MOUNTAIN: Uncommon transient, seven at Bear Head Lake on 10 Sep 1981 (Hatch and Weidl).

DUNLIN *Calidris alpina*

Uncommon transient.

At Melville, 22 May 1940 (LTMcK). One or two seen daily at Leech Lake from 24 May through 7 Jun 1945 (WHC, Priestly 1945b). Only two records at Sheho: 18 May 1950 and 15 May 1959 (WN). Two at Leech Lake, 24 May 1973 (AS). At GSL, 31 May 1989 (WA). One at Little Quill Lake, 27 May 1994 (DH).

The only fall record is of an adult among a large flock of Sanderlings at GSL on 5 Sep 2000 (WA,JA, Bain and Shanahan 2000c).

STILT SANDPIPER *Calidris himantopus*

Uncommon to common transient.

Sixty-seven at Foam Lake marsh on 10 Jul 1974 (AS). Thirty-eight on marsh beside Yorkton Trailer Park, 24 May 1982 (DH).

At Little Quill Lake on 23 May 1990, Paul Chytyk counted 12,600 Stilts among 91,500 shorebirds (Koes and Taylor 1990b); there were 500 there on 30 May 1998 (DH).

ESA: 14 May 1954 (WA) and 22 May 1940 (LTMcK). **LSD**: 1 Jun 1954 and 27 May 1962 (WA). **EFA**: 17 Jul 1946 at Leech Lake (Farley Mowat, CSH) and 20 Aug 1966 (WA). **LFD**: 6 Sep 1969 (WA).

Banded 1,454 (Dickson) + pre-1955, 1 (at Leech Lake, CSH).

Recoveries, banded in area, 2 [0.1%] (MT 1, [died of botulism at Medicine Lake NWR one day short of five years later]; Venezuela 1).

BUFF-BREASTED SANDPIPER *Tryngites subruficollis*

Rare transient.

Two at Sheho, 20 May 1955 (WN). One at GSL, 29 Aug 2000 (Bain and Shanahan 2000c).

A remarkable 44 were seen at Little Quill Lake on 14 Aug 1989 (Koes and Taylor 1990a)—[the 70 seen on 14 Aug 1991 were on the west side of the lake, outside of the YDM area (PC, Koes and Taylor 1992a; ARS, pers. comm.)].

SHORT-BILLED DOWITCHER *Limnodromus griseus*

Uncommon transient.

One of three singles at Leech Lake was collected on 18 Jul 1946 (Farley Mowat, unpub. field notes 1946). Seen twice at GSL: 22 June 1989 and 18 May 2000 (WA).

ESA: 18 May 2000.

Banded 102 (Dickson).

DUCK MOUNTAIN: Uncommon transient; 20 at Bear Head Lake on 10 Sep 1981 (Hatch and Weidl). At sewage lagoon: one on 24 Aug 1989 and two on 4 Jul 1990 (TH).

LONG-BILLED DOWITCHER *Limnodromus scolopaceus*

Fairly common transient.

At Foam Lake marsh, 44 on 25 May 1974 (AS), and one on 17 Jun 1989 (WA,FAS).

A flock of 3,500 dowitchers (species not separable) was counted on Little Quill Lake on 14 Aug 1989 (Koes and Taylor 1990a). Fourteen fall sightings at GSL.

ESA: 13 May 1962 (WA) and 15 May 1940 (IMP). **LSD:** 7 Jun 1942. **EFA:** Group of six at Leech Lake, 17 Jul 1946 (Farley Mowat) and four at Leech Lake, 17 Jul 1973 (AS). **LFD:** 19 Oct 1958.

Banded 99 (Dickson).

WILSON'S SNIPE *Gallinago delicata*

Fairly common summer resident. Breeds.

Raine collected three sets of 4 eggs, at Crescent Lake and the Saltcoats marshes, 4 and 5 Jun 1901, WFVZ.

Its "winnowing" call, made by its flight feathers, can be heard over larger marshes, including Lower Rousay Lake and Horseshoe Lake (CSH,DH). Disappeared by the dry years of 1988 and 1989 at Salt Lake, north of Sheho (WN). In 1989, heard at Churchbridge on 11 Jun and at Ituna on 22 Jun (CSH,MIH).

ESA(45): 7 Apr 1958, Av 23 Apr. **LFD**(30): 2 Nov 1958, Av 11 Oct.

Breeding: At GSL, early nest, 4 eggs, 28 May 1958; 4 eggs, 20 Jun 1987; two tiny young, 24 Jun 1956; two young, 13 Jun 1997 (WA). Nest, 4 eggs, found by Herb Upshall, 7 miles north of Saltcoats about 1990 (JRJ).

BBS: overall mean, 4.9. Theodore, mean 5.3 (1970s, 5.4; 1980s, 3.2; 1990s, 8.5; n=25).

Banded 12 (Dickson) + pre-1955, 6 (CSH 4, Blakeley 2)

Langenburg: Common, winnowing over sloughs, 1990 (TH,BL).

DUCK MOUNTAIN: SMNH 1926. SNHS 1956. Seen in 1976 (JAS). Uncommon summer resident, heard winnowing at four places through June and July 1981 (Hatch and Weidl). Two at sewage lagoon, 24 Aug 1989 (TH).

[AMERICAN WOODCOCK *Scolopax minor* HYPOTHETICAL

Vagrant.

One sighted at Hazel Dell on 13 May 1984 by Curtis and Marilyn Pollock (Pollock 1986a). One, 11 miles south of Preeceville, May 1987 (M. Weiman in Smith 1996).

DUCK MOUNTAIN: One heard but not seen on 21 May 1988 (RK, Gollop 1988a).]

WILSON'S PHALAROPE *Phalaropus tricolor*

Fairly common summer resident on sloughs. Breeds.

In 1973, 10 groups of five birds each at Crescent marsh on 26 Jul (AS). Eighty-one at Foam Lake marsh on 23-25 May 1974 (AS).

ESA(31): 26 Apr 1945 (RB) at Saltcoats, 8 May 1944 at Sheho (WN), and 1955 at GSL (WA), Av 21 May. **LFD**: 31 Aug 1982.

Breeding: Early nests with 4 eggs near Saltcoats on 22 May 1958; another had 4 eggs on 28 Jun 1959 (BH). Nests at the Rousay lakes: 4 eggs on 6 Jun 1943 (IMP,CSH); 3 eggs on 28 May 1948 were a complete clutch of 4 eggs on 29 May (CSH). Nest with 4 eggs at Rokeby marsh, 31 May 1945 (WHC).

Good-sized flocks appeared at Salt Lake, north of Sheho, each spring. Small flocks often flew over Niven's nearby farmyard, "just skimming the ground." One night about dusk a flock came over, "diving like bullets," and crashed into the telephone wires. Over a dozen were knocked to the ground, eight were killed outright, and one had its wing sheared from its body. Others were just stunned and ran away through the grass (Niven 1946).

At GSL, three clutches of 4: on 28 May 1958 (this nest had four young on 17 June), 10 Jun 1962, and 23 Jun 1967 (WA). In 1973, nests with 4 eggs on 29 May and with 3 eggs on 2 Jun (AS).

PNRS: at Stornoway on 8 Jun 1960, four incomplete clutches of 2, 2, 3 and 3 eggs (SZ), and on 6 Jun 1962, a clutch of 4 (RD).

BBS: overall mean, 0.3. Theodore, 4 years; highest, 3, 1975, 1976, mean 0.3.

Banded 127 (Dickson).

Langenburg: Uncommon and local in groups of two to six, 1990 (TH,BL).

DUCK MOUNTAIN: Rare summer resident, seen on 2 Jun along west park boundary and on small island near Miles Island, 18 Jun 1981 (Hatch and Weidl).

RED-NECKED PHALAROPE *Phalaropus lobatus*

Common, sometimes abundant, transient.

May migrate in large flocks. Over 500 on 30 May 1945 at Leech Lake (WHC). Eighty-five at Foam Lake marsh on 29 May 1974 (AS). Over 3,000 at GSL on 2 Jun 1989 (JA,WA, Gollop 1989b).

At Little Quill Lake on 10 Aug 1989, Paul Chytyk counted 6,000 (Koes and Taylor 1990c); on 23 May 1990 there were 26,000 Red-necked Phalaropes among 91,500 shorebirds (Koes and Taylor 1990b).

ESA(7): 18 May 1940. **LSD**: six on 6 Jun 1973 (AS). **EFA**: 9 on 30 Jul and 120 on 11 Aug 1961, also 40 from 31 Jul through 7 Aug 1960 at GSL (WA, Krause 1961; Nero 1961). **LFD**: a specimen brought by Judy Lynn and Shirley Pfeifer, Yorkton, 14 Sep 1959.

Banded 42 (Dickson).

RED PHALAROPE *Phalaropus fulicariaus*

Vagrant.

A single individual at Little Quill Lake, 18-22 Jul 1990, was seen first by S. Alexander (ARS,TH,RK,BL; Smith 1996; ARS, pers. comm.; Koes and Taylor 1990c).

FRANKLIN'S GULL *Larus pipixcan*

Common summer resident, decreasing. Breeds.

Flocks of 500 to 1,000 have appeared at GSL during grasshopper infestations (WA). Hundreds in exceptionally heavy migration at Sheho, 29 Apr through end of May 1957 (WN, Gammell 1957b). In 1973, nearly 200 present on Crescent marsh on 26 May and over 20 were seen at Upper Rousay Lake on 24 Jun (AS). There were 361 at Foam Lake marsh on 6-8 Jun 1974 (AS). Flock of more than 50 on Salt Lake north of Sheho on 17 Jun 1989 (WA,FAS).

ESA(48): 10 Apr 1986, Av 25 Apr. **LFD**(22): 30 Sep 1962, Av 13 Sep.

Breeding: In the 1960s, a pilot from Yorkton Flying Services flew low over the Foam Lake marsh and reported a large colony of Franklin's Gulls, but I have mislaid the year. Since at least 1990, there has been a colony of 100 or more pairs in the freshwater marsh maintained by the DU dam where Milligan Creek enters the east arm of Little Quill Lake (Chuck Deschamps, pers. comm.).

BBS: overall mean, 2.9. Theodore, 5 years; highest, 141, 1992, mean 4.6.

DUCK MOUNTAIN: SMNH 1926. SMNH 1951. SNHS 1956. Three at Ministik Beach, 20 Jun 1969 (HWRC). Seen in 1974, 1976, 1977 (JAS). Fairly common, 10 over East Bay on 4 Jun, several over Whitethroat Trail on 15 Jun, a few over Spruce Island on 18 Jun 1981 (Hatch and Weidl). SNHS 1989.

BONAPARTE'S GULL *Larus philadelphia*

Uncommon transient and summer resident. Breeds at Duck Mountain.

Only occasionally identified in spring migration, such as late May and early Jun 1972 (WA). In 1973, between 20 and 50 seen on four occasions, 1 through 14 May, and up to 10 non-breeding yearlings were seen at York Lake from 30 May through 6 Jul (AS).

Fairly regular in fall migration on GSL where, for example, 120 were seen on 3 Sep 1959 (Krause 1960a). In 1960, they were present daily from 4 through 26 Sep (WA, Krause 1961). In 1961, a year when grasshoppers were prevalent, Bonaparte's Gulls appeared on 27 Jul, increased to 70 by 5 Aug, and stayed until 17 Sep (WA, Nero 1961; Nero 1962a). In 1962, 40 arrived on 29 Jul, with the last 24 seen on 14 Oct (WA, Nero 1962c; Nero 1963). In 1967 there were 14 on 26 Aug, 30 on 30 Aug and the last eight on 24 Oct (WA). Three adults at Foam Lake marsh on 7-8 May 1974 (AS). They stop also at smaller lakes, including McOuat Lake west of Gorlitz on 28 Aug 1977 (Morgotch 1978).

ESA(27): 24 Apr 1977, 1989, 1994, 2000, Av 5 May. **LFD**(37): 8 Nov 1999, Av 10 Oct.

BBS: Theodore, 1, 1988, mean 0.03.

Breeding: DMPP only.

DUCK MOUNTAIN: SMNH 1926: study skin female, #1893. SMNH 1951. Two adults and five yearlings at Ministik Beach on 20 Jun 1969 (Copland 1969). Seen in 1977 (JAS). Nest in Black Spruce, southeast corner of park, 17 Jun 1981 (Weidl 1983). High count of 45 at Ministik Beach, 17 Aug 1990 (TH). Uncommon summer resident; adult over Madge Lake on 18 Jun; in migration, 14 on 9 Sep and 10 on 10 Sep 1981 (Hatch and Weidl). SNHS 1989.

RING-BILLED GULL *Larus delawarensis*

Common summer resident. Breeds at Fishing Lake and Duck Mountain.

At York Lake, a night roost on the south shore contained over 100 birds on 12 May 1973 (AS), with a similar number on 11 Aug 1991 (DH). Eight at Foam Lake marsh on 31 May 1974 (AS). A flock of over 150 at Salt Lake, north of Sheho, on 17 Jun 1989 (WA,FAS).

ESA(38): 31 Mar 1997, Av 14 Apr. **LFD**(27): 19 Nov 1983, Av 28 Oct.

Breeding: Two nests on gravelly island in Leech Lake, early July 1944, and one nest on same island, June 1945 (JHW,WHC). Colony of several hundred nests on island in Fishing Lake, 28 Jun 1963 (CSH). Five nests with 1 to 3 eggs on small island in Stonewall Lake south of Invermay on 8 Jun 2001 (AS).

BBS: overall mean, 1.5. Theodore, 14 years; highest, 32, 2001, mean 2.0.

Banded 236 (CSH at Fishing Lake in 1963).

Recoveries, banded in area, 9 [3.8%] (SK 4, SD 1, TX 1, Mexico 3).

Recoveries, banded outside area, 10 (SK 7, five at Last Mountain, one at Redberry and one at Big Quill; MB 1, north end of Lake Winnipegosis; ND 2, Chase Lake, Stutsman County).

Langenburg: Fifteen at Langenburg town dump, 1990 (TH,BL).

DUCK MOUNTAIN: SMNH 1926. SMNH 1951. SNHS 1956. Two at Ministik Beach, 20 Jun 1969 (HWRC). Seen in 1974, 1975, 1976, 1977 (JAS). Common in summer, regular over Madge Lake and at the garbage dump. Two nests, each with 2 eggs, on small islands on 18 Jun. Late fall concentration of 40 on 5 November 1981 (Hatch and Weidl). SNHS 1989.

CALIFORNIA GULL *Larus californicus*

Uncommon but probably regular transient. One breeding record.

Increased numbers in spring migration at Sheho in 1949 (WN, Gammell and Gammell 1949b). At Leech Lake on 2 May and 6 Jun and at York Lake on 5 May 1973 (AS). An immature flew over Upper Rousay Lake on 13 Aug 1974 (CSH,MIH). One seen at Hyas on 21 Apr 1985 (CP). Twenty at Langenburg town dump, 1990, one with a dyed red patch on its right cheek; its legs were submerged and thus hiding any band it might have carried (TH,BL). Five at York Lake, 11 Aug 1991 (DH).

ESA: 22 Apr 1991 (WA).

Breeding: at least two nests present on island in Fishing Lake, 28 Jun 1963; two young were banded (CSH).

Banded 2 (CSH).

DUCK MOUNTAIN: SMNH 1926, immature male, #1891. Seen in 1975, 1976 (JAS). Common in summer, regular over Madge Lake and at the garbage dump (Hatch and Weidl 1981). Two at Ministik Beach, 10 Jul 1990 (TH).

HERRING GULL *Larus argentatus*

Fairly common transient.

One at Fishing Lake gull colony on 6 Jun 1974 (AS).

ESA(14): 5 Apr 1943 at Yorkton (CSH) and 15 Apr 1985 at GSL (WA), Av 13 Apr at Yorkton and 28 Apr at GSL. **EFA:** 27 Sep 1941 and 1942. **LFD**(4): 5 Nov 2000.

Recoveries, banded outside area, 1 (MI).

Langenburg: One immature at Langenburg town dump, 1990 (TH,BL).

DUCK MOUNTAIN: Two at garbage dump on 15 Jun 1981 (Hatch and Weidl). One, 22 May 1989 (TH).

SABINE'S GULL *Xema sabini*

Vagrant.

DUCK MOUNTAIN: One feeding, flying, diving, and perching at Ministik Beach on 20 Jun 1969 (Copland 1969, RLS, HWRC, Hatch 1969).

[BLACK-LEGGED KITTIWAKE *Rissa tridactyla*

A first-winter bird sighted 26 Oct 1990 and photographed west of Dubuc, or 5 miles south and one mile east of Lemberg, a few miles south of the south boundary of the Yorkton area, but not cited in Hooper's update of Callin (L. Rowell in Smith 1996, and in Koes and Taylor 1992a).]

CASPIAN TERN *Sterna caspia*

Rare transient.

Spring: at GSL: one on 19 May 1998 (WA).

Summer and fall: at GSL, one on 12 July 1991 and one on 15 Aug 1979 (WA, Harris 1980); one flying over Lake of the Prairies, south of Togo, on 6 Aug 1976 (CSH,MIH); 10 on 4 Sep 1996 (WA, Bain and Holder 1996); one on 16 Sep 1989 (WA).

DUCK MOUNTAIN: An injured adult male collected at Madge Lake on 4 Jun 1985 by Nigel Caulkett (SMNH).

COMMON TERN *Sterna hirundo*

Fairly common transient and uncommon summer resident. Breeds at Duck Mountain but rarely elsewhere.

Fairly common at GSL, 23 Jun 1936 (JDS). Near Yorkton, 18 Aug 1943 (IMP). At York Lake, 17 Jul 1947 (CSH). In 1973, occasionally present at York Lake, eight on 13 Jun, four on 19 Jun, two on 28 Jun, one on 4 Jul (AS). One flew over Horseshoe Lake on 9 Jul 1975 (DH). Four at Little Quill Lake on 30 May 1998 (DH).

ESA(36): 21 Apr 1942 (WN) and 24 April 1997 and 2000 (WA), Av 9 May. **LFD**(23): 11 Oct 1983 (WA, Harris 1984), Av 10 Sep.

Breeding: At GSL in 1952, a new nesting colony consisted of at least six nests with 1 to 3 eggs at Burgis Beach, GSL; 24 young were banded on 16 Jul (Houston 1952). In 1953, only four nests were located and these were flooded out twice in succession with rising water levels (Gammell et al. 1953). In 1955, with the lake flooded, the terns and Piping Plovers moved to a weedy mud flat in a field 2 miles southeast of the lake; 70 adults were overhead but only 2 eggs and two downy young were found on 2 July and one half-grown young on 21 July (Anaka 1955). In 1959, terns were carrying minnows to a large marsh adjoining Horseshoe Lake; I waded into the marsh on 15 June but found no nests or young (WA).

At the east end of Crescent Lake in July 1958, Common Terns were thronging around the island, clearly attending to nestlings; I did not have a boat with me at the time and could not approach any closer (CSH).

BBS: overall mean, 0.1. Willowbrook, 5. Ituna, 1.

Banded 1 (WA) + pre-1955, 24 (CSH), as above.

Langenburg: Flying in Cutarm valley, 11 Jun 1989 (CSH,MIH).

DUCK MOUNTAIN: SMNH 1926: study skin female, #1892. SMNH 1951. SNHS 1956. Seen in 1974, 1975, and 1977; young in 1976 (JAS). Common breeder on Madge Lake with 59 nests on one island and 32 on another on 18 Jun 1981 (Hatch and Weidl). SNHS 1989.

PNRS: On 14 Jun 1963, on an island in Madge Lake, 22 nests had 3 eggs, four had 2, and two had 1 (LM).

FORSTER'S TERN *Sterna forsteri*

Rare transient. One breeding record.

Identified only occasionally in spring, e.g., 15 at York Lake on 5 May and four at the Rousay lakes on 27 May 1973 (AS). One seen by Anaka on 6 Sep 1959 (Krause 1960a). One pair, sitting on a muskrat house at Horseshoe Lake in June 1975, raised the possibility of breeding (DH). Five at Little Quill Lake on 10 Jul 1976 (DH). Present at DU's Foam Lake marsh complex, 14 Jun, and at York Lake, 22 Jun 1989 (CSH,MIH).

ESA(6): 10 May 2001. **LFD**: 6 Sep 1965.

Breeding: A small colony in a marsh in a valley at the southern edge of the area, 1 mile north and 2 miles east of Yarbo, 1990 (TH,BL).

Langenburg: One flying over open prairie near Cutarm Creek (TH,BL).

DUCK MOUNTAIN: Seen in 1970 (RJL) and 1973 (BL). Seen on 9-10 Jun 1985 (Pollock 1986c). One at Kamsack Beach, 9 Aug 1989 (TH).

BLACK TERN *Chlidonias niger*

Very common summer resident in marshy areas with bulrush. Breeds.

Three fresh eggs collected at Foam Lake on 18 Jun 1911 (H.C. Grose, SMNH).

At Horseshoe Lake in 1975, 23 on 3 Jun and 20 on 9 Jul (DH). In a dry year, 1989, a tour of the old haunts nearest Yorkton resulted in only one sighting of this species, 20 in one marsh beside Highway 10 near Sturdee, but farther distant there were 80 at Crescent Lake on 20 Jun and groups of 10 and 40 west of Churchbridge on 21 Jun (CSH,MIH).

ESA(48): 10 May 1959, Av 18 May. **LFD**(34): 19 Sep 1965, Av 23 Aug.

Breeding: Early nests with 2 and 1 eggs, 5 Jun 1955. By 26 Jun that year there were 11 nests, seven with 3 eggs, one with 2 eggs, two with 1 egg, and one with one downy young. A second colony had 10 nests on 26 Jun 1955, seven with 3 eggs, two with 2 eggs, and one with 1 egg. A late nest with 3 eggs, 22 Jul 1954 (WA). Nest, 3 eggs in Ravine Reserve (one of the adult terns struck my head!), 8 Jun 1946. Early hatching, downy young in nest at Lower Rousay Lake, 1 Jul 1946 (CSH).

In 1973, two nests, each with 3 eggs, 11 Jun; nest, 3 eggs, 20 Jun; and nest, two young and 1 egg, 8 Jul (AS).

PNRS: at Melville in 1960, two nests with 1 egg each on 10 June, one with 3 eggs on 25 June, and one with an egg and newly hatched young on 7 Jul (VS); at Saltcoats, 1, 3, and 3 eggs on 13 Jun 1958 and 1 and 2 eggs on 19 Jun 1959 (BH); at Stornoway, five nests had 1, 1, 1, and 2 eggs on 6 Jun 1962, but had 2, 3, 3, and 3 eggs on 12 Jun, when additional nests held 3, 3, 2, 1, and 1 eggs (RD); there were eight nests in 1960 and two in 1963 (SZ).

BBS: overall mean, 19.3. Theodore, mean 23.5; decreasing (1970s, 33.3; 1980s, 16.4; 1990s, 16.1; n=31).

Banded 3 (Dickson) + pre-1955, 1 (Blakeley).

Langenburg: Very common, up to 80 and 90 around large marshes and two to four around half-acre potholes, 1990 (TH,BL).

DUCK MOUNTAIN: SMNH 1926: 2 eggs. SMNH 1951. SNHS 1956. Fairly common, one to three seen on 3, 17, 18, and 25 Jun and 10 Jul 1981 (Hatch and Weidl).

ROCK PIGEON [formerly Rock Dove] *Columba livia*

Common year-round resident. Breeds.

Three eggs collected north of Saltcoats (NJ,PMA).

One reason for tearing down the old grain elevators at Sheho in 1977 was the hope of eliminating the Rock Doves! (WN, Serr 1978a). Several pairs moved into an old barn on the Niven farm in 1979 and 1981 (WN).

Breeding: Nest, two young less than a week old, 21 Apr 1968, in abandoned farm at Horseshoe Lake (WA).

BBS: overall mean, 3.8. Theodore, 13 years, including 11 of first 14 years, but only twice since; highest, 11 in 1980, 1989, mean 2.2.

CBC: Even though Rock Doves were not counted on CBCs in North America until 1966, and hence were omitted for the first 14 years in Saskatchewan, they nevertheless remain in eighth place overall. Rank #8. 87 counts, 3,302 birds. Highest: 328, Yorkton 1988.

Langenburg: present, 1990 (TH,BL).

DUCK MOUNTAIN: Seen in 1970 (RJL). SNHS 1989.

BAND-TAILED PIGEON *Columba fasciata*

Vagrant.

One sighted 1 mile west of Hyas in 1966 or later (CP in Smith 1996). One seen at GSL on 28 Aug 1985 (Anaka 1986; Harris 1986).

MOURNING DOVE *Zenaida macroura*

Very common summer resident. Breeds.

R.P. Rooke noted this species only on rare occasions in the 1890s (Rooke 1947). Doves spread northwestward with settlement, numbers increasing slowly but steadily over the first 80 years of the 20th century. Unknown at Sheho in 1910; Niven's first record was on 30 May 1920 and they steadily increased in numbers thereafter.

At Yorkton, increased between 1925 and 1944 (IMP), to become common. At GSL, numbers have been stable since about 1970 (WA).

ESA(48): 5 Apr 1974 at Sheho (WN) and 12 Apr 1974 at GSL (WA), Av 24 Apr. **LFD**(37): 30 Oct 1939 (IMP) and 20 Oct 1957 (WA), Av 24 Sep.

Breeding: This species has an unusually prolonged nesting season. Early nest, 2 eggs, 10 May 1952, and another 25 May 1953 (PPa,CSH). Another early nest had two young, 29 May 1960 (CSH). Late nest, 5 Sep 1964, had 1 egg and one newly hatched young, which fell from the nest and perished; the remaining egg was then deserted (WA, Lister 1965).

At Saltcoats in 1959, nest with 1 egg on 28 May had 2 eggs on 30 May; another nest had 2 eggs on 17 May (BH). Nests with 2 eggs: at York Lake on 7 Jul 1944; 3 miles north of Yorkton on 23 May 1948; at Sheho 20 May 1968; at Yorkton on 18 May 1981; along Isabel Priestly Nature Trail, York Lake, 31 May 1986 (DH); south of Kelliher on 16 May 1992 (CSH). An unusual nest, with 1 egg on 15 May 1989, was on the seat of a long-discarded binder with a large willow growing through and around it to provide cover (WA). Nests commonly found near Sheho, including one close to farm house in 1976 (WN).

Also breeds at Kamsack, Rhein, Togo, Jasmin (ARS, *Atlas*).

BBS: overall mean, 14.6. Theodore, mean 14.4 (1970s, 14.6; 1980s, 14.0; 1990s, 16.9; n=33).

Banded 7 (CSH 3, WA 4) + pre-1955, 1 (CSH).

Langenburg: Seen and heard throughout the area in shelterbelts and wooded valleys, 1990 (TH,BL).

DUCK MOUNTAIN: SMNH 1951. SNHS 1956. Seen in 1974, 1976, 1977 (JAS). Common summer resident, regularly seen in park core area, along Highway 57, and at Little Boggy Creek. Nest, 2 eggs, at latter site on 16 Jun 1981 (Hatch and Weidl).

Point counts: 1 individual at 1 stop (ARS).

Aspen count totals: OA 2, YA 3, MA 2 (TH).

PASSENGER PIGEON *Ectopistes migratorius*

Extinct.

Originally it ranged west beyond Crystal Lake; Daniel W. Harmon (1820:91) listed it when he resided during the winter of 1800-1801 at the North West Company trading post of Alexandria on the Assiniboine, 10 miles north and 2 miles east of the present site of Canora. Charles A.

Boultbee took pigeons as far north as Fort Pelly in the fall of 1874 (Atkinson 1905). W.J. McLean last encountered large flocks of them passing northwesterly from the Whitesand River near Fort Pelly about the middle of May 1877. In 1881 John Macoun saw some in early September on the Swan River northwest of Livingstone, and shot "large numbers for the pot" on the upper Assiniboine near present-day Kamsack. While firing at the pigeons in a tree, Macoun's nephew, Davie, tumbled out of his boat into the river from the recoil of the gun. The low flats along the river were covered with Red Osier Dogwood; the pigeons were feeding on the ripe berries (Houston 1972b). After Ernest E. Thompson [Seton] homesteaded just over a mile west of present Runnymede, 7-22 Jun 1884, he wrote: "Common summer resident, probably all over, as it was noted on Riding Mountain, along the Assiniboine...as well as all over the Red River valley" (Seton 1886, Houston 1972b).

BLACK-BILLED CUCKOO *Coccyzus erythropthalmus*

Common summer resident, especially in years of Forest Tent Caterpillar invasions. Breeds.

In 1927, H.M. Rayner, then a schoolteacher, boarded with a farm family near Jasmin. A cuckoo built its nest and raised a brood within 25 feet of the children's swing; it became so tame Rayner could approach within an arm's reach without flushing the bird off its nest.

Rayner did not see another cuckoo until 7 Jun 1957, when a cuckoo was killed by flying into his kitchen window in Ituna.

Invasions of Forest Tent Caterpillars occurred in 1940-42, 1971-72, 1976 and 2002. At Sheho, cuckoos were especially common in 1940-42, but were present most years until 1944; then none was seen in 1945 or 1946 and only one in 1947. They were present from 1948 through 1952 (Niven 1953), absent from 1953 through 1955, present from 1956 through 1967, absent in 1968 and 1969 and common once again in 1971 (Houston 1971c) and 1972 (Niven 1973).

Three sightings in the wetlands south of Yorkton in 1973 (AS). One at GSL, 27 May 1978 (WA, D.D. Hayward, Serr 1978b). One was killed when it flew into a picture window on the Bill Elaschuk farm near Wroxton on 27 Jun 1978 (LM). Six and two at York Lake, 30-31 May 1986 (MDG, Doris Ihrig, DH). One heard near Milligan Creek, northwest of Foam Lake, 22 Jun 1989 (CSH,MIH). One heard near Preeceville and another near Pelly, 24 Jun 1991 (MIC,MIH,CSH).

ESA(24): 7 May 1949 at Saltcoats (RB), 12 May 1923 at Sheho (WN), and at GSL on 17 May 1987 (WA), Av 2 Jun. **LFD**(8): 29 Aug 1984, Av 14 Aug.

Breeding: Nest, 3 eggs hatching, 28 Jun 1940 (Philip Rosentrator, Vernon Barnes, CSH, mentioned in "Chickadee Notes"). Nest, four young, at George Byblow's yard within Yorkton city limits, 13 Jul 1941 (CSH).

PNRS: at Melville, 2 eggs and one young on 8 Jul 1961 (VS): at Stornoway, nest with 2 eggs on 13 Jun 1961 and 3 eggs on 12 Jul 1962 (RD).

BBS: overall mean, 3.1. Theodore, mean 3.8; decreasing (1970s, 6.3; 1980s, 3.7; 1990s, 2.5; n=25).

DUCK MOUNTAIN: SMNH 1926. Seen in 1976 (JAS). One along road to Batka Lake, 21 May 1988 (RK,CP,RE,TH). Rare; heard or seen along west side of park on 2 Jun, near Ministik Beach on 4 Jun, and along road to Bear Head Lake on 25 Jun 1981 (Hatch and Weidl). One on Highway 57, just south of park gate, 9 Jun 1989 (TH). Two seen and heard near ski hill, 18 Jun 2000 (PSa,DV,BL).

BARN OWL *Tyto alba*

Vagrant.

One found dead 1 mile west of Hyas, 1979 (CP in Smith 1996).

EASTERN SCREECH-OWL *Megascops asio*

Rare year-round resident. No breeding records.

One seen at Dulmage's farm at Ravine Reserve, 22 Feb 1941, and one in Yorkton on 18 Feb 1942 (IMP). One, captured in the garden of Constable Hatch on Fifth Avenue in Yorkton, was banded on 15 Nov 1945 (CSH). Gray-phase individuals were found dead on 1 and 19 Feb 1951 (Shaw 1954). One red-phase bird wintered, December 1985 through February 1986, and was photographed in a spruce tree in a farmyard near Dunleath, section 3-26-2W2, in December (Hjertaas 1986). Gray-phase bird photographed in tree beside a house in Canora on 8 Mar 1986 (Pollock 1986b; R. Krukoff and C. Pollock, *Canora Courier*, 12 March 1986). One brought injured to veterinarian Kenn Wood in early January or February 1986. One "whinny" call in response to tape on Cussed Creek, NW15-27-5W2, 4 miles east of Springside in 1986 (WH,

L. Lepp). One found injured on Highway 47 north of Springside on 14 Mar 1980, and taken the next day to the Moose Jaw Wild Animal Park (T. Kenway and WA in Adam 1989). One seen 14 Mar 1980 and one heard 9 Mar 1992 at GSL (WA,JA).

CBC: Rank #57. 3 counts, 3 birds. Yorkton 1985, 1995; Kamsack 1989.

Banded pre-1955, 1 (CSH).

Langenburg: One wintered near Langenburg, 1991-92 (Dennis Petracek, David Weiss, Koes and Taylor 1992b).

DUCK MOUNTAIN: Heard near Lac la Course, 27 May 1942 (JDS). Heard calling from an aspen-White Spruce stand near Batka Lake, 3 Jun 1981 (Weidl 1983). Heard by Nigel Caulkett at entry gate and at golf course, 16 and 20 Jun 1986 (Adam 1989). Heard calling after midnight, 23 May 1987 (RK,CP,RE,TH). Heard at Kamsack Beach on 9 Aug 1990 (RK).

GREAT HORNED OWL *Bubo virginianus*

A fairly common year-round resident, increasing at the peak of each 10-year snowshoe hare cycle (e.g., 1960). *Breeds.*

Nesting along the Whitesand River on 5 May 1884, with remains of 20 Northern Pocket Gophers in the nest (Christy 1885; Seton 1909b, 1:571). Duck Mountains and Assessippi; a light-coloured *arcticus* shot near the Duck Mountains in the fall of 1893 [Seton 1909a].

At Sheho, with clearing of aspen bush, there was a decrease in Great

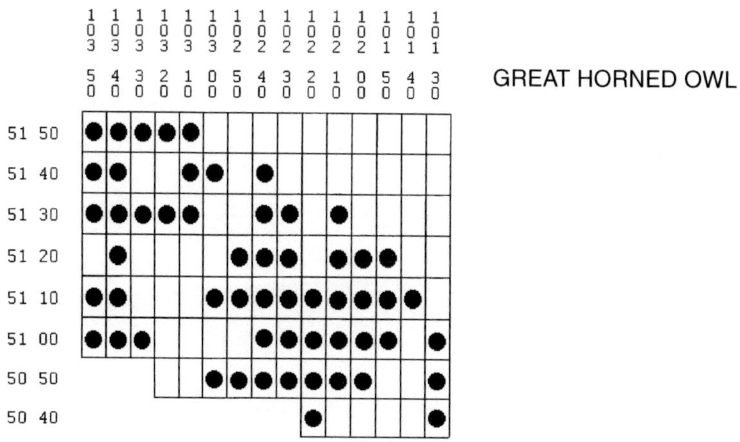

Horned Owls (WN). Present near bird study plots at GSL in 1969 (Flack 1976).

Breeding: Early nests, 1 egg at Crescent Lake on 6 Mar 1945 (J. Hagel); 3 eggs, 14 Mar 1954; and 2 eggs, 15 Mar 1954 (WA). Brood size of owls in the Yorkton-Saltcoats area was 22 from 10 nests in 1959 and 40 from 20 nests in 1960 (Houston 1971d). Early hatching, Lower Rousay Lake, 21 Mar 1946, of an egg that must have been laid about 19 Feb, when the temperatures dipped below 0 degrees F. every night (E.S. Wood).

On 18 Apr 1954, with three downy young in the nest, the female gave a distraction display by walking awkwardly on a branch nearby, then

My Great Horned Owl banding project received a big impetus from an unpredicted source, a weekly television show [see the results below and in the Boreal Owl account]. Bill Horseman found owl nests while trapping muskrats near Saltcoats in the spring of 1958; this was reported in the Changing Seasons section of Audubon Field Notes as follows: "A fine example of field work was the locating of 10 Great Horned Owl nests and the banding of 22 individuals" (Bull 1958). This spurred us on to band 67 young in 29 nests in 1959 (Houston 1971d). Then, in 1960, the Yorkton TV station asked me to do a 10-week series of programs during spring migration, "Watching Parkland Birds with Stu," sponsored by Brooke Bond to publicize the Roger Tory Peterson bird cards in every package of their tea and coffee. We offered one Peterson's Field Guide to the Birds to the farm lad finding the most owl nests for banding, but on the final program on 17 May we presented three guides: one to 16-year-old George Chopping of Dubuc, who had found 16 nests and found another four later in the month; one to 11-year-old Myles Ferrie of Invermay, who also found 16 nests; one to Ronald Hilderman of MacNutt, who found 10 nests. Each lad had a companion who had tramped with him during the nest search: Doug Lambie of Dubuc, Eddie Shepherd of Invermay and Tim Wendell of MacNutt; each of them received a year's subscription to Blue Jay. That year, at the peak of the hare cycle, owl nesting density averaged one pair for each 2 mi^2. Chopping had walked 31 miles one day to find four nests, and 25 miles another day to find three nests (Houston 1960b).

dropping to a lower branch, then tumbling to the ground (Anaka 1954). A downy young, blown from its nest in a severe wind- and rainstorm, 10 May 1986, was placed 6 feet above the ground in a box nailed to the same tree; it was raised successfully by its parents. Density of 0.4 pairs per 10 km^2 at Yorkton (Sargeant et al. 1993).

BBS: overall mean, 0.9. Theodore, 10 years; highest, 2 in 1984, 1990, 1992, mean 0.4.

CBC: Rank #26. 102 counts, 154 birds. Highest: 9, Saltcoats 1954.

Banded 952 (CSH 894, WA 27, RLS 19, Harris 1, Gerard 9, Anweiler/Fox 2) + pre-1955, 17 (CSH 9, Wotherspoon 3, Wickencamp 5). Owls were banded in all 56 10-minute blocks where a search for a nest was made; the highest number of 147 was in the block surrounding the farm of the octogenarian nest-finder, Ian Lochtie, south of Kelliher.

Recoveries, banded in area, 81 [8.4%] (SK 60, MB 10, MN 5, ND 1, NE 1, SD 4; see map in Houston 1978b). Anaka's three recoveries were all within 2 miles of where banded, but Scott's three recoveries within Saskatchewan moved from 30 to 220 miles, one from Duff to be shot at Canwood over a year later. One banded as a nestling near Melville on 17 May 1959 was caught in a trap north of Roblin on 4 Dec 1972; for a year or more this bird held the North American longevity record of 13 years 7 months (CSH).

Recoveries, banded outside area, 6 (all SK).

DUCK MOUNTAIN: SMNH 1926. SMNH 1951. Seen in 1976 (JAS). Fairly common, adults with two and three young, respectively, on 2 Jun and 18 Jun. Adult heard calling near Yellow Lake on 1 Jun 1981 (Hatch and Weidl). SNHS 1989.

SNOWY OWL *Bubo scandiacus*

Fairly common fall and winter transient.

Numbers vary considerably, and are much lower than in central and western Saskatchewan. Unusually plentiful in the winters of 1939-40 and 1942-43. Absent at Sheho and GSL during winter of 1975-76 (WN, WA, Harris and Lamont 1976), also absent in 1982-83 (WN). Present every winter from 1979 to 1989 at GSL, but seen only in 6 of 14 winters since (WA). Absent some winters at Melville (AS).

EFA(17): 8 Oct 1993, Av 8 Nov. **LSD**(12): 23 Apr 1991, Av 4 Apr.

CBC: Rank #29. 51 counts, 86 birds. Highest: 9, Yorkton 1954—the top number of individuals for the continent that year.

Banded pre-1955, 1 (CSH).

DUCK MOUNTAIN: Uncommon in winter, one on rocky island off Midway subdivision on 5 Nov 1981 (Hatch and Weidl).

NORTHERN HAWK OWL *Surnia ulula*

Rare in winter. One 1899 breeding record.

Four eggs were collected by Frank Baines at Crescent Lake on 6 June 1899 from a nest 9 feet up in a willow. The parent bird was collected, so the identification is unquestioned (Macoun and Macoun 1909:316). A specimen collected 20 miles east of Melville by Charles Barr about 15 March 1927 is in SMNH.

Two were found dead on western edge of Yorkton in the winter of 1938 (Brother Clarence). One seen at Sheho, 3 May 1946 (WN). One at Manitoba boundary east of Yorkton, 23 Mar 1968 (R.J. Walker, Hatch 1968b). One was present 2 miles south of Saltcoats from 5 through 19 Dec 1982 (BH). One seen near Norquay and another near Stenen, both on 23 Jan 1984 (CP). One near Wroxton on 13 Feb 1992 (WH,JRJ). During the irruption winters of 1990-91 and 1991-92, there was one record each year from the YDM area (James 1994). During the pre-irruption year of 1999-2000, one was sighted 6 miles southeast of Theodore on 24 Jan (AS); another perched in Ketchen on the late spring departure date of 6 May 2000; it allowed close approach for photography (DF).

Seven records at GSL: 27 Apr 1954, 8 Nov 1956, 26 Dec 1973 (WA, Houston 1974), 14 Apr 1978, 28 Dec 1993, 3 Jan 1994, 23 Nov 2000.

Breeding: Mary Brennan reported two pairs nesting on her farm near Ituna in 1962 (Brennan 1965a).

CBC: Rank #52. 4 counts, 4 birds. (GSL, 1953; at GSL from 23 Dec 1955 through 1 Jan 1956 (JG); 1973; 1993.)

BURROWING OWL *Athene cunicularia*

ENDANGERED, COSEWIC, MAY 2000

Formerly a rare summer resident of open prairie at the extreme northern edge of its range in eastern Saskatchewan; then breeding; locally extirpated since the 1980s.

None seen between Yorkton and Saskatoon in 1903 (Atkinson in Macoun and Macoun 1909:317).

Formerly bred for at least two decades in the 1930s and 1940s. In answer to Mrs Priestly's query in 1944, pairs were reported from Bredenbury, Saltcoats, Melville, and Yorkton (Priestly 1945a), the northeast limit of its maximum range. Strangely, the first sighting on the Thompson farm at Bangor was in 1952 (Thompson 1953). None seen in 1973 (AS).

Breeding: Farthest north nest known within the YDM area, 3 miles southwest of Stenen, had four young banded on 20 Jul 1939 (Emma Wickencamp). Three pairs nested near Melville in the late 1930s and early 1940s; one pair was in a hole in the bank of the reservoir (LTMcK). Nested 3 miles south and 2 miles west of Saltcoats for several years in succession, about 1938 (Frank Baines). Nested 5 miles east and 2 miles north of Yorkton, in a small pasture in front of Cope's farm home near Anemone school, from 1940 to 1945 (CSH). Nest with five young (four banded), 3 miles west of Melville, 9 Jul 1947 (CSH). Nest, with three young on mound beside nest near GSL on 21 Jul 1950 (WA). Allan Frederickson, since deceased, reported a "ground owl" in his alfalfa field, 3 miles north and 5 miles east of Theodore in 1982 (Burrowing or Short-eared?), but the last record of an unequivocal Burrowing Owl in a presumed nesting situation was one perched on a telephone cable in Melville near three burrows with whitewash and two pellets, 4 May 1984 (AS).

Banded pre-1955, 8 (Wickencamp, 4; Houston, 4).

BARRED OWL *Strix varia*

Rare summer resident at Duck Mountain. Breeds.

CBC: 1 count, 2 birds. Kamsack 1995 (1 Jan 1996).

DUCK MOUNTAIN: On 21 Jul and 29 Aug 1970, Robert J. Long heard calls frequently, even in midday. At 1:30 a.m. on 29 Jul he was awakened by loud calling near his cabin. He recorded on tape an interaction between two individuals (Long 1970); nest with young in 1975; seen in 1976 (JAS). Two seen along Red Squirrel Trail, 21 May 1988 (RK,CP,RE,TH) and on 27 May 1994 (ARS, R. Edmonds). SNHS 1989. One heard and seen along Pelly Point Trail on 20 Jun 1993 (BL,RK).

GREAT GRAY OWL *Strix nebulosa*

Rare fall and winter transient.

One shot in error by Harry Anaka, 27 Mar 1954, at GSL, and sent to SMNH (Carmichael 1954; Houston 1957g). In 1961, an individual was present from midsummer into early autumn, with sightings on 7 and 8 Jul, 28 Aug, 8, 14, and 19 Sep 1961 at GSL (Anaka 1961b; Nero 1961; Nero 1962a).

One present during Christmas week 1968 at Saltcoats. Another at GSL on 6 Feb and 12 Mar 1974 was part of a widespread invasion that winter (WA, Houston 1974). One seen at Hyas from 29 Dec 1985 through Mar 1986 (CP). One seen 4 miles north of Willowbrook on 27 Jan 1997 (AS) and one a mile east of Dunleath, 23 Feb 1997 (DH). Saskatchewan's largest-ever reported influx of Great Grays, at the southern edge of the mixed forest and extending regularly 10 or 20 miles south of that, especially around feedlots where mice were plentiful, was during the winter of 2000-2001; this was most marked near Ketchen and Preeceville (DF). One in aspen beside Whitesand Lake, 3 miles south of Margo, on 29 Nov 2000, and one 6 miles southeast of Parkerview on 27 Dec 2000; a single bird at the latter location in Jan 2003 (AS). One sighted in an aspen bluff, 2 miles southwest of Calder, 23 Nov 2002, by Jeff Morari.

CBC: Rank #53. 4 counts, 4 birds. Bangor 1988. Kamsack 1991. GSL 2000. Preeceville 2000.

DUCK MOUNTAIN: Heard calling 30 May 1997 (RK, Bain and Holder 1997c).

LONG-EARED OWL *Asio otus*

Fairly common summer resident in aspen "bluffs." Breeds.

Two specimens procured at Fort Pelly, 16 September 1881 (Macoun in Seton 1891). One egg collected north of Saltcoats (NJ,PMA). Two sets of 4 eggs in the WFVZ, misidentified by Walter Raine as being those of the Northern Hawk Owl [Lloyd Kiff], were collected from a nest in a willow 9 feet above ground at Crescent Lake on 6 May 1899, and another, in a former crow nest 25 feet up in an aspen north of Yorkton, on 1 June 1901. On the latter occasion, the alternate name of "Long-tailed Owl" was given.

Considerably more evident in years when mice are numerous, especially in 1960. If one sees one of these birds in June, it almost always

When I was about six years old, our farmyard was surrounded by mature aspen and willow woodland on three sides. One day I wandered into the bush alone and was met by the most horrible screeching and caterwauling as an owl (as it proved to be) dropped to the ground in apparent distress, pretending it was injured. Fear lent wings to my feet as I made a mad dash for home. My father went with me back to the site, where we found four or five recently-fledged Long-eared Owls sitting in a row on a stout branch. The adult repeated its performance in an attempt to lead us away from its young (WA).

has young in a nearby crow or magpie nest. William Niven thinks their decrease in numbers since 1950 has been related to the increasing clearing of aspen bush. Less common at GSL since 1967.

ESA: female on nest with 4 eggs, 19 Apr 1959 (WA). **LFD:** 11 Sep 1954.

Breeding: Early nests, 3 eggs, near Yorkton, 8 May 1946 (CSH), and Melville, 3 eggs, 11 May 1947 (GAA). Another nest had five young, the youngest hatched that day, on 19 May 1946 (CSH). Five nestlings were banded near the edge of York Lake on 25 Jun 1952 (CSH). At Sheho, nest with 5 eggs, 23 May 1954; another nest was found on 31 May 1982 (WN).

In 1958, five young were found by Bill Horseman west of Saltcoats on 22 Jun. In 1959 there were four nests: three young banded with Stanley Zazelenchuk at Stornoway on 24 May, two at Saltcoats with Horseman on 26 May, five at Donald McKen's farm at Orcadia on 8 Jun (CSH), while Bill Anaka had a nest at GSL with 1 young on 31 May. Numbers then exploded in 1960, after hordes of mice prospered while eating much of the wheat lying in the swath under the snow. The first nest visited, with five young, was 6 miles north of Melville on 30 May. The next nest had four young on 8 Jun. Several pairs raised five young. Other localities in that peak year included Otthon, Fenwood, Rhein, Kamsack, and Invermay; the season total of banded nestlings in the Yorkton-Duck Mountain area reached 74 (CSH). Bill Anaka also banded 4 young on 22 Jun 1964, and 3 young and 2 young on 19 Jun 1967.

Scattered nests in subsequent years included a nest found by Larry Morgotch east of Springside and one by David Ashdown south of Okla, each with three young on 18 Jun 1966; an incubating female north of Foam Lake on 21 May 1981 and 31 May 1982, and two nests in 1985 (WN); a nest in a crow nest in an aspen west of Goodeve on 2 May 1986 (AS); and a nest near GSL with two young on 11 Jun 1990, the only nest found since 1967 (WA). Wayne C. Harris banded three nestlings near Wadena on 12 Jun 1990.

PNRS: 18 nests near Stornoway: Stanley Zazelenchuk had two with 4 eggs and two young in 1959; seven with 2, 5, 5, 6, and 6 eggs and five and six young in 1960; one with four young in 1961 and one with 5 eggs in 1962. In 1962, Ross Derkatch had one nest with 5 eggs and in 1963, nests with 4 and 5 eggs and the other four with five young each.

Also breeds at Wadena, Margo, Jasmin (ARS, *Atlas*).

CBC: 1 count, 1 birds. Kamsack 1993.

Banded 98 (WA 10, Harris 4, CSH 84) + pre-1955, 6 (CSH).

Recoveries, banded in area, 2 [1.9%] (Mexico 2). In January 1965, Glen Peterson of Orland, California, was in the village of Mitla, Oaxaca, Mexico, when an elderly Mexican took the banded foot of an owl from his pocket. A boy had found the owl dead and hanging in a tree nearly five years previously, three days' walk to the south, near Ejutla, Oaxaca (Houston 1966a). Anaka's owl, after banding in the nest 7 miles east of Otthon, went to Monterey, Nuevo Leon, Mexico.

DUCK MOUNTAIN: Rare, two calling near Batka Lake on 17 Jun. A roadkill on Highway 8, outside and southwest of the park on 14 Aug 1981 (Hatch and Weidl). One heard at Kamsack Beach subdivision #1 on 9 Aug 1990 (RK).

SHORT-EARED OWL *Asio flammeus*

SPECIAL CONCERN, COSEWIC, APRIL 1994

Rare summer resident, formerly uncommon. Breeds.

Two specimens procured on the Assiniboine and Shell rivers, 24 September 1881 (Macoun in Seton 1891).

Twelve each day at Foam Lake marsh on 14 and 25 May 1974, with further sightings in June (AS). One sighted near Horseshoe Lake on 4 Jun 1975 (DH) and Upper Rousay Lake on 13 Jun 1989 (CSH,MIH). One sighted near Togo, 21 Jun 1989 (CSH,MIH).

ESA(9): 7 Apr 1962 and 8 Apr 1955 at Sheho (WN), and 10 Apr 1975 at GSL (WA), Av 21 Apr. **LFD**(12): 28 Nov 1974 at Salt Lake and 4 Nov 1983 at GSL (WA), Av 17 Sep.

On 4 Jun we visited a Short-eared Owl nest in stubble on the farm of Joe Gillis near Theodore; the previous evening there had been six young in the nest. One adult was killed by flying into a nearby high-tension power line during the night. When we arrived, the remaining adult was dismembering the three smallest nestlings and feeding them to the three oldest. Wildly and unforgivably anthropomorphizing, we remarked that her arithmetic was excellent: if two adults can raise six young, one adult might raise only three! (CSH).

Breeding: Prior to 1960, we had known only of nests on the flats southeast of Crescent marsh (F. Baines) and of one early nest with 3 eggs at Tonkin, 8 May 1952 (Robert Pearce).

In years when mice were numerous, many of these nomadic owls stayed to breed, particularly on upland stubble fields, in place of their more usual habitat locally in damp meadows. In 1960, "the year of the owls," numerous farmers reported Short-eared Owls nesting on their stubble fields. Most had not noticed this species before. Ed Ball, 7 miles south of Willowbrook, had three nests on 320 acres of stubble. Early nests, all on stubble fields: with 7 eggs, northeast of Yorkton, 4 May 1960 (Clarence Fick); 1 egg, May 8 (Mervin Mroske); 6 eggs at Barvas and 6 eggs at Saltcoats, 13 May (Jack Jowsey and Jim Trowell). We captured an adult female on her nest with 7 eggs, 1 mile south of Dubuc on 15 May. The first three young were banded 13 miles east and 2 miles north of Yorkton on 28 May, but five smaller young and 1 egg were left unbanded in the same nest. Altogether, 68 young were banded in 21 nests, though a number of smaller owlets weren't large enough to hold a band at the time of the visit (Houston 1960b).

Has nested in grassy meadows at Sheho (WN). At GSL, neighbours have reported young on their fields in late summer (WA). Present at Barvas marsh (Kornder and Belcher 1996; Wapple 1999).

PNRS: at Stornoway in 1960, nests with 2 and 9 eggs and seven and nine young (SZ).

Also breeds at Wadena (ARS, *Atlas*).

BBS: overall mean, 0.06. Churchbridge, 1. Ituna, 2.

CBC: Rank #49. 4 counts, 10 birds. Yorkton 6, 1953.

Winter: One in Jan 1940 near Togo (J.R. Foreman) and another 20

Jan 1956 at GSL (WA, Anaka 1952).

Banded 56 (CSH).

DUCK MOUNTAIN: Uncommon. One, immediately outside the park, south of Little Boggy Creek, on 18 Jun 1981 (Hatch and Weidl).

BOREAL OWL *Aegolius funereus*

Rare and unpredictable winter transient.

Fairly common in early 1960. One was shot 9 miles north of Melville on 28 Jan 1956 by Alf McKay. A massive and unprecedented influx of at least 39 different individuals, 25 of which were banded, occurred in the winter of 1959-60, beginning with one caught in a weasel trap at Saltcoats on 19 Dec 1959 by Bill Horseman, and one snared near Melville by Gary Anweiler on 7 Jan 1960. By 22 Jan, CSH had banded four. We agreed to show Gary's fifth owl on CKOS-TV before it was banded and released; that day Martin Busch brought in another from Calder. Other localities included Kuroki, Springside, and Churchbridge.

CBC: 1 count, one bird. Bangor, 27 Dec 1952.

One seen at Sheho, early February 1985 (WN). One was found dead

During a special 15-minute TV program devoted exclusively to a discussion of Boreal Owls, Gary Anweiler and I each had an owl perched on one wrist, secured with falconers' jesses built by Gary. Exposed to the bright TV lights, these spunky little owls snapped their beaks and flapped their wings–and became a household name throughout the viewing area. Those who were out of range or not tuned in, heard about the owls from their neighbours. Immediately after the program, Steve Shebeluk of Calder went out to his machine shed and caught the owl he had seen while doing evening chores. That weekend, Gary and I drove more than 300 miles to pick up owls from Fenwood in the west to Arran and Inglis, Manitoba, in the east, with the added bonus of a Saw-whet Owl at Hazelcliffe. The influx finally ended with one caught in a magpie nest on 27 Apr by Horseman at Saltcoats (Anweiler 1960; Krause 1960b). The final banding total of 25 Boreal Owls was a new North American record. It led to a 10-week show, "Watching Parkland Birds with Stu," through the spring migration that year.

in a haystack near Hyas in Mar 1986 (R. Bellows). In 1988, one was found dead at Churchbridge in February and another in March at Canora (AS). Heard calling from a farm spruce shelterbelt near Hyas on 10 Apr 1998 (CP, Bain and Shanahan 1998a).

Banded 24 (CSH).

Recoveries, banded in area, 3 [12.5%] (SK, all froze or starved close to the place of banding).

NORTHERN SAW-WHET OWL *Aegolius acadicus*

Uncommon transient and summer resident. Rare in winter.

Present at Fort Pelly, 1881 (Macoun in Seton 1891). One caught at Leross during winter of 1920-21 by Thomas Boston (Natural History Club column in [Montreal] *Family Herald and Weekly Star*, 9 Mar 1921). One was shot by George Markham near Yorkton about 1930; for years the mounted specimen was in the possession of Ted Groom of Yorkton.

One at GSL on 4 and 6 Mar 1955, heard calling later (Gunn 1956a). Two heard at Horseshoe Lake, 21 May 1988 (RK,CP,RE,TH). A pair present in aspen woods near Hyas, heard and seen through the summer, from 1984 through 1988 (R. Bellows). Heard calling each spring in aspen, 6 miles north of Saltcoats, since 1984 (JRJ). One present in spruce trees in Freimarks' yard, 1.5 miles east of Horseshoe Lake, on 6 May 1986 and from 21 Nov to 17 Dec 2000 (WA).

BBS: Theodore, 1 in 2001, mean 0.03.

CBC: 1 count, 1 bird. Saltcoats 1942, and specimen collected, 31 Dec 1942 (Jim Rogerson).

Winter: One noted at Sheho from 26 Jan through 11 Feb 1943 (WN). One 5 miles east of Saltcoats, 28 Jan 1948 (RB). One appeared in Dallas Fairburn's yard at Ketchen for five minutes, 25 Dec 2000 (DF).

Banded 3 (CSH 2 [caught north of Hazelcliffe on 30 Jan and at Waldron on 11 Feb (CSH)], WA 1).

DUCK MOUNTAIN: SNHS 1989. Three heard along 6 miles of Batka Lake Road, from 10 to 11 p.m., 2 May 1992 (TH).

COMMON NIGHTHAWK *Chordeiles minor*

Once fairly common to common, now a rare summer resident. Breeds.

Two eggs, Crescent Lake, 4 Jun 1901, W. Raine, AMNH #9131.

At Sheho, nighthawks disappeared as a nesting species by the 1950s, and thereafter were seen only in migration: one, 22 May 1975; none in

In my youth the "mosquito hawk" was widespread, nesting on bare or stony ground and on flat gravel roofs where available. Numbers of this wide-gaped bird have decreased appreciably since the 1940s, coincident with chemical spraying to control mosquitoes in Yorkton and major towns, and with the increase in ground predators such as coyotes and foxes. Do these apparent associations represent cause-and-effect or mere coincidence? (CSH).

1976 (for the first time); one on 27 Aug 1977; one on 24 May 1978; two in 1979; one on 12 Aug 1980; one in 1982; none in 1986; one on 21 Jul 1988; and none in 1989 (WN). In 1973, in the wetlands south of Yorkton, only four sightings all summer (AS). Largest numbers in fall migration in 2002 were 50 flying over Ketchen from 5:45 to 6:15 p.m. on 27 Aug (DF).

ESA(43): 10 May 1949 at Sheho (WN), 15 May 2000 at GSL (WA), and 19 May 1939 (WN), Av 26 May. **LFD**(30): 17 Sep 1959, Av 31 Aug. Usually departs well before the first frosts reduce the number of insects in the air.

Breeding: On the flat gravel roof of Victoria School, Fourth Avenue at Duncan Street, there was 1 egg, 28 Jun 1940, and two young on the same roof on 23 Jul 1946 (CSH). Nest, 2 eggs on bare ground of baseball field in Yorkton, 1944, and another on stone chips on a grave in the Yorkton cemetery, 19 Jul 1945 (IMP).

Banded pre-1955, 3 (CSH, see above).

DUCK MOUNTAIN: SMNH 1926. SMNH 1951. Seen in 1977 (JAS). Fairly common in open areas, at park core area and near Little Boggy Creek (Hatch and Weidl 1981).

WHIP-POOR-WILL *Caprimulgus vociferus*

Rare, sporadic summer resident in northern part of YDM area.

Regular in aspen bush and wasteland along gulches and on the south side of the sand hills at the south end of GSL for about 20 years from the time the Gunns first settled there in 1887. Heard calling only once more, in July 1943, by John A. Gunn. Heard calling each night, from a bush nearly 2 miles north of the lake, by Robert Rousay in his first year at Upper Rousay Lake in 1889. Heard by Frank Baines on three separate occasions before 1900 at Crescent Lake. Heard at Browns' farm east of Theodore, where a small creek runs into the Whitesand River, in 1915 and 1916 (Pearl Guest, letter of 19 April 1950).

This retiring species is rarely seen—yet since it is named for its distinctive onomatopoeic call, there is a better chance of knowing its presence than for most other rare species. Heard in the Assiniboine River valley, east of Yorkton, by J.R. Foreman in 1938 and another heard west of his home on Wallace Avenue, Yorkton, in 1938. One heard by Mrs M.J. Culver near Yorkton in July 1941. Seen and heard each evening, 4 through 15 Jun 1958, at GSL (Anaka 1958).

DUCK MOUNTAIN: One heard at Lac la Course, 6 miles west of the northwest corner of the park, 27 May 1942 (JDS).

CHIMNEY SWIFT *Chaetura pelagica*

Vagrant.

A recently dead, well preserved male was found and photographed during routine cleaning of an active chimney at the G. Switzer farm, 8 miles east and 7 miles south of Yorkton, on 15 Jul 1967 (Switzer 1970).

Langenburg: One pair seen, 1 Jun 1946 (JDS).

RUBY-THROATED HUMMINGBIRD *Archilochus colubris*

Uncommon to fairly common summer resident. Breeds.

One of the latest birds to arrive in spring, frequenting hollyhocks and delphiniums in flower gardens, and an increasing number of hummingbird feeders in gardens throughout the area. Up to seven at a time feeding on a small tree exuding sap in August 1956 (Joyce Gunn). Under-reported on BBS, due to small size and weak vocalization.

ESA(40): 9 May 2000 (WA, Bain and Shanahan 2000a), Av 23 May. **LFD**(28): 23 Sep 1942 (CSH) and 11 Sep 1984 (WA), Av 31 Aug.

Breeding: A nest was found in 1943 at Mamornitz School ground near Buchanan by Peter Bodiuk, a grade three student; the children watched the nest each day, noting the tiny eggs. The nestlings grew surprisingly quickly and were "fully feathered" at four days (Priestly 1943c). Nest at Sheho in 1975 (WN). Nest, 2 eggs, on horizontal branch of Black Poplar, 5 feet above ground, in the yard of Tom Yates, Ituna, 11 Jul 1983 (Brennan 1984).

BBS: overall mean, 0.06. Churchbridge, 1. Willowbrook, 1. Theodore, 1.

Banded pre-1955, 3 (CSH).

Langenburg: one male on 1 Jun 1976, in the Assiniboine valley north of Marchwell (DH).

DUCK MOUNTAIN: SMNH 1926: Two nests with 2 eggs, 18 Jun, #1870a, group, #1871, study skin male, #1898, "much needed for museum collection." SMNH 1951. SNHS 1956. Seen in 1976, 1977 (JAS). Fairly common, seen at old sawmill site near west boundary on 3 Jun; at Little Boggy Creek on 16 Jun; along Red Squirrel Nature Trail on 16 Jul; and at four other sites in August 1981 (Hatch and Weidl). SNHS 1989. Late fall date, 8 Sep 2002, Pickerel Point (TH).

PNRS: Nest on horizontal branch of birch, 13 feet above ground, had 2 eggs on 22 Jun; one hatched on 23 Jun and the second the next day (HWRC).

Point counts: 1 individual at 1 stop (ARS).

Aspen count totals: OA 2, MA 1 (TH).

ANNA'S HUMMINGBIRD *Calypte anna*

Vagrant.

One was present at Regina Aye's feeder, on the south side of Highway 16 opposite Springside, from 20 to 29 Sep 1997. She did not notice it until the Ruby-throats had left and she had removed the hummingbird feeders, but she promptly replaced the feeders and the bird fed repeatedly. This was Saskatchewan's second record, but the first fully documented. The bird was docile and sang a variety of squeaks and rasps while perched. On 6 Oct the feeders froze; Regina enticed the bird into her house with a blooming fuchsia plant in the open bedroom window. The bird was clearly going to die in the oncoming Saskatchewan winter, so was picked up by Bob Kreba from the Royal Saskatchewan Museum on 10 Oct and flown to Vancouver on 11 Oct; it died just before it could be released (Bain and Holder 1997d, 1997e).

RUFOUS HUMMINGBIRD *Selasphorus rufus*

Vagrant.

One appeared at the new hummingbird feeder at the Hilton farm near Bangor, 31 Jul to 7 Aug 1984. It was the "boss hummingbird"; the Ruby-throats could feed only if they came in twos and threes, one feeding while the others combatted the Rufous. When feeding, the Rufous held its bill sloping downwards into the feeder, while the local hummingbirds held their bills parallel to the ground (Hilton and Hilton 1985).

BELTED KINGFISHER *Ceryle alcyon*

Uncommon summer resident. Breeds.

Very abundant along the Red Deer, Swan, and Assiniboine rivers in 1881 (Macoun in Seton 1891). Half a dozen pairs were present all summer at Crescent Lake from about 1909 to 1923, when fish were plentiful (Frank Baines). John A. Gunn noted them as resident at the south end of GSL from 1887 until 1927, when the minnows disappeared from the lake. Kingfishers were most often noted in 1922, a year of high water levels, when water ran in the creek leading out from the south side of the lake.

Present sparingly in summer at the Rousay and York lakes. Ten seen from a canoe paddled along 2 miles of York Lake shoreline, 7 Oct 1951 (Houston 1951; Gammell and Gammell 1952a). In 1973, seven sightings of one to four individuals between 30 Apr and 28 Jun (AS). Five seen while canoeing 2 miles of the Whitesand River, beginning at Highway 9, on 11 Aug 1988 (DH). The number of migrant kingfishers decreased at GSL after 1990.

ESA(22): 19 Apr 1952 at Sheho (WN) and 1976 and 1977 at GSL (WA), Av 3 May. **LFD**(30): 22 Oct 1983, Av 2 Oct.

Breeding: About 1924, Niven found an active nest hole along a creek bank just east of Sheho.

BBS: overall mean, 0.09. Willowbrook, 1. Theodore, 3 years, including 2 in 1972.

DUCK MOUNTAIN: SMNH 1951. Seen in 1975, 1976, 1977 (JAS). Nest, Batka Lake. Fairly common, especially along Little Boggy Creek and a Black Spruce bog near the southeast corner of the park in Jun and Jul, and at two other sites in Aug 1981 (Hatch and Weidl). SNHS 1989. Large number, eight, in migration, 7 Sep 2002, upper Assiniboine River (TH).

LEWIS'S WOODPECKER *Melanerpes lewis*

SPECIAL CONCERN, COSEWIC, NOVEMBER 2001

Vagrant.

One shot by Frank Baines, 3 miles south and 6 miles west of Saltcoats, about 1920 (Houston 1949).

RED-HEADED WOODPECKER *Melanerpes erythrocephalus*

SPECIAL CONCERN, COSEWIC, APRIL 1996

Rare summer resident. Breeds.

One at Yorkton in June 1936 (IMP). One at Sheho, 6 Jul 1946 (WN). One at Saltcoats, 25 and 28 Jun 1950 (Baines 1950). One at GSL, 24 May 1982 (WA,JA, documented, Wedgwood 1982); a pair resided in the L. Freimark yard, 1.5 miles east of Horseshoe Lake from 13 Jun to mid-Jul 1998, among numerous dead aspen trees and snags, but neither nest nor young were located (WA [this sighting was given an erroneous date of 20 Sep 1998 by Bain and Shanahan 1998c]).

Breeding: Pair carrying food to a hole 30 feet above ground in a Black Poplar, 6 miles north and 2 miles east of Pelly, 24 Jun 1991 (MIC,MIH,CSH, Koes and Taylor 1991c). A pair was observed feeding four nearly-fledged young on 1 Jul 1993 (RK, pers. comm.)

RED-BELLIED WOODPECKER *Melanerpes carolinus*

Vagrant.

A young male Red-bellied Woodpecker was seen on the Brygider farm, 2 miles south of Saltcoats on 25 and 26 May 1991 (Jowsey 1992).

WILLIAMSON'S SAPSUCKER *Sphyrapicus thyroideus*

Vagrant.

Sighted 5 miles east of Saltcoats, 26 Apr 1948 and 30 Apr 1947 (RB). Two at GSL, 3 Sep 2003 (Daryl Ford).

YELLOW-BELLIED SAPSUCKER *Sphyrapicus varius*

Fairly common summer resident. Breeds at Duck Mountain and sparingly in heavy aspen to the south.

Brother S. Matthias reports it was common in the big timber of the Beaver Hills, south and west of Theodore, in the early 1920s. Four in one of two mature aspen study plots at GSL, 1969 (Flack 1976).

ESA(45): 12 Apr 1986, Av 24 Apr. **LFD**(42):26 Oct 1981, Av 28 Sep.

Breeding: At GSL: a nest found in 1938 (IMP). In 1957, an adult was working inside a nest hole on 2 Jun and young were calling inside the cavity on 27 Jun; three young fledged from another nest on 8 Jul 1963 (WA). Two nests on 22 Jun 1969, 10 and 15 feet above ground, one with

One morning just after sunrise a woman, who was nervous about staying in a cabin at Good Spirit Lake while her husband worked all week in Yorkton, woke to what sounded like someone knocking at her door. She ran to the door but found no one there. She thought someone was harassing her until she saw a sapsucker tapping away on the stovepipe jutting above the roof of the adjacent cabin–the noise she had thought was someone knocking (JA).

young squeaking (JADF).

At Saltcoats, nest with young, 27 Jun 1957 (BH).

PNRS: at Saltcoats, young could be heard in cavity on 7 Jun 1969 and five fledged on 6 Jul (JRJ).

Also breeds at Kamsack, Yorkton (ARS, *Atlas*).

BBS: overall mean, 1.2. Churchbridge, 2. Ituna, 1. Willowbrook, 7. Theodore, mean 1.5 (1970s, 1.7; 1980s, 2.1; 1990s, 1.2; n=23).

CBC: 1 count, 1 bird. Kamsack 1993.

Winter: One at the feeder of J. Popoff, Springside, 26 Jan 2003 (Bain 2003).

Banded 4 (CSH) + pre-1955, 3 (CSH 1, Wotherspoon 2).

Langenburg: One male, Assiniboine valley, north of Marchwell, 1 Jun 1976 (DH). Three separate sightings of single males in coulees and valleys, 1990 (TH,BL).

DUCK MOUNTAIN: SMNH 1926: nest on 20 Jun, #1995; study skin male, #1922. SMNH 1951. SNHS 1956. Seen in 1974, 1977; nest with four young, 1975; nest with young almost fledged, 1976 (JAS). Common throughout park. Nest with young in aspen along Pelly Point Trail on 24 Jun 1981 (Hatch and Weidl). SNHS 1989.

Point counts: Rank #48, 2 individuals at 2 stops (ARS).

Aspen count totals: OA 8, MA 1 (TH).

In mid-July 1958, sapsuckers bored a series of holes in a willow near the Gunn store at GSL; the sap ran freely and attracted hornets, wasps, and seven Ruby-throated Hummingbirds in daytime and moths at night–a virtual feeding station. The tree was never free from birds for more than five minutes (Gunn 1956c).

DOWNY WOODPECKER *Picoides pubescens*

Fairly common year-round resident, but more apparent in winter at feeding stations which offer suet. Breeds.

Has decreased at Sheho since the 1940s, due to clearing of most of the heavy aspen bush. Two in one mature aspen study plot at GSL, 1969 (Flack 1976). A male at Horseshoe Lake, 30 Jan 1976, flew from one cowdung patty to another, pecking each, presumably for food (DH).

Breeding: Adult feeding one young, 29 Jun 1989; recently fledged young are often brought to feeders to eat rolled oats (WA).

PNRS: at Stornoway, six young each at two nests on 18 Jun and 3 Jul 1961, four young on 2 Jul 1962, and five young on 23 Jun 1963 (RD).

BBS: overall mean, 0.5. Theodore, 14 years, all with either 1 or 2, mean 0.6.

CBC: Rank #15. 208 counts, 942 birds. Highest: Kamsack 33, 1995.

Banded 26 (CSH 3, WA 23) + pre-1955, 2 (Wickencamp 1, Wotherspoon 1)

Returns, 23 retrapped on 31 occasions after a year or more (WA). Two males and one female retrapped after one year, two males and a female after three years, a male after six years, a female after eight years, and a male with his band worn thin after nine years and one month (WA). At the time, this was the oldest Downy Woodpecker recorded in Canada (Brewer et al. 2000); the North American longevity record is 11 years, 5 months.

Recoveries, banded outside area, 0 [retraction of an erroneous published record; a bird banded by Mrs Emma Wickencamp on 24 Feb 1938 was reported as "band only" two years and 11 months later, on 10 January 1941. The erroneous listing on the computerized printout is apparently due to a clerical transcription error which retrospectively gave the latitude and longitude of Imperial (51° 20, 105° 20), well outside the Yorkton area's west boundary, as the place of banding, instead of 51° 50, 102° 20, which is Stenen, where Mrs Wickencamp did all of her banding, and where the copy of the original banding schedule in my possession indicates unequivocally that the bird was banded. This error was repeated in Brewer et al. (2000), who chose this as one of three examples of interesting Downy Woodpecker dispersal. In truth, the bird was recovered at the same locality where it had been banded.]

DUCK MOUNTAIN: SMNH 1926. SMNH 1951. Three in 1959 CBC. Seen in 1974, 1975, 1976, 1977 (JAS). Fairly common throughout park (Hatch and Weidl 1981).

Point counts: 1 individual at 1 stop (ARS).

HAIRY WOODPECKER *Picoides villosus*

Fairly common year-round resident, but more apparent in winter at feeding stations which offer suet. Breeds.

Five eggs, 12 feet up in an aspen near Yorkton, 6 Jun 1898 (WFVZ). Four eggs, in hole in tree 10 feet above ground, and 6 eggs, 15 feet above ground in a poplar, 31 May 1901 at Yorkton (W. Raine in Macoun and Macoun 1909:327).

Has decreased at Sheho since about 1950, due to clearing of most of the heavy aspen bush. Two in one mature aspen study plot at GSL, 1969 (Flack 1976).

Breeding: At GSL, young were calling in nest, 8 Jun 1955, and fledged 21 Jun. Nest with young in hole in aspen, 8 feet above ground, Horseshoe Lake, 5 Jun 1975 (DH). A pair began excavating a hole, 8 feet above ground in a dead aspen, 5 Apr 1986. Recently fledged young are often brought to feeders to eat rolled oats (WA).

BBS: overall mean, 0.5. Theodore, 17 years; highest, 4, 1980, mean 0.8.

CBC: Rank #16. 182 counts, 738 birds. Highest: 24, Kamsack 1996 and 2001.

Banded 21 (CSH 2, WA 19) + pre-1955, 2 (Wotherspoon).

Recoveries, banded in area, 1 [4.3%](SK 1, Wotherspoon bird found dead at Lintlaw 7 years, 1 month after banding, as reported in Brewer et al. [2000]).

Returns, 2 (both retrapped a year after banding, WA).

Recoveries, banded outside area, 1 (MB 1, a female, banded by Ed Robinson at East Bay, Manitoba, on 13 Jan 1935, was turned in by a neighbour to John A. Gunn at GSL on 15 Aug 1938).

Langenburg: A male drumming on a dead tree, 1 Jun 1976, Assiniboine valley north of Marchwell (DH). A single bird carrying food in Cutarm valley, 1990 (TH,BL).

DUCK MOUNTAIN: SMNH 1926. SMNH 1951. SNHS 1956. One in 1959 CBC. Seen in 1976, 1977 (JAS). Fairly common throughout park (Hatch and Weidl 1981). SNHS 1989.

Point counts: Rank #37, 4 individuals at 4 stops (ARS).

Aspen count totals: OA 1.

AMERICAN THREE-TOED WOODPECKER *Picoides dorsalis*

Rare in fall and winter. Breeds at Duck Mountain.

One sighting at Yorkton on 15 Sep 1942 (CSH). Male found dead in Yorkton on 28 Oct 1958 by grade four Simpson School pupils of Miss Pauline Summers, Georgina Demchuk, Janet Parno, and Caroline Crandall (Houston 1959a).

Two sightings at GSL: a female in the Anaka yard, 19 Oct 1982 and one on the 23 Dec 1997 CBC.

CBC: Rank #54. 3 counts, 4 birds. Highest: 2, Canora 1977.

DUCK MOUNTAIN: Year-round resident. SMNH 1951. SNHS 1956. Pair feeding young at nesting hole, 21 May 1988 (CP), and in core area on 24 Jun 1990 (ARS). Uncommon in winter, rare in summer, one along the Red Squirrel Nature Trail on 14 Aug 1981 (Hatch and Weidl). One, 21 May 1988, along road to Batka Lake (RK,CP,RE,TH); four along Red Squirrel Trail, 20 Aug 1989; three behind golf course, 27 Dec 1989 (TH).

BLACK-BACKED WOODPECKER *Picoides arcticus*

Vagrant.

At GSL: one visited Joyce Gunn's feeding station on 1 Apr 1954 (Gunn 1954a); one, 24 Oct 1965; one on 1 Jan 1967 CBC (WA). One foraging east of Parkerview on 8 Nov 1997 (Eric Schmidt, AS).

CBC: 2 counts, 2 birds. Kamsack 1992 and GSL 1996 (1 Jan 1997).

DUCK MOUNTAIN: A pair observed along the Red Squirrel Nature Trail on 10 Sep 1981 (Hatch and Weidl). One female, 26 Mar 1989, at Kamsack Beach; one on 20 May 1989 near garbage dump on road to Batka Lake; three behind golf course on 27 Dec 1989 (TH).

NORTHERN FLICKER *Colaptes auratus*

Common summer resident. Breeds.

Along the Assiniboine in 1881, wherever there were trees (Macoun in Seton 1891). Four eggs, Crescent Lake, 29 May 1890, F. Baines, ROMZ #2337. Three eggs collected north of Saltcoats (NJ,PMA).

Has decreased at Sheho since about 1950, due to clearing of most of the heavy aspen bush.

ESA(48): 5 Apr 1986 and 1987, Av 18 Apr. **LFD**(43): 11 Oct 1972, Av 29 Sep.

Breeding: Early nesting, 12 May 1946, with young from this nest able to fly on 26 Jun. Five nestlings banded on 23 Jul 1947 near Yorkton (CSH). Late working on nest hole, GSL, 26 May 1957 (WA).

Oology: one egg near Canora in 1965-66 (H. Cymbalisty, SMNH).

PNRS: four nests at Stornoway with 3, 3, 5, and 5 eggs, 1959-62 (SZ,RD); nest west of Canora had 10 eggs on 25 May and seven young on 23 Jun 1959 (WC); four nests at Saltcoats in 1958 and four in 1959, with 6, 6, 6, 7, 7, and 7 eggs and two with six young (BH); 4 eggs on 12 Jun and four young west of Canora on 20 Jun 1969 (EK).

BBS: overall mean, 4.9. Theodore, mean 5.0 (1970s, 5.8; 1980s, 4.4; 1990s, 4.0; n=32).

CBC: 1 count, 1 bird. Yorkton 1997.

Banded 2 (CSH) + pre-1955, 118 (CSH 10, Patrick 9, Blakeley 5, Briggs 63, Wickencamp 25, Wotherspoon 6).

Recoveries, banded in area, 1 [1.0%] (SK, banded at Colmer siding west of Melville, by J.A. Briggs, 10 Jun 1932, found dead in Melville, 1 Jun 1933).

Langenburg: Six sighted, 1990 (TH,BL).

DUCK MOUNTAIN: SMNH 1926. SMNH 1951. SNHS 1956. Seen in 1974, 1975, 1976, 1977 (JAS). Fairly common around park core area, Jubilee subdivision, and Spruce Island. Fair numbers migrate through in September (Hatch and Weidl 1981). SNHS 1989.

PILEATED WOODPECKER *Dryocopus pileatus*

Uncommon summer resident, resident at GSL only since 1968, and increasing. Breeds.

One noted on the Swan River, 2 September 1881 (Macoun in Seton 1891).

Occasionally wanders south into, and even more rarely breeds in, the southern part of the area: John A. Gunn noted only two at GSL in 60 years, one in autumn of about 1936, and one in the spring of about 1942. Mike Heintz saw one on a telephone pole in front of the old collegiate on Darlington Street, Yorkton, in the springs of 1940 and 1941. One seen in heavy aspen forest, 3 miles north of Jedburgh, in the 1950s. One at Saltcoats on 9 Nov 1964 stayed about two weeks (BH). Two in one mature aspen study plot at GSL, 1969 (Flack 1976). One seen 3 miles west and 3 miles north of Ebenezer, 27-28 Oct and again on 8 Dec 1982 (PPa). One on Whitesand River north of Theodore golf course, August

1990; one at Norm Quinton's on Cussed Creek, 24 Nov 1990 (Len and Nan Owen; Gilroy 1990). One seen at Horseshoe Lake, 7 Oct 1993, one at GSL on 10 Mar 1995, and two on 15 Nov 1999 (DH).

Breeding: a) three or four young, 18 feet above ground in hollow tree, GSL, 1968; Joyce had been seeing this species since 1 Apr 1967 (Gunn 1968; WA and Joyce Gunn, Hatch 1968d; b) a pair at nest hole, GSL, 19 May 1969; and c) female working on nest cavity on 8 May and incubating by 20 May 1988, GSL; d) active nest in L. Freimark yard, NW 31-28-4W2, 1.5 miles east of Horseshoe Lake, 3 Jun 1990 (WA); e) freshly excavated hole, with distinctive large chips beneath, near Lake of the Prairies, 25 Jun 1991 (Murray and Allan Bowes); f) nest high in a dead Black Poplar in Cutarm Creek valley south of Saltcoats, 1995 (BH); g) east of Parkerview in 2001, freshly excavated hole, 18 feet up in aspen on 18 May and two young nearly fully feathered on 23 June; less than a mile distant a pair defended an aspen with a recently excavated hole 21 feet above ground on 18 April 2003 (AS).

BBS: Theodore, 2 years (1, 1988 and 1990), mean 0.06.

Winter: One present in an aspen bluff with a lot of deadfall on the farm of Morley Smithson near Sturdee from January through March 1991 (Dale Kitcko, cited in Gilroy 1991, #1349). Has wintered regularly at the Herb and Mary Upshall farm, 4 miles south of Tonkin, since about 1980 and at the Jowsey farm, 6 miles north of Saltcoats, since 1990; birds have remained as late as 7 and 8 May (JRJ). Came almost daily for suet to a feeder in Ketchen in the winters of 1999-2000 and 2000-2001 (DF).

CBC: Rank #30. 35 counts, 85 birds. Highest: Kamsack 11, 1984.

DUCK MOUNTAIN: A regular, year-round resident of the mixed forest at Crystal Lake and Duck Mountain. SMNH 1926: breeding. SMNH 1951. SNHS 1956. Fairly common year-round: along west park boundary on 2 Jun, at Jubilee subdivision on 4 Jun and Little Boggy Creek on 16 Jun 1981 (Hatch and Weidl, Weidl 1983). SNHS 1989.

Point counts: Rank #37, 4 individuals at 4 stops (ARS).

Aspen count totals: OA 4 (TH).

OLIVE-SIDED FLYCATCHER *Contopus cooperi.*

Uncommon transient. Breeds at Duck Mountain.

Common in the Duck Mountains, male shot 12 June 1884 (Seton 1891).

One singing in the Assiniboine valley, 23 Jun 1940 (IMP). One singing on 5 and 6 Jun 1943 in Ravine Reserve (IMP).

ESA: 22 May 1938 at York Lake (IMP) and 24 May 1983 at GSL (WA). **LFD**: 11 Sep 1960.

Breeding: see DMPP.

DUCK MOUNTAIN: SMNH 1926. SMNH 1951. Seen in 1975, 1977 (JAS). Uncommon, seen at Black Spruce bog at southeast corner of park on 17 June, along road to Bear Head Lake on 25 Jun, and at Batka Lake on 22 Jul 1981 (Hatch and Weidl). Pair feeding young at nest at Boomerang Lake, 3 miles northeast of Runnymede, 3 Jun 1988 (CP). SNHS 1989. Seen 25 May 1991 (TH).

WESTERN WOOD-PEWEE *Contopus sordidulus*

Fairly common summer resident. Breeds.

Common in the more open woods and groves in the Duck Mountains [different habitat than that of the Eastern Wood-Pewee]; male shot 12 June 1884 (Seton 1891).

Rare summer resident in southern part of area. Niven has arrival dates only for 1939, 1940, and 1942. One called daily from a tree in the old hospital grounds facing Tupper Avenue, Yorkton, from 19 through 23 Jun 1957 (CSH). Two were present in one study area in old-growth aspen forest at GSL in 1969 (Flack 1976). A few pairs remain all summer at GSL. Two singing in coulees leading to Lake of the Prairies, 21 Jun 1989, and one near York Lake the next day (CSH,MIH). Heard near Pelly, 24 Jun 1991 (MIC,MIH,CSH).

ESA(26): 20 May 1945 at Yorkton (IMP) and 21 May 1958 at GSL(WA), Av 2 Jun. **LFD**(12): 11 Sep 1956, Av 11 Aug.

Breeding: Two nest records. A pair nested in the aspen bush near Niven's north field in 1948, arriving on 3 Jun; young were still present with their parents near the nest on 10 Aug (Niven 1973). Adults feeding three half-grown young in the nest on a horizontal aspen branch 14 feet above ground east of Parkerview on 3 Aug 1990 (AS).

BBS: Theodore, 4 years, including 2 in 1988, mean 0.2.

Banded pre-1955, 19 (Wotherspoon).

Langenburg: Four singing males, 1990 (TH,BL).

DUCK MOUNTAIN: SMNH 1926: study skin of male, #1902, nest and 4 eggs, Jun 18, #1954. SMNH 1951. SNHS 1956. Seen in 1974, 1976, and 1977; nest with young 1975 (JAS). Common throughout park, including boundary trails, nature trails, Kipp Lake, and

Little Boggy Creek (Hatch and Weidl 1981). SNHS 1989.

Point counts: Rank #43, 3 individuals at 3 stops (ARS).

Aspen count totals: OA 1, YA 2, MA 1 (TH).

EASTERN WOOD-PEWEE *Contopus virens*

Uncommon summer resident at Duck Mountain and GSL. No breeding records.

Very common in depths of thick woods, Duck Mountains, 14 June 1884 (Seton 1891).

Occasional sightings at GSL, where the two species of wood-pewee overlap. Song of one taped and another heard during 11-12 Jun 1977 SNHS summer meeting (WCH,SL, Bob Godwin, Kathy Meeres, Ron Jensen, Don Weidl, and many observers). Two at GSL, 27 May 1983 (WA, Harris 1983), and one on 3 Jul 1985 (WA,JA, Harris 1985b).

ESA: 24 May 1986.

BBS: Theodore, 4 years, 1 each, mean 0.1.

DUCK MOUNTAIN: One heard in aspen forest on hillside in southeast corner of park, 17 Jun 1993 (TH), and another heard 23 Jun 1998 (RK, Bain and Shanahan 1998b).

Aspen count total: MA 1 (TH).

YELLOW-BELLIED FLYCATCHER *Empidonax flaviventris*

Uncommon transient.

Noted at Duck Mountain (Seton 1909a).

Close-up view obtained at the Ravine Reserve on 3 Jun 1943 (IMP).

DUCK MOUNTAIN: Recorded by R.J. Long, Jul-Aug 1970. A possible territorial bird was observed in a Black Spruce bog near the ski hill. It was calling from a typical perch about 6 feet up a 9-foot Tamarack sapling, near a Nashville Warbler territory. [In Prince Albert National Park these flycatchers prefer bogs with sphagnum hummocks, not sedge, where there is an open canopy. Nashville Warblers sometimes use the same habitat] (BL).

ALDER FLYCATCHER *Empidonax alnorum*

Common summer resident in damp areas with willows. No breeding records.

Collected at Duck Mountain, 10 and 12 June 1884 (Seton 1891, 1909a).

Noted along Little Whitesand River, probably breeding, 2 Jul 1942 (IMP). Sighted at SNHS summer meet, GSL, 11-12 Jun 1977 (MB). Present where Highway 9 crosses the Assiniboine River south of Sturgis, 29 Jul 1981 (DH). Heard near Preeceville and Pelly, 24 Jun 1991 (MIC,MIH,CSH).

ESA(34): 24 May 1990, Av 2 Jun. **LFD**(17): 13 Aug 1986, Av 8 Aug.

BBS: overall mean, 1.4. Calder, 7. Churchbridge, 8. Ituna, 1. Willowbrook, 33. Theodore, 14 years; highest, 5, 1991, mean 0.8.

Langenburg: Nine singing in aspen stands in or near coulees, 1990 (TH,BL).

DUCK MOUNTAIN: SMNH 1926: study skin of male, #1901. SMNH 1951. SNHS 1956. Seen in 1975 (JAS). Common throughout park; adults scolding along Whitethroat Nature Trail on 16 Jul and adult with fledged young along Pelly Point Trail on 20 Jul 1981 (Hatch and Weidl). SNHS 1989.

Point counts: Rank #48, 2 individuals at 2 stops (ARS).

Aspen count totals: OA 1, YA 3 (TH).

LEAST FLYCATCHER *Empidonax minimus*

Very common summer resident. Breeds.

Abundant in 1969, the commonest species in two aspen study plots at GSL, with 6 and 30 present (Flack 1976).

ESA(47): 5 May 1999, Av 15 May. **LFD**(38): 19 Sep 1954, Av 2 Sep.

Breeding: Early nest building, 31 May 1986 at York Lake, and 1 Jun 1976 near Langenburg (DH). Nest, 2 eggs, 15 Jun 1953 (CCS). Nest, 2 eggs, 29 Jun 1952, and nest, three young, 28 Jul 1959 (CSH). Nest, 4 eggs, near Saltcoats, 22 Jun 1958 (BH,CSH).

At GSL, nest completed 6 Jun had 3 eggs on 12 Jun and three young on 27 Jun 1954. Nest, 2 eggs and 2 cowbird eggs on 6 Jun 1962. Nests with four young on 19 Jun 1961 and 17 Jun 1963 (WA). Nest, 4 eggs, 4 feet above ground in Beaked Hazelnut at GSL, 22 Jun 1969 (JADF).

PNRS: at Saltcoats, 3 eggs on 4 Jun 1958 and 4 eggs each on 22 Jun 1958 and on 26 and 30 Jun 1959 (BH); at Melville, 4 eggs on 4 Jul 1960 and 1 egg on 30 Jun 1961 (VS).

BBS: overall mean, 13.9. Theodore, mean 16.3; increasing (1970s, 10.5; 1980s, 18.0; 1990s, 23.9; n=33).

Banded 8 (CSH 8) + pre-1955, 1 (CSH).

Langenburg: Very common, at least one singing in each aspen bluff, 1990 (TH,BL). Six pairs seen, one nest under construction, 1 Jun 1976 (DH).

DUCK MOUNTAIN: SMNH 1926: 5 eggs on 16 Jun. SMNH 1951. SNHS 1956. Seen in 1974, 1975, 1976, 1977 (JAS). Fairly common throughout park, in core area and on boundary roads and trails (Hatch and Weidl 1981). SNHS 1989.

Point counts: Rank #18, 12 individuals at 10 stops (ARS).

Aspen count totals: OA 6, YA 8, MA 2 (TH).

EASTERN PHOEBE *Sayornis phoebe*

Common summer resident. Breeds.

Has increased at Sheho since the 1940s, now nesting in and around deserted farm buildings. One seen at Little Whitesand Creek east of Yorkton on 17 May 1980 (DH).

ESA(48): 6 Apr 1988, Av 20 Apr. **LFD**(32): 19 Sep 1983 and 1986, Av 4 Sep.

Breeding: Nesting beneath the concrete highway bridge at Sheho each year (WN). Nest at York Lake, 24 May 1945. Adult banded at nest, Jmaeff farm, 6 miles south of Kamsack, 25 May 1958 (CSH). Nest, three young and 1 egg near Margo on 9 Jul 1989 (ARS,PC).

At GSL, newly built nest without eggs, 11 May 1952. Early nest, 5 eggs on 15 May, then 3 eggs and two young on 29 May, five young on 31 May, fledged on 14 Jun 1958; in that year we had eight nests (WA). Late nest, 1 egg on 6 Jul, 4 eggs on 9 Jul and three young on 26 Jul, fledged on 11 Aug 1959 (WA).

PNRS: at Saltcoats, 2 eggs on 15 May 1958, 4 eggs on 1 Jun 1959, and four young on 25 Jun 1959 (BH); at Stornoway, 14 nests with 1, 2, 4, 4, 4, 4, 5, 5, 5, 5, 5, 5, and 5 eggs, and four young in another nest on 14 Jun (SZ,RD).

BBS: overall mean, 1.0. Churchbridge, 4. Willowbrook, 7. Theodore, mean 1.3 (1970s, 1.5; 1980s, 1.1; 1990s, 1.6; n=20).

Banded 1 (CSH) + pre-1955, 3 (Blakeley 2, Wickencamp 1).

Langenburg: One pair each beside two bridges in coulees, 1990 (TH,BL).

DUCK MOUNTAIN: SMNH 1951. SNHS 1956. Seen in 1974 and 1977; nest with young, 1975; nest, 5 eggs, 1976 (JAS). Early singing of seven males, 22 Apr 1989 (TH). Fairly common along roadsides and open clearings. Nest, Pickerel Point Campground office building, 3 Jun; nest, 4 eggs, old cabin at Little Boggy Creek, 16 Jun; adult with three fledged young at west park boundary, 12 Aug 1981 (Hatch and Weidl). SNHS 1989.

Point counts: Rank #33, 5 individuals at 5 stops (ARS).

SAY'S PHOEBE *Sayornis saya*

Rare transient.

Two records at Sheho: one in a flock of Mountain Bluebirds on 27 Aug 1947; a spring record for 20 Apr 1962 (Niven 1973).

GREAT CRESTED FLYCATCHER *Myiarchus crinitus*

Common summer resident in aspen forest. Breeds.

Has increased in numbers and spread west since about 1940. Flack postulated that this might be explained by the maturing of aspen, and availability of appropriate nest holes, only after fires were brought under control with Caucasian settlement. Among Flack's five study sites in mature aspen, from Paynton, Saskatchewan, to southwest Manitoba, the largest number of Great Cresteds (four) was encountered at one of the mature aspen study plots at GSL (Flack 1976).

None seen at Sheho prior to 23 Jun 1956; subsequently seen in 1959, 1963, and in all but three years from 1967 through 1988, with adults feeding young in 1972 (WN, Niven 1973; Houston 1972a). One pair and three singles at York Lake on 30 May 1986 (MDG). One seen east of Parkerview on 18 Jun 1989 (AS). Summer resident in the Jowsey yard 6 miles north of Saltcoats in 1995 (JRJ, SJ). Regularly seen and heard at GSL.

Saskatchewan nests seem not to contain snake skins, as did all 46 nests examined in Florida and 75% of nests examined by A.C. Bent (Lanyon 1997).

ESA(45): 11 May 1975, Av 25 May. **LFD**(30): 13 Sep 1983, Av 25 Aug.

Breeding: Nest with young, 8 feet from ground in hollow maple stump, GSL, 26 Jun 1952 (Houston 1952; Gammell and Gammell 1952c). A pair was feeding nearly full grown young in 1972 at Sheho

(WN, Houston 1972a). Nine sighted at GSL, 11-12 Jun 1977 (WCH,SL). One seen near Togo on 25 Jun 1991 (MIC,MIH,CSH).

BBS: overall mean, 2.9. Willowbrook, 26. Theodore, mean 3.8; possibly increasing (1970s, 2.2; 1980s, 5.2; 1990s, 4.6; n=30).

Langenburg: Eight singing males in heavy aspen, 1990 (TH,BL).

DUCK MOUNTAIN: SMNH 1926: Surprising numbers, though scarce elsewhere in province; mounted male, #1884, study skin male, #1899, female, #1900. SMNH 1951. SNHS 1956. Seen in 1974, 1975, 1976 (JAS). Fairly common, one along road to Bear Head Lake on 1 Jun, several along south boundary road on 2 Jun, one near boat marina on 4 Jun, at Little Boggy Creek on 18 Jun, near Ranger Bay on 23 Jun, near Pickerel Point Beach on 14 Jul, and twice in August 1981 (Hatch and Weidl). In 1994, seen on 27 May (ARS, R. Edmonds) and on 19 Jul (ARS).

Point counts: 1 individual at 1 stop (ARS).

Aspen count totals: OA 3, YA 1 (TH).

WESTERN KINGBIRD *Tyrannus verticalis*

Fairly common summer resident from Yorkton, Theodore, and Sheho, south, but rare at GSL. Breeds.

Reached Melville in 1921 and for four years running, a pair nested behind a can on an electric light pole. At the time, trees were small and this was the only pair in Melville (McKim 1926).

A new arrival to the southern part of the area as the annual prairie fires were controlled, and trees were planted in farmsteads (Houston 1979). One of two species showing a rapid increase in numbers after the first nests were found in Yorkton and in Saltcoats in June 1940; by 1943 they were "quite common...a familiar nesting species in the boulevard maples" in residential areas of Yorkton. One seen near Togo on 25 Jun 1991 (MIC,MIH,CSH).

ESA(9): 18 May 1994, Av 31 May. Fall dates: 22 Aug 1943 (IMP), 23 Aug 1942 (CSH), and 2 Sep 1994 (WA).

Breeding: One pair nested on the farm of Robert Mess, 5 miles west of Bredenbury, in 1937. First nests found in Yorkton and Saltcoats in 1940. First reached Sheho on 7 Jun 1942, with one or two seen most summers since, and with one nesting in a spruce in a farm shelterbelt 3 miles northeast of the Niven farm; a pair nested on the Niven homestead for the first time on 11 Jul 1988 (WN).

> The 30,000 abandoned farmsteads throughout southern Saskatchewan "make a significant contribution" to Western Kingbird populations; the Western nests high in the canopy, and prefers domesticated poplar to maple stands (Smith 1972).

At Yorkton in 1943, a pair built their nest in the lowest branch of a spruce just outside the front door of the William Pickering home on Agricultural Avenue. Mrs Pickering put out strips of raw meat along the branch near the nest; these were quickly fed to the young birds (Priestly 1943c).

In 1973, sighted at Crescent marsh, Lower Rousay Lake and Upper Rousay Lake (AS). Sighted at SNHS summer meet at GSL, 11-12 Jun 1977 (MB). One seen at Hyas on 17 Jun 1983 (CP). Two near Upper Rousay Lake on 13 Jun 1989 (CSH,MIH). In 1994, six sighted between Leech Lake and Rokeby marsh on 18 and 23 Jul (DH).

BBS: overall mean, 0.7. Theodore, 17 years; highest, 5, 1988, mean 0.9.

Banded: pre-1955, 7 (Briggs).

Langenburg: One seen, 1990 (TH,BL).

DUCK MOUNTAIN: Present, although not expected. SNHS 1956.

Aspen count total: YA 1 (TH).

EASTERN KINGBIRD *Tyrannus tyrannus*

Common summer resident. Breeds.

Duck Mountains, 17 Jun 1884 (ETS). Three kingbird eggs collected north of Saltcoats (NJ,PMA).

Range extends north barely to Pelly and Arran (Wotherspoon 1976). Two in one mature aspen study plot at GSL, 1969 (Flack 1976).

ESA(47): 9 May 1959 and 1981, Av 20 May. **LFD**(37): 12 Sep 1956, Av 30 Aug. One family of young did not leave the nest until 28 Aug, and did not migrate until 7 Sep 1983 (WA).

Breeding: Nests found in city and in farmyards. Nest on artificial shelf placed for robins at Ituna, 1943 (Rayner 1944). Early nest, 2 eggs, 18 Jun and 4 eggs on 2 Jul 1961 (WA). Nest at GSL, 30 feet above ground in major fork of a live tree, feeding young, 22 Jun 1969 (JADF). Two immatures at York Lake on 15 Aug 1973 (AS). Two young at Ravine

> The kingbird is aptly named. We have a Purple Martin house in our yard. A dominant male had threatened everything and everybody. Neither Joyce nor I could walk near the martin house without being harassed--until a pair of Eastern Kingbirds moved in. A day or two later there was a low level flight across the length of the yard, the kingbird riding the martin's back with feathers flying, and the martin screaming its head off. We had a very well-behaved martin for the rest of that summer (WA).

Reserve on 20 Jul 1985 (DH). Nest near Sheho, 11 Jun 1988 (WN). One pair built a nest in the eavestrough of our porch roof, and had 4 eggs on 26 June 1999; on 9 July the unwise placement of the nest became evident when it was flooded in a heavy rain, causing abandonment (WA).

PNRS: at Stornoway, five young in nest on 23 Jun 1960 (SZ) and nest with three young on 10 Aug 1960 (RD).

Also breeds at Duff (ARS, *Atlas*).

BBS: overall mean, 5.9. Theodore, mean 4.7; possibly decreasing (1970s, 6.4; 1980s, 5.0; 1990s, 2.8; n=31).

Banded pre-1955, 9 (Briggs 6, CSH 1, Wotherspoon 2).

Langenburg: Common, 10 pairs seen, 31 May 1976 (DH). Several pairs seen, 1990 (TH,BL).

DUCK MOUNTAIN: SMNH 1926. SMNH 1951. SNHS 1956. Seen in 1976, 1977 (JAS). Fairly common, seen at Little Boggy Creek on 17 Jul and 13 Aug and Kipp Lake on 12 Aug 1981 (Hatch and Weidl). SNHS 1989.

Aspen count total: YA 1 (TH).

LOGGERHEAD SHRIKE *Lanius ludovicianus*

THREATENED, COSEWIC, APRIL 1986

Formerly common, now an uncommon summer resident bordering on rare. Breeds.

A nest with 6 eggs, 5 feet above the ground, at Crescent Lake, 11 June 1901, attributed to the Northern Shrike, and another with 5 eggs on 1 Jun 1902, were almost certainly Loggerhead Shrikes, although the parent was collected by Frank Baines. [The coloration of the western *L. l. excubitorides* was different from the better-known and minimally smaller eastern subspecies, so Raine's misidentification might conceivably have been an honest mistake coloured by wishful thinking, not necessarily a dishonest attempt to extort a higher price for the much more rare eggs of the Northern Shrike!-CSH]. The Baines brothers also collected a number of shrike clutches at Crescent Lake in 1890 (Raine 1892:141). Five eggs, Crescent Lake, 2 Jun 1894, AMNH.

Numbers have decreased greatly since the 1940s. At GSL, John A. Gunn felt there had already been an appreciable decline between about 1940 and 1948. As a youngster in the 1940s, I remember frogs and grasshoppers being impaled on barbed wire fences by shrikes; seen at GSL in only seven years since 1950! (WA).

Only one was seen in 1975, one on 4 May 1976, and none was seen around the Niven farm north of Sheho, 1977-1989 (WN). Studies in 1987, 1993, and 1998 confirmed the low numbers of 0.1 to 0.2 breeding pairs per 100 km^2 (Johns et al. 2002). On 25 Jun 1988, a pair was acting territorial on a farm at the southwest corner of Yorkton (AS).

ESA(4): 28 Apr 1943 (WN), 29 Apr 1962 (WA) and 30 Apr (RB). **LFD**: 19 Sep 1944 (CSH).

Breeding: Early nest, in a spruce across from the planing mill in Yorkton, 17 May 1942 (IMP). Nest, 7 eggs, Melville, 21 May 1957 (GAA). Nest, 7 eggs, Pearce farm near Tonkin, 30 May 1955 (CSH). A nest with 1 egg on 28 May 1945 had 6 eggs on 4 Jun (WHC).

PNRS: Three young banded on 26 Jun 1971 south of Kelliher (WCH); at Saltcoats, 7 eggs on 7 Jun 1958 and four young on 25 Jun 1959 (BH); at Melville, 3 eggs on 30 May 1969 (GAA) and 7 eggs, 31 May 1960 (VS).

BBS: overall mean, 0.1. Churchbridge, 1. Ituna, 1. Willowbrook, 1. Theodore, 1 in 1976 only, mean 0.03.

Banded pre-1955, 31 (Briggs 6, Blakeley 5, Wotherspoon 20). [Three nestlings listed as banded by subpermittee WCH south of Leross in late June 1971 represent a probable banding office error in location entry; they were banded near Raymore].

Langenburg: one on 31 May 1976 (DH), but none sighted by Herriot and Luterbach.

DUCK MOUNTAIN: Sighted by Bob and Susie Kohlmeier at SNHS spring meet, 10 Jun 1989.

NORTHERN SHRIKE *Lanius excubitor*

Uncommon in winter.

EFA(30): 8 Oct 1986 and 2002, Av 24 Oct. **LSD**(10): 20 Apr 1958, Av 5 Apr.

CBC: Rank #36. 22 counts, 27 birds. Highest: Kamsack 3, 1987.

One seen during Saltcoats CBC in 1953 and one in Christmas week at Sheho the same year. Also seen at Sheho on 4 Apr 1964, 23 Mar 1968, 7 Nov 1974, and 15 Oct 1976 (WN). One individual seen on 26 Dec 1956, the only time recorded in the first 19 Yorkton CBCs. Noted in only 2 (1967 and 1968) of 22 CBCs at GSL. One sighted at Horseshoe Lake, 30 Jan 1976 (DH).

YELLOW-THROATED VIREO *Vireo flavifrons*

Rare summer resident at Duck Mountain and along the Assiniboine River valley. No breeding records.

"Taken at Yorkton by W. Raine" (Seton 1909a:221; Macoun and Macoun 1909:605).

Langenburg: One singing on Assiniboine valley slope north of Marchwell, 1 Jun 1976, during a heavy Forest Tent Caterpillar invasion; the bird flew back and forth across the cut marking the Manitoba boundary (DH).

DUCK MOUNTAIN: Present. SMNH 1951: Male specimen, the museum's first for Saskatchewan, 4 Jun 1951 (FWL,SMNH). SNHS 1956.

BLUE-HEADED VIREO *Vireo solitarius*

Rare transient and summer resident. Breeds.

Seen in the Duck Mountains, 10 Jun 1884, apparently breeding (Seton 1891).

Rare transient at Yorkton. One seen in May 1936 (IMP) and along Isabel Priestly Nature Trail, York Lake, by Paule Hjertaas and Jean Brochu, 31 May 1986 (DH). Five sightings at GSL: 23 May 1954, 3 May 1955, 7 Sep 1958, 31 Aug, and 16 Sep 1994 (WA).

LFD at Melville: 20 Sep 1941 and 24 Sep 1943 (LTMcK).

Breeding: Nest with eggs in flimsy aspen sapling, 10 feet above ground, in the Anaka yard at GSL, 2 July 2002. The nest would have been well hidden had the Forest Tent Caterpillars not eaten all the leaves, leaving the nest fully exposed; the young fledged successfully (WA).

DUCK MOUNTAIN: SMNH 1926: study skin male, #1910, female, #1911, the first known Saskatchewan specimens; nest, 3 eggs, #1957. Sighted 30 Jul 1939 (IMP). SMNH 1951. SNHS 1956. "Probably nesting" in 1976 (JAS). Uncommon; pair building a nest in a birch along the Pelly Point Nature Trail on 24 Jun 1981; another pair seen next day along Red Squirrel Nature Trail (Hatch and Weidl). SNHS 1989. Several heard, 18 Jun 2000 (BL).

Point counts: Rank #24, 7 individuals at 7 stops (ARS).

WARBLING VIREO *Vireo gilvus*

Very common summer resident. Breeds.

Seen on west side of Duck Mountain in 1884 (Seton 1891).

Eight and 12 in the two mature aspen study plots at GSL, 1969 (Flack 1976).

ESA(45): 13 May 1992, Av 22 May. **LFD**(31): 16 Sep 2000, Av 2 Sep.

Breeding: On several occasions, including 14 Jun 1942, I have followed the voice of a singing male, to find him singing on his nest, a habit common in this species. It is rarely possible to look into the pendant nest, hanging much like an oriole's from a slender branch, so the eggs often cannot be counted. Another nest, 14 Jun 1958. A Warbling Vireo nest in a willow 6 feet above ground near Gorlitz on 2 Jul 1967 contained 1 vireo egg and 4 cowbird eggs (CSH).

At GSL, nest, 4 feet out on branch, 15 feet above ground in poplar, 22 Jun 1969 (JADF); nest with small young, in Balsam Poplar thicket, 21 Jul 1987, but presumably predated by weasels later that day (WA).

PNRS: near Saltcoats, on 14 Jun 1959, a nest held 3 vireo eggs and 2 cowbird eggs (BH); at Stornoway, 4 eggs on 20 Jun 1959 (SZ), and 3 eggs on 4 Jun 1961 (RD).

BBS: overall mean, 11.9. Theodore, mean 13.8 (1970s, 12.1; 1980s, 15.1; 1990s, 15.4; n=33).

Langenburg: Very common, singing in every shelterbelt and aspen bluff, 1990 (TH,BL).

DUCK MOUNTAIN: SMNH 1951. Uncommon, sightings on 3 Jun and 14 and 15 Jul 1981 (Hatch and Weidl).

PHILADELPHIA VIREO *Vireo philadelphicus*

Fairly common summer resident locally among closely packed, tall, thin aspen at GSL and Duck Mountain. Breeds.

The first nest documented for this species anywhere in North America was found by Ernest Thompson Seton on the upper Assiniboine, south of Fort Pelly, on 9 June 1884. The exact location was between 1 and 2 miles west of the present hamlet of Runnymede. "The nest hung from a forked twig about 8 feet from the ground, in a willow which was the reverse of dense, as it grew in the shade of a poplar grove. The nest was pensile, as usual with the genus, formed of fine grass and birch bark." Four days later the female was shot and the 4 eggs collected (Seton 1885, 1909a).

One was found dead and identified at Yorkton, 5 Jun 1945 (Priestly 1945b). One seen at GSL on 5 Aug 1971 (BL). One at Yorkton, 24 May 1971 (Houston 1971c), and along the Isabel Priestly Nature Trail at York Lake, 31 May 1986 (DH). One singing near Togo on 25 Jun 1991 (Murray and Allan Bowes).

ESA(6): 19 May 1987, Av 5 Jun. **LFD**(6): 7 Sep 1958, Av 28 Aug.

Breeding: see above.

BBS: overall mean, 0.2. Churchbridge, 1. Theodore, 3 years; highest, 4, 1998, mean 0.2.

DUCK MOUNTAIN: SMNH 1926: Specimen male, #1879, female, #1880, study skin male, #1909, new for collection. SMNH 1951. SNHS 1956. Seen in 1976 (JAS). Uncommon, sightings of individuals on 12 and 13 Aug 1981 (Hatch and Weidl). SNHS 1989.

Aspen count totals: OA 4, MA 2 (TH).

RED-EYED VIREO *Vireo olivaceus*

Common summer resident in heavy aspen woods, such as near GSL. Breeds.

Eight in one of two mature aspen study plots at GSL, 1969 (Flack 1976).

ESA(38): 13 May 1941 (IMP) and 17 May 1964 (WA), Av 26 May. **LFD**(32): 18 Sep 1942 and 17 Sep 1943 at Yorkton (IMP) and 9 Sep 1956 and 1962 at GSL (WA), Av 29 Aug.

Breeding: Nest, 2 eggs and 1 cowbird egg, in Chokecherry 3 feet above ground, 11 Jul 1959; after cowbird egg was removed, one young fledged 26 Jul 1959. Late nest had two young, 1 Sep 1986 (WA).

BBS: overall mean, 6.6. Theodore, mean 9.6 (1970s, 9.6; 1980s, 10.7; 1990s, 10.6; n=33).

Banded 1 (CSH).

Langenburg: Common, especially in coulees and valleys, 1990 (TH,BL).

DUCK MOUNTAIN: The monotonous three-syllable monologue is the most obvious bird call throughout Duck Mountain. SMNH 1926: 2 and 3 eggs, #1955 and #1956. SMNH 1951. SNHS 1956. Seen in 1974, 1975, 1976, 1977 (JAS). Common throughout park (Hatch and Weidl 1981). SNHS 1989.

Point counts: Rank #1, 108 individuals at 74 stops (ARS).

Aspen count totals: OA 17, YA 12, MA 17 (TH).

GRAY JAY *Perisoreus canadensis*

Rare winter visitor in parkland. Fairly common year-round resident at Duck Mountain, where it breeds.

Young following parents in Duck Mountain, 11 June 1884 (Seton 1891). In the 1880s and 1890s, so common in fall and winter as to be a nuisance to the trappers at GSL (JAG). Also common at Crescent Lake every winter, with one or two most winters from 1883 until 1935 (Frank Baines).

Not present in the early years of settlement at Sheho, but became fairly common throughout the 1920s, when six or 10 moved about the district each fall and winter. Thereafter, a few individuals might stop by once or twice, but not every winter, through 1946, when two were present from the early date of 10 Aug through 10 Dec (WN, Shaw 1946b). Between 1946 and 1972, most winters had none, with a rare single individual on a few occasions (Niven 1973), and one in the Niven yard on 22 Oct 1976 (WN).

At Yorkton, two on 18 Feb 1937 and three on 7 Mar 1942 (Ethel Lloyd), and two during the winter of 1945-46, making numerous visits to J.R. Foreman's feeding station on Wallace Avenue. One on Yorkton CBC, 1945, 1946. Now irregular and sporadic in winter at Yorkton and GSL, though it missed five years 1958-62 at the lake (WA, Lister 1965), and then one wintered, 1972-73 (WA, Houston 1973b). "Hundreds" on 16 Sep 1976, with seven daily until 15 Nov (WA, Serr 1977a). Part of minor southward irruption at GSL, fall 1984 (WA, Harris 1985a). Fewer seen at GSL since about 1990.

EFA(12): 28 Aug 1955, Av 5 Oct. **LSD**(11): 30 May 1956, Av 19 Apr.

CBC: Rank #28. 37 counts, 133 birds. Highest: 17, Kamsack 1992. Yorkton sightings were in 1945, 1946, 1947, 1949 (2), 1952, and 1976, with none since.

Banded 5 (on 4 and 9 Jan 1956, 21 Feb 1959 and two on 17 Mar 1968, WA) + pre-1955, 1 (CSH, at feeding tray, at 102 Second Avenue North, Yorkton, on 7 Dec 1952).

Recoveries, banded in area, 1 [16.75%] (SK, injured a few miles south of GSL, two months later, WA).

DUCK MOUNTAIN: SMNH 1926: Fairly common year-round resident. Breeding. SMNH 1951. Seen in 1975, 1976, 1977 (JAS). Six in 1959 CBC. Fairly common in mixed woods year-round; two adults and an immature near Selwood Campground on 15 Jun 1981 (Hatch and Weidl). Young, SNHS 1989.

Point counts: Rank #48, 2 individuals at 2 stops (ARS).

Aspen count total: OA 1 (TH).

STELLER'S JAY *Cyanocitta stelleri*

Vagrant.

Two records. Two spent the summer on the Baines farm at Crescent Lake, 4 miles south and 8 miles west of Saltcoats, 28-22-2W2, in the 1920s (Frank Baines in Houston 1949). Miss Pauline Summers got a close-up view of this midnight-blue jay at the sandpit in Yorkton in the spring of 1944 (Priestly 1944).

BLUE JAY *Cyanocitta cristata*

Uncommon year-round resident, most evident during fall wanderings and at feeding stations in winter. Breeds.

Swan River, 2 September 1881 (Macoun in Seton 1891). Several pairs were present at GSL for a few years about 1937 (JAG).

On 23 Aug 1949, Merle Wershler, a member of Pauline Summers' Simpson School Nature Club, watched a Blue Jay swoop down and carry off a flicker (Wershler 1949). A flock of 20 at Foam Lake on 18 Sep 1984 was the largest group seen (WN).

Breeding: Nest, three young, in the Baines' yard in the town of Saltcoats, early June 1944. Nested in a spruce 10 feet from Mrs Quaid's back door on Assiniboia Avenue in Yorkton, May 1946. Nest in a White Spruce in L. Friemark's yard, 1.5 miles east of Horseshoe Lake, in May 2002, fledged in June.

BBS: overall mean, 0.1. Willowbrook, 2. Theodore, 5 years; highest, 9, 1999, mean 0.1.

CBC: Rank #14. 121 counts, 1,118 birds. Highest 100, Kamsack 2000.

Banded 15 (CSH 1, WA 14) + pre-1955, 1 (Wickencamp).

Return, 1, retrapped one year after banding (WA).

Recoveries, banded outside area, 1 (banded east of Watertown, SD, on 24 Dec 1958, caught in trap at Sturgis before 27 Mar 1964).

DUCK MOUNTAIN: SMNH 1926. SMNH 1951. SNHS 1956. Seen in 1975, 1976, 1977 (JAS). Fairly common year-round throughout park (Hatch and Weidl 1981). SNHS 1989.

Point counts: Rank #33, 5 individuals at 5 stops (ARS).

Aspen count totals: YA 4, MA 1 (TH).

BLACK-BILLED MAGPIE *Pica hudsonia*

Very common year-round resident. Breeds.

Upper Assiniboine, below the Shell [south of the YDM area], 27 September 1881 (Macoun in Seton 1891). Common at GSL for the first 10 years after John A. Gunn arrived in 1887. It then disappeared for about 30 years.

Numbers began to increase during the late 1930s, with winter sightings predominating. At Crescent Lake, they first re-appeared some time after 1920 (Baines 1956). At Sheho, none was seen between 1910 and 1925; a few wandering individuals were noted between 1926 and 1937. A noticeable increase in numbers in 1938 (DU "Kee-men" reported 6 to 15 per square mile during the last week of July 1938) was followed by a population explosion in 1939 when magpies became for the first time a common year-round resident.

At Yorkton, the magpie was still an uncommon species in 1940, when my first sighting for the year was not until 22 Jun! Numbers increased quickly during the 1940s (Houston 1977b), yet the largest group seen through 1960 was only 20 (CSH). By 1970, there was a flock of 150 at GSL on 9 Sep (JG, Houston 1971a).

Breeding: I did not find my first nest until 1953, when six young were present on 31 May and banded on 7 Jun (CSH). An early nest on

12 Apr 1955 had 2 eggs, with 1 egg added daily for the next three days (Brother Halward). A sudden increase in magpies became evident in 1958 and seemed associated with above-average productivity; Horseman found 12 nests with 1, 3, 4, 5, 5, 6, 6, 7, 8, and 9 eggs, and two and six young; survival was high and nesting was early, with 26 banded in five nests and seven in another three nests on 1 Jun 1958; another 11 were banded in two nests on 11 Jun and four on 22 Jun (CSH). In 1959, Horseman found nine nests with 5 and 6 eggs and three, three, three, five, five, six, and seven young (BH). In 1959, Anweiler found 12 nests near Melville, including nests with 8 eggs on 29 Apr and on 1 May; from that first nest all eight hatched but only four survived to 26 May (GAA). In 1959 at GSL, 32 young fledged from five nests, an unusually high average of 6.4 per nest (WA). In June 1959, 14 were banded in five nests at Melville and three in one nest at Saltcoats (CSH).

A flock of 22, mostly short-tailed immatures, was at York Lake on 28 Jun 1973 (AS). In the 1960s and 1970s, magpies moved into cities and towns for the first time. Sargeant et al. (1993) found two nests.

Oology: One egg collected north of Saltcoats, probably after 1945 (NJ,PMA).

PNRS: at Stornoway, one nest in 1959 with 7 eggs, and three nests in 1960, with 2, 4, and 9 eggs (SZ).

BBS: overall mean, 17.2. Theodore, mean 21.6 (1970s, 23.9; 1980s, 17.1; 1990s, 23.0; n=33).

CBC: Rank #9. 226 counts, 2,824 birds. Highest: 87, Yorkton 2000. Four of the first 10 CBCs at Yorkton had only one magpie counted, with a maximum of eight; in 1951 there was a sudden increase to 24. By the winter of 1958-1959, numbers were higher than at any time in the century (Krause 1959b).

Banded 181 (CSH 67, WA 108, RLS 6) + pre-1955, 17 (Bard 11, CSH 6). Most banding was in nests found by Bill Horseman at Saltcoats (46) and west of Saltcoats (5, CSH), and at GSL (108; 59 in the 10-minute block west of Canora and 49 in the block south of Buchanan, WA).

Recoveries, banded in area, 9 [4.5%] (SK 8, MB 1). One of Bard's magpies was caught in a trap and one was shot, both the same year, and two Houston birds were shot locally the same year. A magpie banded at Wadena in Oct 2000 was caught by hand and released near Kenaston, 24 Mar 2001 (CSH). Anaka's four were shot half a mile east of the banding site; trapped near Mikado the same year; trapped near Hyas, SK; and shot at Moorepark, MB, the last two a year after banding (WA).

Recoveries, banded outside area, 1 (banded by J.A. Briggs at Biggar on 15 Jul 1939, shot at Theodore in early November that year).

Langenburg: Common, 1990 (TH,BL).

DUCK MOUNTAIN: Absent in 1926 and 1951 (SMNH). Uncommon year-round in open areas and roadsides (Hatch and Weidl 1981).

AMERICAN CROW *Corvus brachyrhynchos*

Abundant summer resident. Breeds.

The numbers of this intelligent, adaptable, omnivorous species increased quickly as land was broken at the turn of the century. It was rare when the Baines family arrived at Crescent Lake in 1883 but gradually increased until 1915, when numbers levelled off (Baines 1956). At Sheho, crows also increased with settlement, thriving in the presence of man and his agriculture, but with clearing of the aspen bush in the 1950s, their numbers showed a slight downward trend (Niven 1973). Five fresh eggs at Foam Lake on 12 May 1911 (H.C. Grose, SMNH). Four eggs collected north of Saltcoats (NJ,PMA).

Large fall flocks appeared as early as 1962, when in early September, on a hayfield near Calder, at least 500 but probably 1,000 or more, descended to eat grasshoppers (Ninowski 1962).

ESA(48): 5 Mar 1981, Av 22 Mar. **LFD**(42): 15 Oct 1957, Av 2 Oct.

Breeding: Early nest, 5 eggs, 7 May 1946. Early hatching, four young, 28 May 1946. Near Saltcoats, 31 young were banded in 13 nests in early Jun 1958, beginning on 1 Jun; in 1959, young from 13 nests, with 1, 3, 4, 4, 4, 4, 5, 5, and 5 eggs, and three, three, three, and five young, went unbanded (BH). At GSL, banded 175 young in 53 nests, an average of 3.3 young per nest; peak year was 1959, with 15 nests. A nest on 10 May was lined with highly odoriferous fur from a Striped Skunk; by 17 May a lining of poplar inner bark had been added (WA). Sargeant et al. (1993) found 10 nests in the Yorkton area.

PNRS: at Stornoway, nests had 6 eggs, and two and two young in 1959; 3, 4, 4, 5, and 5 eggs in 1960; and 4 eggs in 1961 (SZ); at Melville in 1959, nests contained 1, 3, 3, and 5 eggs; at Kamsack, a nest had four young on 18 May 1958 (JHJ); at Mikado, 4 eggs were hatching on 31 May 1966 (EK).

BBS: overall mean, 70.4. Theodore, mean 72.5 (1970s, 87.0; 1980s,

65.4; 1990s, 63.6; n=33).

Winter: An early record of a single wintering crow on 26 and 27 Dec 1932 at Invermay (Thomas W. Condon, Lawrence 1932).

CBC: Rank #55. 4 counts, 4 birds. Saltcoats 1953, Melville 1989, Yorkton 1996, Bangor 2000.

Banded 219 (CSH 44, WA 175) + pre-1955, 44 (CSH 8, Bard 13, Wotherspoon 23).

Recoveries, banded in area, 33 [12.5%] (SK 22, KS 1, NE 4, OK 5, TX 1). A Patrick crow from Yorkton was shot in Oklahoma; eight Wotherspoon crows from Hyas were shot and another three found dead, two in Nebraska and one in Oklahoma. Anaka crows from GSL were shot during their first southward migration, at Seward, KS, and Waller, TX, while another was shot during its second winter at Albert, OK (WA). Of 13 crows released in the Yorkton area as part of Fred Bard's anti-crow campaign (Houston 1969d), five were shot within the area, that fall or the subsequent spring, and one banded near Dubuc on 27 May 1937 was shot 2 Jan 1939 at Cement, OK. Houston crows were shot at one year and one was found dead at two years; shot the same fall at Blair, OK, and Grand Island, NE; shot after two years at Hastings, NE (Houston 1969d). A crow banded as a nestling at Saltcoats on 2 Jun 1958 was shot a few miles south of there at 12 years of age, on 7 Oct 1970, by Peter Chobotar, who confirmed the exact day he shot it (CSH). At the time it was the second oldest crow in the banding files for North America, although the current record is now 14 years, 7 months. Local recoveries near the banding site include three shot at three years of age within 2 miles of the nest in which they had hatched. Although a small sample, this suggests shorter dispersal than is evident for most other species.

Recoveries, banded outside area, 4 (SK 1, AB 1, OK 2).

Langenburg: Common, 1990 (TH,BL).

DUCK MOUNTAIN: SMNH 1926. SMNH 1951. SNHS 1956. Seen in 1974, 1975, 1976, 1977 (JAS). Common in all woods and open areas, but especially at garbage dump; young birds along Pelly Point Nature Trail on 15 Jul and on Spruce Island on 16 Jul 1981 (Hatch and Weidl). SNHS 1989.

Point counts: Rank #9, 25 individuals at 23 stops (ARS).

COMMON RAVEN *Corvus corax*

A fairly common year-round resident, increasing. Breeds.

Quite common at GSL in winter, between 1887 and the settlement of the district at the turn of the century (John A. Gunn). They then retreated northward.

The raven re-invasion of the YDM area, a phenomenon well documented, began with rare fall and winter sightings: one at Cussed Creek, 9 Oct 1938 (IMP); one at GSL, Oct 1939 (John A. Gunn); one present regularly at Leech Lake for three weeks in Aug 1944, feeding on carrion such as dead shorebirds; one noted at Yorkton by Gordon Betker, 26 Jan 1947; one 4 miles north of Yorkton, 8 Aug 1956 (CSH); one, with a 4-foot, 2-inch wing span, was shot at Saltcoats on 18 Oct 1956 (BH).

It was not until 23 Sep 1947 that William Niven saw his first wandering raven; several stayed during Oct and Nov 1954 (Gammell and Huenecke 1955). Since then, one or two appeared at some time between late fall and early spring nearly every year, until a more widespread southward incursion occurred in 1972; 19 were seen on 7 Nov and others were present through the end of December (WN, Houston 1973a). In 1974, 17 Sep was the earliest arrival to that date and by 1978, ravens were present at Sheho from 20 Oct through 15 Nov 1978, and then off and on throughout the winter. The first summer date was 7 Jun 1981, when one was chased by crows (WN).

> The raven is an incredible bird, intelligent, versatile, bold, displaying a seemingly incredible sense of humour. Along the beach at GSL, ravens will play with any object that attracts their attention–a piece of plastic or paper (minnow bucket lids are favourite toys), or bits of driftwood or twigs. It becomes a game with two or three tossing the desired object around until tiring of the game. Snow bathing is another sport when conditions are right. On 26 Oct 1991 at 1700 hours, three ravens were observed diving into 6 inches of new-fallen snow, then lying with wings flapping, twisting, and turning before flying off. On a bitterly cold day, with temperature approaching minus 40, and a wind of 30 mph, ravens will face the wind and do cartwheels and stalls, obviously enjoying themselves. When a coyote is feeding at a deer kill, ravens will brazenly sneak in and attempt to steal morsels (WA).

Seen once in the Yorkton area in 1985 (Sargeant et al. 1993). Eight ravens were feasting on a deer kill near Yorkton on 23 Nov 1975 (WA, Serr 1976). In 1978, one at the edge of Yorkton on 11 Mar (DH), and one present on 22 Sep and from 9 through 26 Nov at GSL (WA). In the fall of 1980, populations increased even more dramatically (WA, Harris 1981a). By 1987, up to 10 could be seen per day at GSL (WA), and 20 between Yorkton and Wynyard on 15 Nov (DH, Harris 1987).

Ravens became regular in winter near Saltcoats about 1980, followed by the first nesting pair in 1982 (BH). Now ravens are one of the commonest winter birds. On 3 Mar 1989 single individuals were seen at Tuffnell, Wroxton, and Stornoway, with clusters of three and seven on the south and east edges of Kamsack, and on 8 Mar there were individuals near Springside and near Sheho (CSH,MIH). There were 50 near Parkerview on 1 Mar 1990; 111 flying south-southeast in a loose string half a mile long near Parkerview on 23 Jan 1999; 265 in a loose flock spread out over half a mile, east of Parkerview on 20 Feb 2000 (AS). A flock of nearly 100 ravens was moving south near Calder, 21 Nov 2002 (Jeff Morari).

EFA: 17 Aug 1956. **LSD**: 27 Mar 1974. [Dates prior to it becoming a year-round resident]

Breeding: At GSL, first nesting was in 1980; this nest had four young on 25 May, but they failed to fledge (WA). Ravens have nested each year since. Carrying nest material 11, 15, and 18 Apr; young banded by CSH on 15 Jun 1986. Two young, 14 May 1988, fledged on 23 May; the same nest was re-used in 1989 (WA). First nest at Saltcoats in 1982 (BH).

Nests in 1989 at Parkerview, with four young barely able to fly on 11 May, and a defended nest west of Goodeve on 22 May (AS). Five young banded in nest at GSL, 20 May 1990 (CSH,WA). Numbers of flying young near Preeceville and Pelly, 24 Jun 1991 (MIC,MIH,CSH). Ian Lochtie reported a pair nesting near the Kelliher nuisance ground in 1992; the next year a pair raised young that were flying by 27 May on his farm south of Kelliher. A raven nest 1 mile east of Saltcoats on 12 Apr 1993, was lined with deer hair, and contained four newly hatched young and 1 egg (BH).

Subsequent nestings were near Hubbard on 17 April 1996; at Wroxton on 25 Apr 1997; west of Goodeve on 30 Apr 1997; under the dome of St George's Ukrainian Catholic Church in Melville on 15 May 1997; and near Calder, Leross, Goodeve, and Parkerview in 1998 (AS). By 1999, it was common for aspen bluffs of more than 40 acres in extent to hold a pair of ravens nesting deep in the center (Larsen 1999, BH).

The first attempted urban nest, on 29 Mar 1998, was on the third floor ledge of a condominium on the east side of Yorkton. One bird did most of the work while the mate supervised. Many sticks fell onto the ground; after two days the remaining sticks were removed by the caretaker (WA).

David Weiman noted three young, flapping around and nearly ready to fly, squawking from the top of a beaver house in a pond in parkland habitat, 5 miles northwest of Preeceville (Weiman 2001).

Also breeds at Goodeve (ARS, *Atlas*).

BBS: overall mean, 0.9. Calder, 1. Willowbrook, 3. Theodore, 9 years; highest, 11, 2001, mean 1.4. The 33-year Theodore BBS gives excellent insight into the gradual summer invasion. One raven was seen in the 1980 BBS, then one in 1989 and nine in 1990, three in 1991, and five, ten, five, and four, respectively, in 1997, 1998, 1999, and 2000.

CBC: Rank #11. 112 counts, 1,765 birds. Highest: 129, Preeceville 2001.

Banded 10 (CSH).

DUCK MOUNTAIN: SMNH 1951. One in 1959 CBC. Seen in 1974, 1975, 1976, 1977 (JAS). Common year-round (Hatch and Weidl 1981). Nineteen at the landfill on 3 Mar 1989 (CSH,MIH). SNHS 1989.

Point counts: Rank #8, 27 individuals at 23 stops (ARS).

Aspen count totals: OA 6, MA 2 (TH).

HORNED LARK *Eremophila alpestris*

Abundant migrant. Until very recently, an abundant summer resident of open fields, now common. Breeds.

The only species that resides regularly within growing grainfields. As a result, their numbers increased as land was cleared at Sheho. There has been a drastic drop in numbers since 1970.

On Highway 47 between Melville and Willowbrook on 10 Mar 1995, 172 migrants were counted (DH). Later, from mid-Apr to mid-May, flocks of 100 or more, of the completely migratory subspecies, en route to the Arctic tundra, chiefly *E.a. hoyti*, pass through (WN).

ESA(45): 9 Feb 1968, Av 25 Feb (mainly resident pairs). **LFD**(35): 4 Nov 1963, Av 16 Oct.

ESA for *E.a. hoyti*: 19 Apr 1964, Av 29 Apr. **LSD:** 21 May 1956 (WN).

Breeding: Many Horned Lark nests on the Niven farm were destroyed by spring cultivation, chiefly in May, but the birds usually renested with success (WN). They breed extremely early, sometimes when snow remains on the ground. Early nests, with 3 eggs at Melville on 2 Apr 1961 (VS), 2 eggs at Saltcoats on 3 Apr 1959 (BH), 3 eggs at Sheho on 9 Apr 1959 (WN), and 16 Apr 1950 near Yorkton (Darcy Wershler). Earliest hatch, at Stornoway, five young and 1 egg on 7 Apr 1961; the eggs were laid before 26 Mar (SZ).

Nest, 4 eggs, Deer Park Golf Course, 6 Jun, had two young hatched on 8 Jun 1953 (CCS). A pair nested successfully and fledged two young in a residential backyard at Melville, even though a lawn mower passed several times over the nest; 1 egg failed to hatch (Anweiler 1965). At GSL, during farming operations I found five nests in 1958, five in 1959, two in 1960, and one each in 1962 and 1963; nests contained 1, 2, 2, 3, 3, 3, 3, 3, 4, 5, 5, and 6 eggs; 3 eggs and one young; and one young alone; early fledging here was of two bob-tailed young, barely able to fly short distances, 6 May 1963 (WA).

PNRS: west of Canora, a nest had 1 egg on 18 Apr 1959, 2 on 20 Apr and 4 on 23 Apr, but the nest failed due to cold weather on 25 Apr (WC); at Melville, nests with 3 lark eggs and 2 cowbird eggs on 4 Jun 1960 and 4 lark eggs on 23 May 1960; at Saltcoats, four young on 24 May 1958 (BH); at Stornoway, nest with three young and 2 eggs on 2 Jul had five young on 7 Jul 1962 (RD).

BBS: overall mean, 53.2. Theodore, mean 55.3; decreasing drastically (1970s, 95.3; 1980s, 49.8; 1990s, 14.2; n=33). BBS Horned Lark numbers peaked at 162 in 1971; they dropped consistently until there were only 5, 10, 5, 4, 4, and 3 in the final six years of the Theodore BBS, 1996-2001. The drop in numbers was coincident with the decrease of "black" summerfallow, the usual farming practice in the 1970s, now replaced by second crops and by chemical summerfallow (Anaka and Switzer 1999). Intuitively, one would have expected that the increased cover of chemical summerfallow versus the open exposure of black tilled summerfallow should have improved nesting success, not the reverse.

CBC: Rank #46. 3 counts, 11 birds. Highest: 6, Melville 1961.

Winter: Horned Larks tend to stay farther north in years with a general paucity of snow cover, including the winter of 1980-81, when they were "quite common" all winter north as far as Yorkton (Harris 1981b). Horned Larks were seen off and on through Dec 1962 and Jan 1963, a winter with little snow (WN). There are a few other December and January records, presented in sequence of date, not year: 6 Dec 1982 near Ebenezer (PPa); two in Niven's farmyard, 31 Dec 1945 (WN); one on road north of Sheho on 7 Jan 1975 (WN); 18 miles south of Yorkton, 15-25 Jan 1945 (C. Stuart Francis); one at Sheho on 24 Jan 1942. Four winter dates at GSL: 23 Nov 1962, 27 Dec 1976, 28 Dec 1978, 29 Jan 1975 (WA).

Banded 1 (WA, an adult male) + pre-1955, 24 (Blakeley 1, Wickencamp 23).

Langenburg: Very common over fields, 1990 (TH,BL).

DUCK MOUNTAIN: Fairly common in open areas immediately outside the park (Hatch and Weidl 1981).

PURPLE MARTIN *Progne subis*

Common summer resident where birdhouses are provided, or old flicker holes in aspen bluffs are available. Breeds.

Martins using flicker holes behind the Niven farmyard at Sheho "began to decrease about 1942 as starlings arrived" (WN, Priestly 1943a; Niven 1973). Later they occupied martin houses erected in both Sheho and Foam Lake (WN). Martins are susceptible to severe spells of cold weather and the consequent lack of insects; several Yorktonites with martin houses had dead adult martins in the houses in 1974 (PPa).

ESA(26): 15 Apr 1933 (Dr D.J. Brass, Lawrence 1933b)[an error?-csh], 22 Apr 1949 (CCS) and 1999 (WA), Av 7 May. **LFD**(28): 17 Sep 1954, Av 25 Aug.

Breeding: Nested in natural cavities in dead aspen on Frank Baines' farm at Crescent Lake from 1938 to 1943. Pairs nested in woodpecker holes in hollow trees in a gully near Melville in July 1941 (LTMcK). Several nests in hollow trees were readily found at Leech Lake; one nest in a dead stub 18 feet high, had five young and 1 egg, a foot below the

> On 27 Apr 1999, a male martin was carried off by a Sharp-shinned Hawk. Responding to the martin's cries, Joyce and I rushed at the hawk, which flew with the martin still in its grasp, only to land a few yards farther on. Was the martin too heavy to carry far? The martin, when released by the hawk, appeared both alert and unharmed; it flew strongly in a beeline back to its nest box (WA).

top opening, on 15 Jul 1946. A few pairs nested in old flicker holes in aspen "bluffs" behind William Niven's farm at Sheho in the 1930s and 1940s—and again in 1972 (Houston 1972a), but most of the few hollow trees remaining had been appropriated by starlings (Niven 1973). Eight to 10 pairs of Purple Martins were feeding young in old tree stumps only 50 yards from an active Cooper's Hawk nest, northwest of Sheho, on 26 Jul 1948 (Street 1949). Pair at nest hole at top of dead aspen in Horseshoe Lake heron colony, 1962, 1963 (23 Jun), 1964, and 1967 (WA).

One pair successfully raised two young in a box 5 feet above the concrete at a gasoline service station at the corner of Fourth Avenue and Broadway in Yorkton for two summers in the 1940s; cars passed within 3 feet of the box, which opened towards the cars. Adult caught on nest, 102 Logan Crescent East, Yorkton, 23 Jun 1957; three of her four young were banded just before they fledged on 30 Jul 1957 (CSH). Now seen more often in cities and towns, and at lakeshore cabins where artificial, multiple-compartment martin houses are erected. Chuck Deschamps estimates a dozen active Purple Martin houses in the town of Wadena, May 2003.

PNRS: two nests at Saltcoats, 1959, 4 eggs and four young (BH).

BBS: overall mean, 1.8. Theodore, mean 2.5; increasing (1970s, 0.2; 1980s, 2.4; 1990s, 5.4; n=20).

Banded 8 (CSH) + pre-1955, 5 (Briggs 1, CSH 1, Wotherspoon 3).

Langenburg: seen on 10 May 1976 (DH).

DUCK MOUNTAIN: SMNH 1926: nest on 21 Jun, #1994. SMNH 1951. Uncommon, nest in a woodpecker hole along west boundary on 3 Jun, and in a tree cavity at Little Boggy Creek on 16 Jun. A pair seen near Rundle Lake, 22 Jul 1981 (Hatch and Weidl). Occupying a martin house, 14 Jun 1990 (MB).

TREE SWALLOW *Tachycineta bicolor*

Very common summer resident. Breeds.

Two in one of two mature aspen study plots at GSL, 1969; the available woodpecker holes were not being used by Tree Swallows or Purple Martins (Flack 1976).

ESA(48): 11 Apr 1977, Av 26 Apr. **LFD**(26): 7 Sep 1942 (CSH) and 1 Sep 1944 (CSH) and 1956 (WA), Av 19 Aug.

Breeding: Swallows used a cavity excavated by another bird, 15 feet above ground in an aspen, GSL, 22 Jun 1969 (JADF). Numerous nest boxes have been placed by the Yorkton Natural History Society, some

During a cold snap on 25 May 1989, more than 150 Tree Swallows, 50 Barn Swallows and 10 Bank Swallows flew over and then landed on the Anaka lawn at GSL (JA, WA, Gollop 1989a). The swallows clumsily hopped around the lawn (Swallows were not designed for walking!), to pick up the almost immobile bugs that were forced down to the ground by the cold and the wind.

along the Isabel Priestly Nature Trail at York Lake.

In the up to 108 houses on Bill Anaka's nest box trail near GSL, 1984-1995, 30 nests failed during cold, wet weather in early July 1993. Twenty swallow nests and 30 bluebird nests were predated by raccoons in 1991, until the houses were replaced by a raccoon-proof design (Anaka 1997).

At Grayson, at the extreme south boundary of our area, John M. Hicke had 312 birdhouses on his trail, 193 occupied by Tree Swallows, 49 by House Wrens and 16 by Mountain Bluebirds (Hicke 1991).

PNRS: at Duff, 5 eggs, 14 Jun 1969, but disappeared by 20 Jun (MM); at Saltcoats, 3, 4, 5, 5 eggs (BH); at Stornoway, 21 nests (SZ,RD).

Also breeds at Bredenbury (ARS, *Atlas*).

BBS: overall mean, 13.1. Theodore, mean 16.0 (1970s, 16.9; 1980s, 14.8; 1990s, 18.8; n=33).

Banded 791 (RLS 756, near Leross, Birmingham, and Duff; WA 35) + pre-1955, 13 (Briggs 7, CSH 2, Wickencamp 4).

Recoveries, banded in area, 2 [0.3%] (SK, two Scott birds died in or near the nest box).

Langenburg: Several pairs at nest boxes, 1990 (TH,BL).

DUCK MOUNTAIN: SMNH 1926, nest 20 Jun, #1993. SMNH 1951. SNHS 1956. Seen in 1975, 1976 (JAS). Common, nests in woodpecker holes at southeast corner of park on 17 Jun and at Little Boggy Creek on 18 Jun 1981 (Hatch and Weidl). SNHS 1989.

Point counts: Rank #48, 2 individuals at 2 stops (ARS).

NORTHERN ROUGH-WINGED SWALLOW *Stelgidopteryx serripennis*

Uncommon summer resident. No breeding records.

Sighted during SNHS summer meet, GSL, 11-12 Jun 1977 (Moray Lewis).

BBS: Ituna, 1 (by Wayne R. Gemmell in 1973).

BANK SWALLOW *Riparia riparia*

Common summer resident. Breeds.

One egg collected north of Saltcoats (NJ,PMA).

In 1973, flocks of over 700 were perched on wires south of Crescent marsh on 2 Aug and again in bulrushes in early morning at Lower Rousay Lake on 10 Aug (AS).

ESA(29): 5 May 1945 (CSH) and 11 May 1963 and 1994, Av 21 May. **LFD**(26): 14 Sep 1968, Av 26 Aug.

Breeding: Bank Swallows make their holes in vertical sand faces, chiefly in gravel pits, and less commonly in temporary gravel piles. One large and stable gravel stockpile, used for two consecutive years, was photographed 11 miles north of Canora (Morgotch 1961a). In 1959, W.S. Richards observed a colony with multiple holes in a gravel stockpile close to the railroad east of Lestock; a few days later, part of the pile collapsed, taking the nests with it (Richards 1961). Another unusual site of two successful nests was in the clay walls of the excavation of Jeff and Ruth Smith's new home on Logan Crescent in Yorkton. House construction was suspended until the young were raised, with two and four banded on 18 Aug 1954, two days before they fledged (CSH). Five nests in bank of Little Whitesand Creek, just east of Yorkton, 17 May 1980 (DH). Fifty active nests in a gravel pit 2 miles east of Preeceville on 19 Jul 1988 (D.F. Hooper, L. Baker). Four nests in bank, 3 miles north and 1 mile west of Pelly, 24 Jun 1991 (MIC,MIH,CSH).

On 5 Jul 1987 there were 125 active nest holes along 0.6 miles of southeastern GSL shoreline, in freshly eroded sand faces 9 to 12 feet high; wave action and high water had caused the sand to slump, leaving a nearly vertical face. "As most faces have slumped in subsequent years, only a few nest holes persist, only 2 feet above ground level and potentially accessible by dogs and natural predators" (WA).

PNRS: at Melville, seven nests with 3, 3, 4, 4, 4, 5, and 5 eggs on 26 Jun 1959 (GAA); at Stornoway, a nest had 1 egg on 12 Jun and 2 eggs on 14 Jun 1961 (RD).

Also breeds at Preeceville, Springside (ARS, *Atlas*).

BBS: overall mean, 5.8. Calder, 4. Theodore, 8 years; high counts in first five years of 22, 50, 36, 51, 25, then only one in 25 consecutive years, and 16 in 1999, mean 6.1.

Banded pre-1955, 21 (CSH 12, Wotherspoon 9).

Langenburg: Colony of 85 birds in dirt-face bank in Cutarm Creek valley, 11 and 21 Jun 1989 (CSH,MIH). Present in same valley in 1990 (TH,BL).

DUCK MOUNTAIN: SMNH 1926. SMNH 1951. Five nests in 1976 (JAS). Fairly common, 44 nest holes in a gravel stockpile at Little Boggy Creek, 17 Jul 1981 (Hatch and Weidl).

Point counts: Rank #24, 7 individuals at 4 stops (ARS).

CLIFF SWALLOW *Petrochelidon pyrrhonota*

Abundant summer resident where bridges cross flowing streams. Breeds.

Very abundant, over 300 pairs at Assessippi, Manitoba, 6 June 1884 (Seton 1891). In the 1890s, nearly every barn west of Saltcoats, including that of Frank Baines, had nesting Cliff Swallows, with between 25 and 40 mud gourd nests on the side of one log barn.

Rare before Caucasian settlement, nesting only in colonies on vertical cliff faces (extremely rare or absent in the Yorkton area), this was the first swallow to adapt to use of newly constructed buildings with eaves. It almost immediately increased to become fairly common, as noted across the Manitoba boundary at Assessippi (Seton 1891). Log barns and new, unpainted barns were disproportionately used; when a new barn received paint, often a year or two after construction, or a fresh painting after the first paint faded, the painters often destroyed the swallow nests. Numbers nesting on barns and sheds gradually decreased until they disappeared near Saltcoats in the late 1930s (FB).

ESA(33): 6 May 1999, Av 21 May. **LFD**(25): 12 Sep 1984, Av 21 Aug.

Breeding: There was a colony on Bill Bennett's farm near Caldervale School in 1919 (Brother S. Matthias). In 1942, there was a colony with 200 adults on the side of a barn, 3 miles west and 3 miles north of Willowbrook (J.R. Foreman). Also in 1942, Foreman counted 600 swallows milling about their nests on the side of the Fraser farm 2 miles north of Sheho; he saw 2,000 adults there in 1946 (though Niven estimated that only a fraction of the initial 2,000 actually nested), but their numbers dropped to 200 in 1947. In spite of the colony only 2 miles

A previous Barn Swallow nest, built in 1955 and used for two summers at GSL, was built up further into the shape of a gourd by Cliff Swallows in 1957; the young had not fledged at the time of the report on 10 Jul (Gunn 1957b).

distant, Niven's first sighting at his farm was not until 29 May 1943.

In 1943, Cliff Swallows daubed mud in 15 or more places on a new stucco home in Melville, partly finished three nests, and then suddenly departed (LTMcK). Twenty nests were built under the eaves of a new, unpainted barn, 10 miles north of Melville in 1945 (Priestly 1945c). Six nests were built on E.S. Woods' barn beside Lower Rousay Lake in 1945, but three pairs were evicted by House Sparrows; the single nest attached to the farmhouse succeeded and fledged young soon after 20 July (CSH).

Now Cliff Swallows are most often found nesting below bridges. From 1970 to 1975, there were 20 nests on a DU structure across Milligan Creek, which flows into Little Quill Lake. In 1989, 20 nests near Sheho; 10 nests near Runnymede on 9 Jun; 10 nests at Springside four days later; a group of 10 nests where Highway 10 crosses the marshy watercourse near Otthon (CSH,MIH). In 1991, 40 pairs nesting under the Swan River bridge, 3 miles north of Pelly, 24 Jun (MIC,MIH,CSH); 50 pairs at bridge across Lake of the Prairies on Highway 10, 25 Jun (MIC,MIH,CSH).

At the two bridges on the Theodore BBS, Bill Anaka and Frank Switzer sent Ian Switzer to jump on each bridge and scare the swallows into flight, resulting in counts, for all 50 stops combined, of 1,031 Cliff Swallows in 1993 and 1,029 in 1994.

PNRS: west of Canora, one young on 2 Jun and six on 7 Jun; two had hatched by 24 Jun 1949 (WC); at Stornoway, three nests in 1959, one of which had 1 egg on 2 Jun, 6 eggs on 7 Jun, first two young hatched on 24 Jun, four young fledged (SZ,RD); four nests on 6 Aug 1961, with 1, 3, 3, 4 eggs; on 6 Jul 1962, early hatch of four young; other nests had 3, 4, and 4 eggs; on 28 Jun 1963, seven nests with 3, 4, 4, 4, 4, 5, and 5 eggs (RD).

Also breeds at Preeceville, Kamsack, Springside (ARS, *Atlas*).

BBS: overall mean, 114.0. Calder, 10. Ituna, 1. Willowbrook, 329. Theodore, mean 172.8; increasing at three bridge nest sites, resulting in the highest total numbers of any species on the Theodore BBS! (1970s, 86.2; 1980s, 103.6; 1990s, 372.5; n=32).

Banded 1 (CSH, a young bird from nest on barn of Fred Harris, 3 miles south of Yorkton, 27 Jul 1958).

Langenburg: About 50 nests in Cutarm Creek valley, 21 Jun 1989 (CSH,MIH), where they were also present in 1990 (TH,BL).

DUCK MOUNTAIN: SMNH 1951.

PNRS: colony at Madge Lake, 18 Jul 1973 (RK).

BARN SWALLOW *Hirundo rustica*

Abundant summer resident. Breeds.

Rare prior to settlement. None present when Frank Baines arrived at Crescent Lake in 1883. It adapted to use of buildings more slowly than the Cliff Swallow, with some years lag between construction of buildings and their use by this species. Barn Swallows first appeared in the late 1890s and increased at a moderate rate over a decade. At Sheho, none was present in 1910; they first appeared in 1916. One egg collected north of Saltcoats (NJ,PMA).

Niven's first observations were of a nesting pair at a deserted house about 1917, and his first recorded arrival date was 23 May 1920.

ESA(46): 23 Apr 1969 at Sheho (WN), 25 Apr 2000 at Ketchen (DF), and 26 Apr 1981 at GSL (WA), Av 7 May. **LFD**(43): 4 Oct 1966, Av 20 Sep.

Breeding: Nests are common, with one to four pairs in almost every barn or shed with an open window. Early nest, partially completed, no eggs, GSL, 24 May 1958 (WA), and York Lake, 25 May 1973 (AS). At York Lake cottage, nests with one and five young on 5 Aug 1977, four young on 17 Aug 1980, and five young on 13 Jul 1989. Nest, 5 eggs, York Lake, 1 Jun 1986 (DH). Two broods from the same nest in Shaws' garage, at 155 Fifth Avenue in Yorkton, had fledging dates of 17 Jul and 7 Sep 1953 (CCS).

Nest, 2 eggs on 29 May 1958, increased to 5 eggs on 2 Jun; first young on 16 Jun, fledging on 6 Jul; total of 34 nests, 1958-61 (JG,WA). Latest nest at GSL, also a second nest, had five young on 4 Sep 1958 (WA). One young at Marchwell on 31 Aug 1980 (MIH).

Barn Swallows began building a nest after a light shower, then waited a month until another light shower produced enough mud to finish the nest; as a result the second brood was still in the nest on 15 Sep (Hruska 1958). Similarly at Sheho in 1980, many nests were not completed due to a lack of sufficient moisture to make mud (WN).

PNRS: at Saltcoats, three nests with 4 eggs and one with five young (BH); at Stornoway, 42 nests (SZ,RD); at Sturgis, 1 egg on 18 Jun 1959, 4 on 21 Jun, and four young on 7 July (Don Buckle).

Also breeds at Togo, Duff, Langenburg (ARS, *Atlas*).

BBS: overall mean, 21.5. Theodore, mean 23.2; possibly decreasing (1970s, 31.1; 1980s, 23.4; 1990s, 17.0; n=33).

Banded 37 (CSH 31, WA 6) + pre-1955, 31 (Briggs 1, CSH 10, Blakeley 1, Wickencamp 19).

Recoveries, banded in area, 0. [Fred G. Bard banded a single flightless Barn Swallow at Davidson (51°10', 105°50') on 6 Jul 1936, but when latitude and longitude were assigned retroactively to each banding site for all banding recoveries, years later, a clerk entered one digit incorrectly: 51° 10' N, 103°50', which placed the bird on the computer list for the Yorkton-Duck Mountain area. The swallow was found dead in a building at Wynyard, 51°40' North and 104°10' West the following mid-July].

Langenburg: Very common, 1990 (TH,BL).

DUCK MOUNTAIN: SMNH 1926. SNHS 1956. Seen in 1974; three nests with young, 1975; nest with eggs, 1976; two nests with young, 1977 (JAS). Common, nests attached to Ministik Beach change house and Ministik Inn (Hatch and Weidl 1981). SNHS 1989.

PNRS: Nests with three young on 24 Jun 1986, three and four young on 4 Jul 1987, and three on 24 Jun 1988 (HWRC).

Point counts: Rank #29, 6 individuals at 4 stops (ARS).

BLACK-CAPPED CHICKADEE *Poecile atricapillus*

Common year-round resident but more evident in winter. Breeds.

Very common along our 1881 route (Macoun in Seton 1891). Frank Baines found two nests in hollow trees on his farm in the 1930s.

At Sheho, few dead trees suitable for nest sites remained as more aspen bush was cleared. Four in one of two mature aspen study plots at GSL, 1969 (Flack 1976).

Breeding: Adult banded on nest with eggs in hollow stump, 3 feet above the ground, 3 miles north of Yorkton, 23 May 1948 (CSH).

At GSL, nest, five young, 25 May 1961 (JG). Fledglings being fed in nest, 22 Jun 1969 (JADF). Pair working on nest hole, 6 feet above ground in a dead aspen, 8 Apr 1984. Young, freshly out of nest, following an adult and begging, 23 Jun 1991. Nest in short rotted stump 2 feet above ground, located 26 May 2002, was intact at 6 a.m. on 17 June; half an hour later the nest had been predated (WA).

Nina Cross reported a chickadee repeatedly entering a wasp nest in February 1993 at Saltcoats until it had eaten all the wasps (Cross 1995).

BBS: overall mean, 2.1. Theodore, mean 3.2; possibly increasing (1970s, 2.9; 1980s, 3.2; 1990s, 4.3; n=27).

CBC: Rank #6. 227 counts, 5,331 birds. Highest: 140, Kamsack 1992.

Langenburg: seen on 27 Apr 1976 near Langenburg and in the

Assiniboine valley north of Marchwell on 31 May 1976 (DH).

Banded 260 (WA 234, CSH 26) + pre-1955, 37 (CSH 18, Wickencamp 5, Wotherspoon 14).

Recoveries, banded in area, 4 [1.3%] (SK 4, Wotherspoon's drowned on the early date of 3 Mar; Wickencamp's at Stenen, caught by a cat at 20 months; Anaka's found dead at two and three years).

Return, 89. The Wotherspoon brothers retrapped birds at one and five years. At GSL, 87 chickadees were recaptured on 189 occasions a year or more after they were banded: 33 after one year, 24 at two, 10 at three, nine at four, three each at five, six, and seven years and one each at eight and nine years. These figures are skewed by the fact that initial bandings will have included older birds, and by the high number in the retraps of one-year-old (13) and two-year-old birds (8) that had been banded in the last three years of banding, 1965-67. However, of 24 banded in the first year, 1956, not a single individual was retrapped (WA). The nine-year-old bird equalled the greatest longevity encountered at Columbia, MO, by Elder and Zimmerman (1983), but is well short of the current North American record, held by a chickadee banded by the late Dr John H. Kennard at Bedford, NH, of 12 years, 5 months (Clapp et al. 1983).

Langenburg: One heard calling in Cutarm valley, 1990 (TH,BL).

DUCK MOUNTAIN: SMNH 1926. SMNH 1951. SNHS 1956. Twenty in 1959 CBC. Seen in 1974, 1975, 1976, 1977 (JAS). Fairly common throughout park; a family group along Little Boggy Creek on 16 Jun 1981 (Hatch and Weidl).

Point counts: Rank #29, 6 individuals at 5 stops (ARS).

Aspen count totals: OA 1 (TH).

BOREAL CHICKADEE *Poecile hudsonica*

Rare in winter in parkland. Year-round resident at Duck Mountain. No breeding records.

In flocks around the Porcupine Mountains in 1881 (Macoun in Seton 1891).

Occasional in winter in southern part of YDM area. One at Ituna CBC, 1 to 3 Jan 1947 (Mrs Isobel Arndt). Seen during seven years at GSL: intermittently from 4 Apr 1973 through 20 Feb 1974 (BA,JA, Houston 1974); 21 days in 1984 in April and between July and October; 13 days in 1985, in Jan, Jun, Aug, Sep, Nov, and Dec; nine days in 1986, in Jan, Sep, Oct, and Dec; 10 days in 1987 in Jan, Mar, Apr, and May;

three sightings between 8 and 24 Oct. At Crystal Lake, Peter and Fran Buryk have this species for much of the year.

CBC: Rank #24. 29 counts, 204 birds. Highest: 25, Kamsack 1969 and Melville 1997.

Banded pre-1955, 1 (One at feeding tray, 102 Second Avenue North, Yorkton, 4 Feb 1953, Houston 1953a).

DUCK MOUNTAIN: Breeding (SMNH 1926). SMNH 1951. SNHS 1956. Fairly common in both White and Black spruce on Spruce Island, Little Boggy Creek, Pickerel Point, Batka Lake, and Red Squirrel Nature Trail (Hatch and Weidl 1981). SNHS 1989.

Point counts: Rank #24, 7 individuals at 6 stops (ARS).

RED-BREASTED NUTHATCH *Sitta canadensis*

Uncommon but increasing year-round resident, seen most frequently during migration and at feeding stations in winter. No breeding records.

Numbers have increased greatly since the 1940s and especially since about 1970. It was seen only rarely during fall migration: The first five records were between 19 Sep 1937 and 28 Oct 1942, then it was present from 4 Sep through 30 Oct 1951 (Houston 1951). Since 1988, there have been 12 summer dates at GSL.

ESA(16): 20 Apr 1994, Av 7 May. **LFD**(33): 4 Nov 1995, Av 6 Oct.

BBS: Theodore, 1 in 1993 and 1995, mean 0.06.

CBC: Rank #23. 41 counts, 224 birds. Highest: 25, Kamsack 1995. Not recorded in any CBC at Yorkton until 1987.

Banded 1 (CSH) + pre-1955, 1 (Wickencamp).

DUCK MOUNTAIN: SMNH 1926. SMNH 1951. Seen in 1974, 1975, 1976 (JAS). Uncommon, seen and heard occasionally along the Red Squirrel and Pelly Point nature trails, at Selwood Campground, and near Batka Lake (Hatch and Weidl 1981).

Point counts: Rank #48, 2 individuals at 2 stops (ARS).

WHITE-BREASTED NUTHATCH *Sitta carolinensis*

Rare year-round resident. Breeds.

Niven's only Sheho record was for 22 May 1920 (WN).

Two at J.R. Foreman's feeding station on Wallace Avenue in Yorkton during CBC, 26 Dec 1945. Seen during 40 of 48 years at GSL; only in extreme eastern Saskatchewan is this species more common than the Red-breasted. At Ketchen, Dallas Fairburn has had a pair resident in his yard since 2000; they bring barely flying young to his feeder.

Breeding: Seven records at GSL: fledged young, 22 Jun 1969 (JADF); a female with three young in early July 1971 (FAS); an adult female feeding young in late July 1971 (WA, Houston 1971c); adults with two young, 29 Jun 1980 (WA, Gollop 1980b); with three young, 23 Jun 1981; with three young, 26 Jun 1982; with two young, 16 Jul 1988 (WA).

Also breeds at Yorkton (ARS, *Atlas*).

CBC: Rank #21. 93 counts, 301 birds. Highest: 30, Kamsack 2001.

Banded 2 (WA).

DUCK MOUNTAIN: SMNH 1951. Seen in 1974, 1975, 1976, 1977 (JAS). In 1981, fairly common, nest in aspen along Whitethroat Nature Trail, 15 Jun, and two adults with one immature near Kipp Lake, 12 Aug (Hatch and Weidl, Weidl 1983). Also seen at Batka Lake and near the south boundary road.

Point counts: 1 individual at 1 stop (ARS).

Aspen count total: OA 1 (TH).

BROWN CREEPER *Certhia americana*

Uncommon transient. Breeds at Duck Mountain.

A well marked migration occurred in Yorkton on 14 May 1942. One summer record: 9 Jul 1943 (IMP). One died when tangled in netting of climbing vines 3 miles southeast of Duff in late fall 2002 (AS).

ESA(8): 3 Apr 1986, Av 17 Apr. **EFA**: 8 Sep 1955. **LFD**: 13 Nov 1985.

Winter: surprisingly, no winter record yet at GSL.

Banded 3, + pre-1955, 1, all CSH, suggesting a fairly consistent timing of fall migration: on 24 Sep 1953; 25 Sep 1956; 22 Sep 1957; 16 Oct 1959 at Melville (CSH).

DUCK MOUNTAIN: Seen in 1976 (JAS). Uncommon; nest found in dead Balsam Fir along Pelly Point Trail, 15 Jul, with other sightings on 4 Jun, 15 Jul, 14 Aug, 9 Sep, and 1 Oct 1981 (Hatch and Weidl 1981, Weidl 1983; Koes and Taylor 1993d). In 1989, sighted 5-9 Jun, 1 Jul, 19-24 Aug (TH).

HOUSE WREN *Troglodytes aedon*

Abundant summer resident. Breeds.

A wren nest with 6 eggs in an old coat sleeve was at Melville on 10 Jun 1920 (H.C. Grose, SMNH). Three eggs collected north of Saltcoats (NJ,PMA).

> Nests in many nooks and crannies in or within 30 yards of a wooded area. The original busybody–every species' business is its business, and every empty or recently fledged nest box must be filled with twigs (WA).

ESA(48): 2 May 2000, Av 15 May. **LFD**(37): 13 Oct 1990, Av 18 Sep. Two and 10 in the two mature aspen study plots at GSL, 1969 (Flack 1976).

Breeding: Early nest with young, at Pearce's farm near Tonkin, 8 Jun 1952. Nested in the basement studs of our new home being constructed on Logan Crescent, Yorkton, July 1954 (CSH). Late nest, three young fledged from Purple Martin house in Anakas' yard at GSL, 20 Aug 1985 (WA).

Unusual nest sites include a previous year's wasp nest under eaves of cabin in the Anaka's yard, June 1967 (WA), and a tire hanging on a fence post on 18 Jul 1973 (AS). A few boxes on the birdhouse trail between Yorkton and Melville have been occupied by wrens.

PRNS: at Saltcoats, 3, 3, 3, 5, 6, 6, 7, and 7 eggs, and six and eight young (BH,JRJ); at Stornoway, 33 nests, 1959-63, including four with 7 eggs, and nests with six and seven young (SZ,RD); near Kamsack in 1958, one nest had 3 eggs on 10 Jun and six young on 29 Jun, while another had 8 eggs on 29 Jun (JHJ).

Also breeds at Canora and Togo (ARS, *Atlas*).

BBS: overall mean, 33.3. Theodore, mean 39.2 (1970s, 36.5; 1980s, 40.5; 1990s, 44.9; n=33).

Banded 4 (CSH) + pre-1955, 127 (CSH 5, Blakeley 19, Wickencamp 36, Wotherspoon 67).

Langenburg: An unusual nest with eggs in a hole in a stump exactly at ground level, Assiniboine River valley north of Marchwell, 1 Jun 1976 (DH). Very common, 1990 (TH,BL).

DUCK MOUNTAIN: SMNH 1951. SNHS 1956. Seen in 1974,

1975; nest with young, 1976, 1977 (JAS). Fairly common, several family groups and fledged young seen (Hatch and Weidl 1981). SNHS 1989.

PNRS: Nest, 6 eggs, 20 Jun 1980 (HWRC).

Point counts: Rank #24, 7 individuals at 7 stops (ARS).

Aspen count total: YA 1 (TH).

WINTER WREN *Troglodytes troglodytes*

Rare summer resident at Duck Mountain. No breeding records.

DUCK MOUNTAIN: Rare; a family group of four was seen and heard along the road to Bear Head Lake, 25 Jun 1981 (Hatch and Weidl, Weidl 1983). One heard 22 May 1991 (ARS). One sporadically vocal at Benito subdivision, 18 Jun 2000 (PSa,DV,BL). A record six heard singing, 2-3 Jun 2001 (RK, Bain 2001).

SEDGE WREN *Cistothorus platensis*

Fairly common summer resident in moist meadows. Breeds.

Abundant resident of grassy, not rushy, sloughs (every little sedgy pool and slough) north to near Cote's Reserve, 1884 (Seton 1891).

Fifty counted at Little Quill Lake on 10 Jul 1976 (DH). Disappeared by the dry years of 1988 and 1989 at Salt Lake, north of Sheho (WN).

ESA(12): 11 May 1990, Av 24 May. **LFD**(9): 29 Sep 1996, Av 10 Sep.

Breeding: A small nesting colony in damp grass bordering Rokeby marsh, 19 Jul 1946 (Farley Mowat,CSH). Dummy nest near GSL, 30 Jul 1967 (WA). Singing male at Horseshoe Lake, 21 May and 25 Jun 1988 (TH). At Upper Rousay Lake, on 13 Jun 1989, one was singing in bulrush on dry former lake bottom, exactly where Marsh Wrens had nested when the lake contained 3 feet of water and live bulrushes in the 1940s. Three seen at DU's Foam Lake marsh complex, 14 Jun 1989 (CSH,MIH). Colony at Rokeby marsh, 22 Jul 1994 (DH).

BBS: overall mean, 0.8. Churchbridge, 1. Theodore, 17 years; highest, 9, 1988, mean 1.3.

DUCK MOUNTAIN: Photographed by F.W. Lahrman at Madge Lake (in Weber 1974). Dummy nest found in 1976 (JAS). Uncommon, regular only at Bear Head Lake (Hatch and Weidl 1981). One heard along Batka Lake road on 20 Jun 1988 (TH). Present in north fringe of park, 24 Jun 1991 (MIC,CSH,MIH).

MARSH WREN *Cistothorus palustris*

Fairly common summer resident in bulrushes. Breeds.

Nine singing males in bulrushes at Horseshoe Lake, 1975 (DH). One sighted in DU's Foam Lake marsh complex, 14 Jun 1989 (CSH,MIH). Heard near Preeceville and Pelly, 24 Jun 1991 (MIC,MIH,CSH).

ESA(12): 11 May 1941 (CSH) and 19 May 1991 (WA), Av 30 May. **LFD**: last week of September 1945 (CSH).

Breeding: Nests, including many more "dummy" nests than occupied nests, have been found at Upper and Lower Rousay lakes, Leech Lake and Rokeby marsh. Early nest, 7 eggs, 11 Jun 1946, Lower Rousay Lake; and 6 eggs in active nest and two empty dummy nests, 26 Jun 1955 (CSH). Nest with 3 eggs in Foam lake marsh on 19 Jul 1974 (AS).

On 6 and 14 Jun 1987, located five colonies for Don Kroodsma's research: near Rhein and Hamton and three near GSL (WA).

PNRS: at Stornoway in 1961, nest, 6 eggs on 13 Jun had six young on 28 Jun (RD).

BBS: overall mean, 0.7. Willowbrook, 8. Theodore, 12 years; highest, 4, 1990, mean 0.8.

Banded 1 (Smith at Little Quill Lake).

Langenburg: Two or three heard singing at a few larger sloughs, 1990 (TH,BL).

DUCK MOUNTAIN: Seen 1970 (RJL). Heard 10 Jul 1990 (TH). Present in north fringe of park, 24 Jun 1991 (MIC,CSH,MIH).

GOLDEN-CROWNED KINGLET *Regulus satrapa*

Uncommon transient, seen chiefly during fall migration.

One found dead on Tupper Avenue, Yorkton, 9 Oct 1951, by Andrew Burant (Houston 1957a; Gammell and Gammell 1952a). At GSL, 31 Oct 1956 (Anaka 1956b). Three at Melville, 10 Nov 1956 (GAA). Small numbers were scattered everywhere through the woods at GSL from 7 through 14 Oct 1961 (Anaka 1962). One near Melville on 21 Sep 1975 (DH). They rarely remain through the winter, with two in Yorkton on 23 Dec 1956, and a late spring departure on 24 Apr 1955 (Houston 1957a).

ESA(4): 29 Mar 1986. **EFA**(27): 22 Sep 1991, Av 8 Oct. **LFD**(27): 19 Nov 1983, Av 30 Oct.

Winter: 24 Dec 1974 through 12 Jan 1975; 22 Dec 1985 (WA).

CBC: Rank #51. 3 counts, 5 birds. Highest: 2, GSL 1974, 1985.

DUCK MOUNTAIN: Seen in 1976 (JAS). Uncommon transient,

one sighting along Red Squirrel Nature Trail on 9 Sep 1981 (Hatch and Weidl).

RUBY-CROWNED KINGLET *Regulus calendula*

Uncommon summer resident. Breeds.

Noted only in spring migration at Yorkton. Summer resident at GSL. Male singing on territory at picnic site at GSL on 12 Jul 1975 (DH). Heard near Preeceville, 24 Jun 1991 (MIC,MIH,CSH).

ESA(39): 7 Apr 1993, Av 24 Apr. **LFD**(22): 14 Oct 1974, Av 2 Oct.

Breeding: On 6 Jun 1998 a female gathered nest material in the Anaka yard, and later four small eggshells were found beneath a large spruce; the male sang until 18 Jun but no young were seen (WA).

BBS: overall mean, 0.1. Churchbridge, 1 (1991, CSH). Ituna, 3 (1973, WRG). Theodore, 2 (1 in 1994 and 1 in 1998, WA,FAS);

Banded 1 (CSH, 5 Oct 1956).

DUCK MOUNTAIN: SMNH 1926. SMNH 1951. SNHS 1956. Seen in 1974, 1975, 1976 (JAS). Common in coniferous woodlands throughout the park (Hatch and Weidl 1981). SNHS 1989. Seven singing on territory, 2 May 1992 (TH).

Point counts: Rank #13, 18 individuals at 18 stops (ARS).

[**NORTHERN WHEATEAR** *Oenanthe oenanthe*

MANITOBA DUCK MOUNTAIN

In a gravel pit near Child's Lake in Manitoba's Duck Mountain Provincial Park, on 9 Sep 1986, Enid Cumming saw two wheatears (Cumming 1988).]

EASTERN BLUEBIRD *Sialia sialis*

Uncommon summer resident. Breeds.

The first Yorkton record was by Jim Rogerson on 18 Apr 1942; the first Sheho records were on 25 Apr 1942 and 24 Apr 1947 (WN). Appeared at the Robert Barnhart farm 5 miles east of Saltcoats on 20 May 1946 and 6 May 1947, and was believed to have nested (RB). Twelve were seen at Crescent Lake on 27 Sep 1964 by Elwin Baines.

1998 was a special "Eastern Bluebird" year in eastern Saskatchewan, including Preeceville (LP).

ESA: 25 Mar 1945 (IMP) and 31 Mar 1946 (WN). One seen at Preeceville, 14 May 1987 (CP).

Breeding: The first nest recorded was in a binder box on the Jack Esson farm, 2 miles north of Cutarm Creek in 1955 (BH). A pair nested 3 miles south of Okla, at the extreme northern margin of this area, with 5 eggs on 6 Jun 1965; the next year the adult female had 2 eggs on 10 Jun and 5 eggs on 13 Jun (Ashdown 1966), and on 20 Jun she was banded (CSH).

The first nest at GSL had four young, 3 Jun 1988; the pair renested with four young in same box, 11 Jul. Subsequently present in 1991 and 1992, then only in even-numbered years through 2000, then each year since. The first year near GSL with two nests was 1991 (WA, Koes and Taylor 1991c). In all, between 1984 and 1995, 42 young fledged from 11 attempts (including three re-nestings; Anaka 1997). On 6 Jun 1998, at a pair of boxes occupied by Tree Swallows and Mountain Bluebirds, respectively, a pair of smaller Eastern Bluebirds was harassing the male Mountain and preventing him from feeding his 10-day-old young in the box. We put up a new, empty, third box nearby. Two days later there were five dead young and no bluebirds of either species evident. On 22 June there were 6 Eastern Bluebird eggs in the new box; later all six young fledged. In 1998, a female Eastern was mated to a Mountain Bluebird male (WA).

Nest in box on Jeff and Sharon Olson's farm east of Parkerview in 2001 raised at least three young.

Also breeds at Margo, Preeceville (ARS, *Atlas*).

BBS: Theodore, 1 in 2001, mean 0.03.

Banded 1 (CSH, see above).

Langenburg: Seen carrying food to nest box within a mile of the Manitoba boundary, 1990 (TH,BL).

DUCK MOUNTAIN: PNRS: Nest, three young, 29 Jun 1989 (HWRC).

MOUNTAIN BLUEBIRD *Sialia currucoides*

Common summer resident along nest box trails. Breeds.

The Mountain Bluebird was an unbelievably rare transient through southern Manitoba and southern Saskatchewan before 1900. Frank Baines saw none, even during migration, before about 1930.

Now common in migration and utilizing nest boxes in summer,

particularly if boxes are placed on fence posts adjacent to sandy pastures. Not seen at Bangor until 1938 (Thompson 1953). Not seen at Sheho, even in migration, between 1910 and 3 Apr 1938. Since then, regular transient, increasing in numbers; large flocks first appeared in fall migration on 11 Sep 1950, with some staying for weeks (Gammell and Gammell 1951). A concentration of 40 at GSL on 10 Sep 1967 was noteworthy for that time (WA, Hatch 1968a).

ESA(47): 6 Mar 1947 at Saltcoats (RB) and 1988 at GSL (WA), Av 30 Mar. **LFD**: 22 Oct 1990.

Nest in a bird box 6 miles south of Yorkton in 1942; in a crack in the side of William Niven's barn north of Sheho in 1946; in Dave Baines' mail box at Crescent Lake in 1947. A common early nest site was in the twine box in a binder. On the Mess farm near Chatsworth School, the men required use of the binder, so moved the nest into an unattached spare binder twine box, and hung it in a nearby tree; the young bluebirds fledged successfully from the new site (Mess 1955).

Breeding: A pair raised two broods in the summer of 1942 in an unusual site: an iron pipe protecting the oblique guy wire of an electric light pole on Third Avenue South in Yorkton (Priestly 1942b). Early nest, 5 eggs, 15 May 1952 (WA). In July 1956, a pair at GSL raised five young in the attic of the henhouse in the Anaka yard, entering through a hole under the eaves. Nests in suitable woodpecker holes at Sheho and in 1976 in a hole in the barn wall (WN).

Ron Bittner began a nest box trail near Abernethy, at the extreme southern edge of the YDM area, in 1987; by 1995 he had 72 houses of 13 designs. In 119 "box-years" of availability of his own experimental box design, Mountain Bluebirds used the houses on 90 occasions, but on only 19 occasions did they use a standard-type box (Bittner 1995b).

The GSL bird box line began with 30 boxes in 1984 and grew to 300, always paired, with two boxes on adjacent fence posts, one for a bluebird and one for a swallow. In good years, 90% of the boxes were occupied. From 1984 through 1995, 3,728 young bluebirds fledged from 929 successful nest attempts (Anaka 1997); more than 4,000 more have fledged since. First nests (577) fledged 2,660 young, Av 4.6; second and late nests (352) fledged 1,068 young, Av 3.0.

The first serious problem on Anakas' nest box trail, beginning in 1990, was predation by raccoons. In 1991, raccoon tracks were present at every box along 3 miles of the nest box trail; we lost eggs or young from 20 Tree Swallow nests and 20 bluebird nests. This forced us to replace all the boxes. The first 116 replacement houses were divided into four groups of 29 each of Bittner B1, Bittner B2 (Bittner 1995b), Brian Shantz (Shantz 1986), and Peterson styles. These new houses had the entrance hole increased from 1.5 to 1.75 inches in diameter. The Shantz design proved the most successful. I added an extra layer of wood to both the exterior and interior aspects of the entrance hole of the original boxes; the extra wood thickness prevented the raccoon paw from plucking out bluebirds or swallows from the box. The larger entrance hole then produced a second problem. European Starlings began usurping the new boxes, increasing from one to five and then in 2000 to 20 boxes. The entrance holes were then reduced back to 1.5 inches in diameter and the starling problem in turn ceased (WA).

PNRS: At Stornoway in 1961, 1 egg on 14 May, 5 on 21 May and 8 on 28 May; a second nest had four young on 26 Jun (SZ); boxes near Duff and Birmingham raised four, five, six, six, and seven young (RLS).

Also breeds at Jasmin, Melville (ARS, *Atlas*).

Oology: One egg collected north of Saltcoats, probably after 1945 (NJ,PMA).

BBS: overall mean, 4.7. Churchbridge, 30. Ituna, 5. Willowbrook, 15. Theodore, mean 6.0; increasing (1970s, 4.6; 1980s, 4.6; 1990s, 9.1; n=32).

Banded 442 (RLS 425, HWRC 11, WA 6).

Recoveries, banded in area, 1 [0.2%] (SK, one banded by RLS near Duff died about 7 miles east, over a month later).

Recoveries, banded outside area, 1 (AB, banded as a nestling south of Bruce on 15 Jun 1982, found dead 18 May 1983 at Melville).

Langenburg: Six pairs at nest boxes, 1990 (TH,BL).

DUCK MOUNTAIN: Seen in 1976 (JAS). Uncommon in open areas; nest in an abandoned woodpecker hole at Little Boggy Creek on 15 Jun 1981 (Hatch and Weidl).

TOWNSEND'S SOLITAIRE *Myadestes townsendi*

Rare wanderer.

Three spring sightings at GSL: 18 Apr 1973 (WA, Renaud 1973), 16 Apr 1979 (WA,JA, Gollop 1979), and 7 Aug 2002 (WA). One fall date at Yorkton, 24 Oct 1940 (IMP), and two at GSL, 19 Oct 1998 (WA,JA, Bain and Shanahan 1998d).

VEERY *Catharus fuscescens*

Common summer resident in heavier aspen bush, especially near ravines and lakes. Breeds.

Abundant in all the willow thickets north to Cote's Reserve in 1881 (Macoun in Seton 1891).

Two in one of two mature aspen study plots at GSL, 1969 (Flack 1976). One singing in coulee leading down to Lake of the Prairies, north of Highway 10, 21 Jun 1989 (CSH,MIH).

ESA(43): 13 May 1940 at Sheho (WN) and 14 May 1990 at GSL (WA), Av 22 May. **LFD**(32): 9 Oct 1956, Av 1 Sep.

Breeding: Early nest, 1 egg, 31 May 1953, had 3 eggs and 1 cowbird egg on 7 Jun (PPa, Houston 1953b; Gammell et al. 1953). Nest, 4 eggs and 1 cowbird egg, 8 Jun 1955 (WA).

Cowbird parasitism can be severe: nest, 2 Veery eggs and 5 cowbird eggs, 5 Jul 1959; on 14 Jul two cowbirds had hatched and 1 cowbird egg remained, but one Veery young had died; on 17 Jul only the two cowbird young had survived. Other nests had 2 Veery and 2 cowbird eggs on 28 Jun 1959; 1 Veery and 2 cowbird eggs on 15 Jul 1962; 1 Veery and 1 cowbird egg, also on 15 Jul 1962 (WA).

PNRS: at Stornoway, nest with 2 Veery eggs and 2 cowbird eggs on 26 Jun 1960 (SZ).

BBS: overall mean, 4.4. Calder, 1. Churchbridge, 7. Willowbrook, 6. Theodore, mean 6.6 (1970s, 7.1; 1980s, 7.0; 1990s, 6.1; n=33).

Banded 6 (WA 4, Smith 2 [at DMPP]).

Langenburg: Five singing between Marchwell and Assiniboine River valley to the north, 31 May and 1 Jun 1978 (DH). Three males heard singing in wooded coulees and valleys, 1990 (TH,BL).

DUCK MOUNTAIN: SMNH 1951. SNHS 1956. Seen in 1974, 1975, 1976 (JAS). Common, heard regularly throughout the park; nest, 4 eggs, on top of a dead Balsam Poplar stump at Spruce Island, 18 Jun 1981 (Hatch and Weidl). One, 15 Jun 1990 (MB).

Aspen count totals: OA 16, YA 6, MA 4 (TH).

GRAY-CHEEKED THRUSH *Catharus minimus*

Fairly common transient.

ESA(12): 12 May 1964, Av 23 May. Migration peaked at GSL on 19 May 1966 (WA, Hatch 1966), and from 16 through 22 May 1974 (WA, Houston and Shadick 1974). Two fall dates at GSL: 13 Sep 1993 and 22 Oct 2002. Sometimes stops over in Yorkton in fall migration, as when nine were banded on 18 Sep 1958 (CSH, Krause 1959a).

Banded 11 (CSH).

DUCK MOUNTAIN: SNHS 1956. One, 20 May 1989 (TH).

SWAINSON'S THRUSH *Catharus ustulatus*

Fairly common to common transient. Uncommon summer resident at Duck Mountain and in heavy aspen forest in parkland. One unconfirmed breeding record.

Breeding in rose thickets a few feet from the ground and in cornel [dogwood] bushes, Swan River, 1881 (Macoun in Seton 1891).

A heavy wave went through Yorkton on 14 May 1940 and another at GSL on 19 May, when 50 were counted following a warm, rainy night (WA, Gammell 1960). Migration peaked at GSL on 19 May 1966 (WA, Hatch 1966); in 1974 they "carpeted the bush" from 16 through 22 May (WA, Houston and Shadick 1974); in 1983, there were 100 on 12 May, the third day of their appearance (WA, Harris 1983). One summer record, but no nest found, 14-19 Jul, at GSL (WA).

ESA(33): 22 Apr 1997, Av 11 May. **LSD:** 29 May 1954 and 2002. **EFA:** 16 Aug 1997. **LFD**(5): 19 Oct 1981.

Breeding: nest with 2 eggs at Stornoway, 7 Jul 1961 (RD, but details of identification not available).

BBS: Theodore, 1 in 1988, mean 0.03.

Banded 17 (CSH 16, Smith 1).

DUCK MOUNTAIN: SMNH 1926: 4 eggs on 10 Jun, #1960 or #1987a, study skin male, #1917, initially labelled as Gray-cheeked even though listed correctly in *Kamsack Times*; identification on study skin but not on egg sets, later changed to Swainson's Thrush (Glen Sutter, pers. comm.). SMNH 1951: 5 eggs, 26 Jun, #5501. SNHS 1956. Seen in 1976 (JAS). Uncommon in mixed woods along nature trails and at Green Lake (Hatch and Weidl 1981). SNHS 1989.

Point counts: Rank #12, 21 individuals at 20 stops (ARS).

Aspen count totals: OA 6, YA 1, MA 2 (TH).

HERMIT THRUSH *Catharus guttatus*

Fairly common transient. Uncommon summer resident at Duck Mountain and in heavy aspen forest in parkland. No breeding records.

William Niven believed they probably nested at Sheho before most of the aspen woods were cleared for farming. Very common in the springs of 1939 and 1940.

One singing at GSL on 4 Jul 1955 (Anaka 1955), and again in 1957, 1959, 1960, 1962, 1963, 1967, 1985, 1986, 1987, and 1989 (WA), and in early July 1971 (FAS, Houston 1971c), in deep aspen forest, but no nest found (WA). One singing in coulee leading down to Lake of the Prairies, north of Highway 10, 21 Jun 1989 (CSH,MIH). As yet there is no proof of breeding, although Smith (1996), in an uncharacteristic error, lists Buchanan as a confirmed breeding site.

ESA(31): 14 Apr 1986, Av 1 May. **LSD**: 31 May 1958. **EFA**: 16 Aug 1997. **LFD**(22): 17 Nov 1978, Av 12 Oct.

BBS: Theodore, 1 each in 1988, 1990, 1991, mean 0.09.

Banded 9 (CSH) + pre-1955, 1 (Wotherspoon).

Langenburg: One unequivocally heard singing in a coulee of the Assiniboine valley, 1990 (TH,BL).

DUCK MOUNTAIN: SMNH 1951. SNHS 1989. Much less common than Swainson's Thrush; one heard and seen, 10 Jul 1990 (TH).

WOOD THRUSH *Hylocichla mustelina*

Vagrant.

One reported in the dunes at GSL on 9 Sep 1985 (Pollock 1986c).

DUCK MOUNTAIN: An adult male seen near turnoff into Benito subdivision on 10 Aug 1990 (RK, in Smith 1996).

AMERICAN ROBIN *Turdus migratorius*

Abundant summer resident. Breeds.

Three eggs collected north of Saltcoats (NJ,PMA).

At Sheho, the robin increased with settlement and became semi-domesticated (Niven 1973). Two in one of two mature aspen study plots at GSL, 1969 (Flack 1976).

Fall migration peaked at 150 on 8 Oct 1967 (WA). Groups of 500 stopped at GSL on 19 Apr 1968 (WA, Hatch 1968c), 100 on 28 Apr 1973 (WA, Renaud 1973), and up to 200 every day from 21 through 26

> Rolled oats is a staple food in our feeders. A female robin, incubating in a spruce at the edge of the lawn, frequently followed this routine: off the nest, to the bird bath for a drink, then to the feeder for a mouthful of rolled oats, then back to the nest–all within 60 seconds. Later she fed rolled oats to her nestlings (WA).

Apr 1989 (JA,WA, Gollop 1989a).

ESA(48): 19 Mar 1938 (CSH) and 21 Mar 1961 and 1991 (WA), Av 4 Apr. **LFD**(47): 14 Nov 1977, Av 28 Oct.

Breeding: Early nests, 2 eggs on 1 May 1944 at Yorkton (CSH) and 4 eggs on 10 May 1919 at Sheho (WN). Late nest, 2 eggs on 29 Jul 1962, two young on 2 Aug, fledged 14 Aug (WA). Nest, 2 eggs, 6 May 1987 and 4 eggs on 8 May, hatched on 20 May and fledged 2 Jun. Nest, 3 eggs, on viewing tower at Little Quill Lake on 30 May 1998 (DH).

PNRS: at Stornoway, 41 nests, 1959-63, including an unusual clutch of five eggs on 14 Jun 1960 and another with five young on 21 Jun 1961, yet only nine nests had four eggs or young when visited (SZ,RD); at Duff, 4 eggs on 11 May and four young on 24 May (MM); at Saltcoats, 14 nests in 1958 and 10 in 1959, 18 of which had clutches of 4 (BH); at Mikado, Kozak had an unusual clutch of 5 eggs on 23 May 1966, but only 4 had hatched by 1 Jun, and I. Yaholnitsky had a clutch of three young on 24 Jun 1969.

Also breeds at Canora (ARS, *Atlas*).

BBS: overall mean, 22.4. Theodore, mean 28.5; increasing (1970s, 20.4; 1980s, 30.2; 1990s, 38.1; n=33).

CBC: Rank #37. 15 counts, 24 birds. Highest: 4, Kamsack 1991 and 1999.

Banded 64 (CSH 52, WA 12) + pre-1955, 221 (Briggs 15, CSH 127, Blakeley 17, Wickencamp 27, Wotherspoon 35).

Recoveries, banded in area, 10 [3.7%] (SK 8, LA 1, TX 1). Wotherspoon at Hyas had recoveries in Louisiana the second winter and Texas the first winter, and one was found injured at Canora at three years. Blakely had a robin killed in a trap at Tadmore nearly three years after banding, and Emma Wickencamp had one found dead locally after a month. Five Yorkton robins, all recovered locally, were killed by a dog the same month; three were found dead at one year; and one at two years after banding (Houston 1968d).

Recoveries, banded outside area, 1 (Rochester, MN, on 8 May 1968, found dead at Sturgis, 2 Jun 1969).

Winter: One wintered successfully at the Ravine Reserve, 1940-41, and another was seen there on 27 and 28 Dec 1944. One stayed at GSL over Christmas 1945, at least until 17 Jan 1946. One at Ituna, 1-3 Jan 1947 (Isobel Arndt). At GSL, 20 Jan 1953; during Christmas week in 1968; after 23 Dec 1970 (WA, Houston 1971b); 4 Dec 1993 (WA). At Ketchen, 3 Dec 2002 (DF).

Langenburg: Common, 1990 (TH,BL).

DUCK MOUNTAIN: SMNH 1926. SMNH 1951. SNHS 1956. Seen in 1974, 1975, 1976 (JAS). Fairly common along highway and trails (Hatch and Weidl 1981). SNHS 1989.

Point counts: Rank #7, 29 individuals at 27 stops (ARS).

Aspen count total: OA 3, YA 1, MA 1 (TH).

VARIED THRUSH *Ixoreus naevius*

Vagrant.

One lingered in Yorkton from 22 Dec 1991 until mid-January 1992 (DH,WH, many observers; Koes and Taylor 1992b). One was present at the farm of Blaise Cooper's parents, 9 miles north of Jasmin, from 23 Dec 1997 until the night of 2 Jan 1998, when it disappeared (ARS).

CBC: 1 count, 1 bird. Yorkton 1991.

GRAY CATBIRD *Dumetella carolinensis*

Common summer resident. Breeds.

Three eggs collected north of Saltcoats (NJ,PMA).

Absent for several years at Sheho but returned in 1987 (WN). Seen near Pelly, 24 Jun 1991 (MIC,MIH,CSH).

ESA(47): 3 May 1951 (RB) and 11 May 1993 (WA), Av 22 May. **LFD**(36): 28 Sep 1942 (CSH) and 27 Sep 1956 (WA), Av 8 Sep.

Breeding: Nest, four young at Sheho, 17 Jun 1919 (WN). Early nest, 1 egg, 31 May 1955, had 4 eggs on 4 Jun and four young on 22 Jun (WA). Nest with young, York Lake, 5 Jun 1957 (CSH). Nest, 2 eggs, York Lake, 7 Jun 1964 (LM).

PNRS: at Saltcoats, 4 eggs each on 15 Jun 1958 and on 4 Jul 1959 (BH); at GSL, 3, 4, 4, 4, 5, and 5 eggs, also three young plus 1 egg deserted, 1958-61 (JG); at Mikado, 4 eggs on 19 Jun 1969, of which two hatched on 24 Jun (IY); at Stornoway, 1960-63, 2, 3, 3, 4, 4, and 5 eggs (SZ,RD).

Also breeds at Jasmin (ARS, *Atlas).*

BBS: overall mean, 5.8. Theodore, mean 5.0 (1970s, 5.4; 1980s, 6.4; 1990s, 3.9; n=31).

Banded 11 (CSH 8, WA 3) + pre-1955, 17 (Briggs 2, CSH 4, Wickencamp 1, Wotherspoon 10).

Langenburg: Very common, 1990 (TH,BL).

DUCK MOUNTAIN: SMNH 1926. SMNH 1951. SNHS 1956. Rare, one seen near Green Lake on 10 Jul 1981 (Hatch and Weidl).

BROWN THRASHER *Toxostoma rufum*

Common summer resident north to Togo, Yorkton, and Sheho; rare farther north at GSL and Duck Mountain. Breeds.

Not common at Swan River, 1881 (Macoun in Seton 1891). Two eggs collected north of Saltcoats (NJ,PMA).

Six singing at York Lake, 1 Jun 1986, but none seen during Horseshoe Lake survey in 1975 (DH).

To our surprise, present in residential area, one block from business section of Yorkton, in the 1940s and 1950s. Numbers dropped appreciably at GSL between 1950 and 1989 (WA). In 1989, noticeably less common than in the 1940s (CSH,MIH).

ESA(32): 1 May 1946 (WN) and 6 May 1972 (WA), Av 18 May. **LFD**(10): 21 Sep 1958, Av 31 Aug.

Breeding: Nest at Yorkton, 4 eggs, 26 Jun 1940 (CSH).

PNRS: at Saltcoats, two nests, each with incomplete clutches of 2 eggs on 12 Jun 1958 (BH); at Stornoway, 1 egg on 4 Jul 1959, 2 thrasher and 2 cowbird eggs on 26 Jun 1960, 3 eggs on 7 Jul 1961, and four young on 26 Jun 1961 (SZ); at GSL, 3 nests each with 2 eggs, on 8 and 26 Jun and 5 Jul 1959, and with only 1 egg on 7 Jun 1963 (WA).

BBS: overall mean, 1.4. Theodore, 15 years; highest, 10, 2001.

Banded 5 (CSH).

Langenburg: One on 12 May 1976 (DH) and one, 1990 (TH,BL).

DUCK MOUNTAIN: SMNH 1951. Seen in 1970 (RJL).

EUROPEAN STARLING *Sturnus vulgaris*

An abundant summer resident. Breeds.

A new arrival since 1938, now very common in summer but only fairly common in winter, since the majority have learned to migrate southwest to fruit farms in Oregon and California (Brewer et al. 2000).

The first recorded appearance in the YDM area was at the Bredenbury Pool Elevator in May 1938. One of two wintering at Crescent Lake was collected as a specimen on 18 Feb 1941 (Frank Baines). A pair wintered at Dave Stearn's farm at Willowbrook in 1941-42 and at least 10 the following winter. One starling joined a flock of blackbirds at Lower Rousay Lake on 1 Sep 1944 (Houston 1957b).

The first starlings appeared north of Sheho on 16 Apr 1944; that year several pairs stayed to nest in holes formerly used by Purple Martins on the Niven farm. These birds migrated, with spring arrival dates of 18 Mar 1945 (a flock of 20, J.R. Foreman); 15 Mar 1946; 24 Mar 1947 (WN). Starlings first appeared on a farm 5 miles east of Saltcoats in 1948 (RB), and at Ituna in 1951 (Rayner 1951).

Starlings remained uncommon until the mid-1950s; on 29 field outings by the YNHS, 1952-1954, not one starling was recorded (Houston 1957b). Numbers peaked in the early 1970s: the Theodore BBS recorded 311 starlings in 50 stops in June 1971 (WA, Houston 1971c), decreasing to only one for the entire route in each of 1983 and 2000. A group of 20 at Horseshoe Lake on 9 Jul 1975 (DH). Present near Pelly, 24 Jun 1991 (MIC,CSH,MIH). A large group of 240 noted at Leech Lake on 27 Jul 1994 (DH).

ESA(43): 5 Mar 1973, Av 3 Apr. **LFD**(22): 31 Oct 1959, Av 15 Oct.

Breeding: The first nesting was in a woodpecker hole in an aspen at Sheho in 1944. Pairs nested each year from 1951 to 1957 in holes, presumably made by woodpeckers, below an overhanging roof in a house on a farm near Wroxton (Osborne 1957). A starling nest in a former woodpecker hole, with 6 eggs, 15 feet above the ground, was found just north of Melville on 10 May 1958, and a nest with four young, half a mile east of Melville, on 4 Jul 1959 (GAA).

At GSL, five young fledged from a woodpecker hole in an aspen, 9 Jun 1957. On 3 Jun 1988, three young were raised in a nest box with a hole of 1.75 inch diameter. In 2000, 20 nest boxes on the GSL trail were occupied by starlings (see Mountain Bluebird account)(WA).

PNRS: at Saltcoats, 1 egg on 3 May 1959 (BH); at Melville in 1959, two nests with 5 eggs each on 2 and 8 Jun (GAA); nest with young, 3 miles west of Togo, 13 Jun 1982 (HWRC).

BBS: overall mean, 23.1. Theodore, mean 31.6; decreasing (1970s, 58.4; 1980s, 17.7; 1990s, 11.1; n=32).

CBC: Rank #20. 33 counts, 310 birds. Highest: 47, Yorkton 1994. The first appearance on a Yorkton CBC was five on 26 Dec 1945. Two were present on 29 and 30 Dec 1946. There were 4 in 1948, 25 in 1953, 4 in 1954, 6 in 1955, 4 in 1957 (6 of the first 19 counts). Seven were seen in the 1958 Saltcoats CBC.

Winter: Only in the late 1960s and the 1970s did small numbers of starlings begin to winter around feedlots at Sheho and near GSL.

Banded 5 (Anweiler/CSH, locals at Melville in 1959).

Langenburg: Only one sighted, 1990 (TH,BL).

DUCK MOUNTAIN: Two in 1959 CBC at Kamsack.

AMERICAN PIPIT *Anthus rubescens*

Uncommon transient, apparently decreasing steadily since the mid-1960s.

At Fort Pelly, 15 Sep 1881 (Macoun in Seton 1891).

Was regular and fairly common at Sheho, especially in spring migration, until about 1966; none noted for the next seven years through 1973 (Niven 1973), but a flock went through on 4 Oct 1983 (WN).

ESA(17): 27 Apr 1944 at Sheho (WN) and 29 Apr 1959 at GSL (WA), Av 12 May. **LSD**: 3 Jun 1982 (WA). **EFA**: 17 Sep 1961. **LFD**(7): 23 Oct 1980, Av 11 Oct.

SPRAGUE'S PIPIT *Anthus spragueii*

THREATENED, COSEWIC, MAY 2000

Common summer resident locally in pastures, now declining steadily. No breeding records since 1902.

Breeding (historic): Nest, 5 eggs, collected at Crescent Lake on 31 May 1890 by Fred Baines (Raine 1892:75). Nest, Crescent Lake, of fine dried grasses, built in the ground at the side of a clump of sod. Egg size .85 x .60 inches. (W. Raine, in Reed 1904:309). Nest, 4 eggs, in the grass on the prairie at Crescent Lake, 15 June 1902 (Macoun and Macoun 1909:689). Three eggs in clump of prairie grass, Crescent Lake, 17 Jun 1902, WFVZ. After most grassland was ploughed in the early 20th century, numbers plummeted.

A further steep population decline resumed in the mid-1970s, without any evidence of reprieve to date. Summer resident of grassy flats at Leech and Lower Rousay lakes, but numbers on pastures near GSL dropped drastically by 1989 (WA). In 1989, one singing over pasture at Upper Rousay Lake, 13 Jun, and two over pastures near Crescent Lake, 20 Jun (CSH,MIH). One seen over prairie on south side of Little Quill Lake on 1 Jun 1990 (DH), and one over pasture near Togo on 25 Jun 1991 (MIC,MIH,CSH).

ESA(23): 22 Apr 1984 and 1989, Av 6 May. **LFD**: 31 Jul 1997.

BBS: overall mean, 1.2. Calder, 1. Churchbridge, 2. Theodore, mean 1.9; probably decreasing (1970s, 1.4; 1980s, 3.9; 1990s, 0.3; n=20).

Summer: [Four pipits sang in the Manitoba Duck Mountains on 9 Jun 1997, Paula Grieef, Caroline Curtis, George Holland, in Koes and Taylor 1998b].

Banded pre-1955, 1 (Wotherspoon).

BOHEMIAN WAXWING *Bombycilla garrulus*

Fairly common to abundant in winter, unpredictable and hence truly "bohemian."

A dozen or more visited H.M. Rayner's garden in Ituna on 25 Oct 1942, the first he had seen since the "famous waxwing winter of 1938-39" (Lawrence 1943). In the 1930s and early 1940s, flocks of 30 or 40 were considered large. After eating every Cotoneaster berry in Yorkton, waxwings would be gone within a week. When Mountain Ash (Rowan) and ornamental crabapple trees came into maturity in many Yorkton yards in the late 1940s, the waxwings began to come in larger numbers and, with plenty of food available, would stay for a goodly part of the winter. "Every year the first spot in Yorkton to receive a visit from the Bohemian Waxwings are some very fine [Siberian] crabapple trees in the Canadian Pacific Railway station gardens. Here, quite unperturbed by the shunting of trains only a few feet away...these birds systematically set to and stripped all the tiny apples from the branches" (Priestly 1943b). In 1953, a small flock stayed in Yorkton from 21 Jan through 22 Feb (MIH).

In smaller towns and villages, they appeared for a few days, ate the available berries, and departed (such as a flock of 100 on 24 Feb 1982 near Sheho and of 60 to 75 in Foam Lake, 30-31 Jan 1989, WN). During some winters, such as 1982-83, few appeared anywhere in Saskatchewan; that winter only singles were seen at GSL in January and early February, but flocks of up to 100 were seen after 13 Feb (WA, Gollop 1983).

An example of commensal feeding occurred at GSL in the winter of 1955-56, when the Chokecherry crop was abundant. Evening Grosbeaks came first, cracking the shell and eating the chokecherry nut inside. They then spit out the pulp, which was eaten by waxwings when they arrived a few days later (WA).

EFA(45): 7 Oct 1990, Av 28 Oct. **LSD**(37): 3 May 1977, Av 8 Apr.

CBC: Rank #3. 20 counts, 15,409 birds. Highest: 5,276, Yorkton 1986.

Banded 7 (CSH) + pre-1955, 26 (CSH).

DUCK MOUNTAIN: Eight, 27 Dec 1989 (TH). Fairly common in winter (Hatch and Weidl 1981).

CEDAR WAXWING *Bombycilla cedrorum*

Common summer resident. Breeds.

Present at Duck Mountain, 1884 (Seton 1891).

In my youth this was one of the last summer species to leave in fall, yet one of the latest to return in spring (CSH). Two and five in the two mature aspen study plots at GSL, 1969 (Flack 1976).

ESA(45): 18 May 1988, Av 29 May. **LFD**(37): 2 Nov 1981, Av 1 Oct.

Breeding: Nests frequently in Anakas' yard at GSL. Early nest, 13 Jun 1989 (WA). Nest at Saltcoats, 3 eggs, 27 Jun 1957 (BH). Nest, filled with sticks, with four large young in open nest in Caragana, north of Saltcoats on 27 Jul 1994 (JRJ). Nest in Fairburn's yard at Ketchen, 4 Jul 1999 (DF). Late nest, three young, 5 Sep 1947 (CSH).

PNRS: at Saltcoats, both with 3 eggs, on 30 Jun 1958 and on 4 Jul 1959 (BH); at Stornoway with 2, 2, 2, 3, 3, 4, and 5 eggs, and three young (SZ,RD).

BBS: overall mean, 4.2. Theodore, mean 5.7 (1970s, 5.8; 1980s, 7.5; 1990s, 4.5; n=32).

Winter: Five Cedar Waxwings were present in the Yorkton CBC in 1948. Wintering Cedars became widespread the next winter, 1948-49, at Yorkton, and at Brooks, Alberta, and Wawanesa, Manitoba (Gammell and Gammell 1949a). This coincided with maturation of more fruit trees, and especially of Mountain Ash (Rowan) trees. Some Cedars began to stay over winter, usually with other Cedars but occasionally in mixed flocks with Bohemians. Ten were seen in residential areas on four occasions through 11 Jan 1949, and 10 on 11 Apr (Shaw 1949). The first winter proximity of the two species was noted on 22 Nov 1952, when there was a small flock of Cedars in maples in Houstons' front yard in Yorkton, while a flock of Bohemians was in the crabapple tree in the backyard (Houston 1953a; Gammell and Gammell 1953). A flock of 15 on 28 Mar 1957 was presumed to have wintered within the province, if not within Yorkton (Houston 1957d).

Wintering groups have increased in size: on 3 Jan 1997, in Yorkton, 120 were with a larger flock of Bohemians (J. Pollock, Koes and Taylor 1997). Cedar Waxwings now winter in the residential subdivisions around GSL.

CBC: Rank #12. 20 counts, 1,151 birds. Highest: 492, Yorkton 1994 (with only one Bohemian Waxwing in that count!).

Banded 9 (CSH, including one flightless young and three immatures).

Langenburg: Six seen, 1990 (TH,BL).

DUCK MOUNTAIN: SMNH 1926. SMNH 1951. SNHS 1956. Seen in 1974, 1975, 1976, 1977 (JAS). Common throughout park (Hatch and Weidl 1981). SNHS 1989.

Point counts: Rank #22, 10 individuals at 7 stops (ARS).

Aspen count totals: OA 2, YA 2, MA 1 (TH).

GOLDEN-WINGED WARBLER *Vermivora chrysoptera*

Rare in summer at Duck Mountain.

Male sighted at Ravine Reserve, on northwest edge of Yorkton, on 30 May 1973 by Bob Luterbach (pers. comm., 5 Dec 1980).

DUCK MOUNTAIN: Individuals seen: on 29-30 May 1986 (Shadick 1986; MDG in Gollop 1986) and throughout most of June, with sightings by ARS on 3, 4, 14, and 21 Jun 1986; 23 May 1987 (RK in Smith 1996); 12 Jun 1988 (CP in Smith 1996); near gate into ski hill area, 30 Jun 1990 (TH, Koes and Taylor 1990c); and 22-24 Jun 1995 (RK, Koes and Taylor 1995c). Two were observed on southern slopes of Duck Mountain in 1998 (Cumming 1998).

TENNESSEE WARBLER *Vermivora peregrina*

Common transient. Fairly common summer resident in heavy aspen forest, common at Duck Mountain. No breeding records.

Noted at Duck Mountain (Seton 1909a).

Seen in migration at Sheho most years (WN).

ESA(45): 9 May 1981, Av 18 May. **EFA:** 18 Aug 1958 (MIH). **LFD:** 15 Sep 1993 (WA).

Summer: Pairs at GSL in 1971, and apparently on territory in aspen woods, 4-21 Jul 1984, 3-12 Jul 1985, 28 Jun-17 Jul 1987, and 13-17 Jun 1988, but no nest found (WA).

BBS: overall mean, 0.2. Theodore, 8 years; highest, 3, 1987.

Banded 13 (CSH).

DUCK MOUNTAIN: Prefers sphagnum bogs and increases with Spruce Budworm infestations (RK). SMNH 1926. SMNH 1951. SNHS 1956. Seen in 1977 (JAS). May 1986 (MDG,SJS). Sightings of 10, SNHS 1989 (JFR). Common throughout park, but especially near wet Balsam Poplar areas in June (Hatch and Weidl 1981). In contrast, on 28-30 June 1990, not one was heard or seen (TH, Koes and Taylor 1990c).

Point counts: Rank #48, 2 individuals at 2 stops (ARS).

ORANGE-CROWNED WARBLER *Vermivora celata*

Uncommon transient and summer resident. No breeding records.

A bird flew into a picture window at 280 Victoria Avenue, Yorkton, on 6 Oct 1957; the specimen was sent to SMNH. Present at GSL six years, 11 Jun 1977 (Bob Godwin, Kathy Meeres), 22 Jun 1991, 1993, 1994, 1996, 1997. One near Togo, 25 Jun 1991 (Murray and Allan Bowes).

ESA(8): 29 Apr 1955, Av 12 May. **EFA:** 1 Sep 2002 (WA) and 18 Sep 1958 (MIH). **LFD**(7): 7 Oct 1993, Av 29 Sep.

BBS: overall mean, 0.13. Theodore, 6 years; highest, 2, 1999, mean 0.2.

Banded 4 (CSH).

Langenburg: One singing male in coulee joining Assiniboine valley, 1990 (TH,BL).

DUCK MOUNTAIN: Typical habitat is the hillsides of Little Boggy Creek valley, where shrubs are mixed with mature deciduous trees with patchy openings (RK). Uncommon in deciduous woodland, sighted near Little Boggy Creek on 16 and 18 Jun 1981 (Hatch and Weidl). Seen in May 1986 (MDG,SJS). In 1988, one singing 21 May and 20 Jun along road to Batka Lake (RK,CP,RE,TH). One sighted, SNHS 1989 (JFR).

NASHVILLE WARBLER *Vermivora ruficapilla*

Uncommon transient. Fairly common summer resident at Duck Mountain. No breeding records.

A male in full song, evidently on territory, 11 June 1884, in Duck Mountains (Seton 1891 and 1909a). 23 June 1936 at GSL (JDS).

ESA: 12 May 1940 (IMP). **LSD**: 3 Jun 1945 (A. McPherson). **EFA**: 29 Aug 1983 at Norquay and at Hyas, 4 Sep 1984 (CP). One fall date at Yorkton (see banding).

BBS: Theodore, 4 years; 1 each in 1972, 1974, 1975, 1976, mean 0.1.

Banded 1, 25 Sep 1955 (CSH).

DUCK MOUNTAIN: SMNH 1951. SNHS 1956. May 1986 (MDG,SJS). This species becomes more spotty in distribution in the mixed forest farther west in Saskatchewan. A territorial male was singing in a Black Spruce bog, with sphagnum hummocks, in June 1973. [Hooper (1992) found them nesting in a Black Spruce bog, the first confirmed nest record for Saskatchewan]. The second habitat is in young aspen on a hillside in dry mixed forest (BL). Fairly common in mixed woods, along south boundary road and the road to Bear Head Lake, at Little Boggy Creek, and along nature trails (Hatch and Weidl 1981). 22 May 1981 (ARS). Three sighted, SNHS 1989 (JFR). Present in north fringe of park, 24 June 1991 (MIC,CSH,MIH). Heard on all three visits to the marl bog in June 1993 (Herriot 1995). Present near ski hill, 18 June 2000 (BL).

Point counts: Rank #33, 5 individuals at 4 stops (ARS).

NORTHERN PARULA *Parula americana*

Vagrant.

One male and three females or immatures near the dying crown of a spruce in Patrick Park, King Street, Yorkton, on 27 Sep 1972 (BL in Smith 1996, letter 5 Dec 1980).

DUCK MOUNTAIN: Seen 25 Jul 1988 (CP in Smith 1996) and 11 Aug 1990 on Pelly Point Trail (RK, Koes and Taylor 1991a).

YELLOW WARBLER *Dendroica petechia*

Very common summer resident. The most widely distributed warbler in cities, towns, and aspen bluffs. Breeds.

One egg collected north of Saltcoats (NJ,PMA).

Four and 12 in two mature aspen study plots at GSL, 1969 (Flack 1976).

ESA(48): 7 May 1988, Av 18 May. **LFD**(28): 19 Sep 1954, Av 28 Aug.

Breeding: Nest with four young, 26 Jun 1940 (IMP). Nest with 3 eggs on 13 Jun 1960 had 2 hatch by 20 Jun; the next year a nest with 3 eggs on 5 Jun 1961 had 5 eggs on 13 Jun (JG). Nest, 1 egg, Saltcoats, 2 Jul 1967 (CSH).

At GSL, earliest nest completed 24 May 1989, had 1 cowbird egg removed on 26 May, 1 warbler egg on 27 May, but nest predated 28 May; the same pair apparently renested in Caragana hedge 6 feet away, with 5 eggs on 14 Jun, five young on 17 Jun, and fledged successfully; another nest had 1 egg with 1 cowbird egg on 4 Jul 1967 (WA).

PNRS: five nests near Leech Lake in 1964, with 4, 4, and 5 eggs, and 3 eggs and one young on 7 Jun, and two young on 19 Jun (LM); at Saltcoats, 2 eggs and two young on 28 Jun 1959 and 2 eggs and 1 cowbird egg on 30 Jun; another nest had 4 eggs on 20 Jul 1959 (BH); at Stornoway, two nests with 3 eggs each on 6 and 10 Jul 1961 (SZ,RD); west of Canora, 1 egg on 7 Jun, 2 on 8 Jun and 3 on 9 Jun, before the nest was destroyed (WC); at Melville, nests with 2, 3, 4, 4, and 5 eggs (GAA,VS). Adults feeding young in Fairburn's yard at Ketchen on 10 Jul 2001 (DF).

BBS: overall mean, 12.8. Theodore, mean 13.3; increasing (1970s, 8.6; 1980s, 11.8; 1990s, 18.9; n=33).

Banded 8 (CSH 5, WA 3) + pre-1955, 8 (CSH 3, Wotherspoon 5).

Langenburg: "abundant," 1990 (TH,BL).

DUCK MOUNTAIN: SMNH 1926. SMNH 1951. SNHS 1956. Uncommon in deciduous woodland, along south boundary road, at Pelly Point, and near Rundle Lake (Hatch and Weidl 1981). May 1986 (MDG,SJS).

Point counts: Rank #37, 4 individuals at 4 stops (ARS).

Aspen count totals: OA 2, YA 1 (TH).

CHESTNUT-SIDED WARBLER *Dendroica pensylvanica*

Rare transient at Yorkton. Fairly common summer resident at Duck Mountain. No breeding records.

Quite common in this wooded region, no doubt breeding, 10 June 1884, Duck Mountain (Seton 1891).

Present all summer at GSL in 1971. One singing in coulee leading down to Lake of the Prairies, north of Highway 10, 21 Jun 1989 (CSH,MIH). One near Togo, 25 Jun 1991 (Murray and Allan Bowes).

ESA: 21 May 1991 at DMPP (ARS), 24 May 1994 (WA), 29 May 1927 and 2 Jun 1945, both at Sheho (WN); and 3 Jun 1945 at Yorkton (A. McPherson).

DUCK MOUNTAIN: Prefers mature aspen with a solid 4- to 8-feet high understory of Beaked Hazelnut or Mountain Maple (RK). SMNH 1926. SMNH 1951. SNHS 1956. May 1986 (MDG,SJS). A surprising 20 sighted, third commonest warbler, SNHS 1989 (JFR). Uncommon in deciduous woodland at Pelly Point, Little Boggy Creek, and southeast corner of park in June (Hatch and Weidl 1981). Regular near the ski hill, 18 Jun 2000 (BL).

Point counts: Rank #43, 3 individuals at 3 stops (ARS).

Aspen count totals: OA 3, YA 5, MA 3 (TH).

MAGNOLIA WARBLER *Dendroica magnolia*

Uncommon transient.

ESA(3): 11 May 1992. Many noted in Yorkton and Sheho during unseasonable snow and cold, wet weather from 26 May to 5 Jun 1945, which detained the migration and gave the best opportunity ever for observing many warblers. One Magnolia picked up dead on 5 Jun was representative of eight species of warblers found weakened or dead throughout the district. One sighted at Sheho on 26 May 1939 (WN) and 22 May 1954 (WN, Gammell et al. 1954b).

EFA: a male at Ketchen on 22 Aug 2000 (DF), at Norquay on 29 Aug 1983 (CP), and at GSL on 4 Sep 1960, 6 Sep 1981, 9 Sep 1956 (WA). **LFD**(5): 19 Sep 1954 (WA) and banded 30 Sep 1959 (CSH).

Banded 1 (CSH).

DUCK MOUNTAIN: SMNH 1926. Three heard, SNHS 1989 (JFR). Uncommon transient (Hatch and Weidl 1981). Present in north fringe of park, 24 Jun 1991 (MIC,CSH,MIH). One seen and heard near Benito Beach subdivision, 16 Jun 1993 (Herriot 1995).

CAPE MAY WARBLER *Dendroica tigrina*

Rare transient.

Spring dates: 11 May 2000, 28 May 2001, and 22-23 May 2002 at Ketchen (DF); 14 May 1955, 16 May 1969, and 1988 at GSL (WA); 17 May 1946 (CCS), 19 May 1945 (IMP), and 21 May 1940 at Yorkton (CSH). **LSD:** 29 May 1947 (Niven 1973; Niven's only record).

Fall dates: at GSL on 18 July 1987; one at Hyas on 16 Aug 1985 (CP); one at GSL, 26 Sep 1997 (DV, Bain and Holder 1997d); and 5 Aug and 1 Sep 2001; and 1 Sep 2002 at Ketchen (DF).

At Ketchen during a spell of cold weather in 2001, a Cape May Warbler sucked at the hummingbird feeder and in 2002 ate at the suet feeder! (DF).

DUCK MOUNTAIN: SMNH 1926. Two on 29 May 1986 (JFR et al.). One heard, SNHS 1989 (JFR). One male heard and seen on Whitethroat Trail, 16 Jun 1993 (Herriot 1995).

BLACK-THROATED BLUE WARBLER *Dendroica caerulescens*

Rare transient.

An adult female was observed at GSL 20 Sep 1997 (WA, Bain and Holder 1997d) and a male on 4 Oct 1998 (WA, Bain and Shanahan 1998d).

YELLOW-RUMPED (MYRTLE) WARBLER *Dendroica coronata*

Uncommon summer resident at GSL, and fairly common at Duck Mountain. Breeds.

The commonest and earliest of the warblers. All observations but one refer to the eastern subspecies, *D. c. coronata.* One singing in coulee leading down to Lake of the Prairies, north of Highway 10, 21 Jun 1989 (CSH,MIH).

ESA(48): 11 Apr 1999 at Ketchen (DF) and 11 Apr 1986 at GSL (WA), Av 22 Apr. **EFA:** 9 Sep 1946 (CSH). **LFD**(45): 26 Oct 1976, Av 8 Oct.

Breeding: At GSL, two young warblers and one cowbird were raised in a spruce, Jul 1956 (Gunn 1956b). In 1957, two young fledged from a nest in a spruce at the Gunns' back door, 18 Jun, and a week later another two young and one cowbird fledged from a second nest (Gunn 1957a). Nest with two young in Balsam Poplar at GSL, 11 Jun 1977 (WCH,SL). Female feeding young cowbird, 7 Aug 1981 and 20 Jul 1986; nest, three nearly fledged young, 10 Jul 1987 (WA).

Also breeds at Buchanan (ARS, *Atlas*).

BBS: Theodore, 1 each in 1987, 1995, mean 0.06.

Banded 35 (CSH) + pre-1955, 3 (CSH).

Langenburg: a migrant near Langenburg on 26 Apr 1976 (DH).

DUCK MOUNTAIN: A bird of mixed forest, always with some spruce (RK). SMNH 1926. SMNH 1951. SNHS 1956. Seen in 1974,

1975, 1976, 1977 (JAS). Fairly common, seen at Spruce Island, Selwood Campground, and along Pelly Point and Whitethroat nature trails (Hatch and Weidl 1981). May 1986 (MDG,SJS). SNHS 1989. Present in north fringe of park, and near Pelly, 24 Jun 1991 (MIC,CSH,MIH).

Point counts: Rank #7, 29 individuals at 27 stops (ARS).

Audubon's subspecies, *D. c. auduboni,* was sighted once near Togo, 18 May 1986, by Stan Shadick (Gollop 1986).

BLACK-THROATED GREEN WARBLER *Dendroica virens*

Rare transient. Uncommon summer resident at Duck Mountain. No breeding records.

Spring dates: 29 May 1927 at Sheho (WN, Niven 1973; this was Niven's only record); 7 May 1945 at Yorkton (IMP); heard in spruce trees south of main beach, GSL, 25 Jun 1988 (TH).

Fall date: at GSL, one on 22 Sep 1996 (WA).

DUCK MOUNTAIN: Prefers mixed woods with mature spruce, often Black Spruce. Forages three-quarters of the way up the trees, but nests nearer the ground (RK). SMNH 1926. SMNH 1951. SNHS 1956. On territory in stands of Black Spruce, 28 Jul 1941 (IMP). In mid-May 1972, among a flock of over 400 migrating warblers in the main cottage area at Madge Lake were a surprising number of Black-throated Greens, "many more than I have seen anywhere before or since; they rarely stop over in Regina or Yorkton, so they either fly over or they follow the southern edge of the mixed forest when travelling northwest" (BL). One heard and seen at Selwood picnic area, 14 Jul 1983, and two along Red Squirrel Nature Trail the next day (DAW). May 1986 (MDG,SJS). Ten heard, SNHS 1989 (JFR). In 1989, four along road to Pelly Point Trail on 5 Jun and five along Woodchuck Ski Trail on 9 Jun (TH, CWS survey). In 1991, 22 May (ARS).

Point counts: Rank #23, 8 individuals at 8 stops (ARS).

Aspen count total: OA 1 (TH).

BLACKBURNIAN WARBLER *Dendroica fusca*

Rare transient. Common resident at Duck Mountain. No breeding records.

Swan Lake and Porcupine Mountains, probably breeding, 1881 (Macoun in Seton 1891).

Spring dates at GSL: J.R. Foreman, 1938; 24 May 1981 (WA); heard singing in spruce at south side of main beach, 25 Jun 1988 (TH). In

Fairburn's yard at Ketchen on 2 Jun 2002 (DF).

Fall dates: at GSL on 29 Aug 1956, 11 Sep 1950, and 15 Oct 1981 (WA).

DUCK MOUNTAIN: Prefers mixed woods with mature White Spruce whose spires break through the canopy of poplar (RK). SMNH 1926: Specimens, male, #1991; female, #1882, study skins male, #1913, #1914, and #1915; these were the first Saskatchewan specimens; Mitchell was surprised by the numbers seen. SMNH 1951: typical warbler species. SNHS 1956. In mid-May 1972, among a flock of over 400 migrating warblers in the main cottage area at Madge Lake were a surprising number of Blackburnians, "many more than I have seen anywhere before or since" (BL). Seen in 1975, 1976 (JAS). May 1986 (MDG,SJS). Commonest warbler, with at least 30 heard or seen, 8-11 Jun, SNHS 1989 (JFR). Fairly common in mixed woods, at Selwood campground and along Pelly Point, Red Squirrel, and Whitethroat nature trails in June (Hatch and Weidl 1981). 22 May 1991 (ARS).

Point counts: Rank #6, 32 individuals at 30 stops (ARS).

PALM WARBLER *Dendroica palmarum*

Fairly common transient in some years, not seen in others.

A heavy migration, 4 to 18 May 1946, was evident at Yorkton, Sheho, and Nipawin. One at Foam Lake marsh on 25 May 1974 (AS). Sixteen sighted at Horseshoe Lake, 19-22 Sep 1975 (DH). On the night of 19-20 Oct 2002, a Palm Warbler sought shelter under a hay bale in Anakas' yard, but with an overnight temperature of -19° C, it was frozen by morning (JA, Bain 2002b).

ESA(25): 29 Apr 1978 at Sheho (WN) and 1 May 1987 at GSL (WA), Av 12 May. **LSD**(9): 25 May 1947 at Sheho (WN), and 26 May 1995 and 2002 at GSL (WA), Av 19 May. **EFA**(27): 12 Aug 1986, Av 10 Sep. **LFD**(32): 5 Nov 1964, Av 29 Sep.

Banded 3 (CSH).

DUCK MOUNTAIN: SMNH 1926, study skin female, #1916. SMNH 1951. Uncommon transient, individuals passed through on 9 Sep and 1 Oct 1981 (Hatch and Weidl).

BAY-BREASTED WARBLER *Dendroica castanea*

Rare transient.

Records of single individuals, all on 26 May, in 1939 and 1945 at Sheho (WN), and 1952 at Yorkton (MIH). At GSL, 23 May 1954, 2 Jun 1982, and 18 May 2000 (WA).

DUCK MOUNTAIN: SMNH 1926: study skin female, #1912. In mid-May 1972, among a flock of over 400 migrating warblers in the main cottage area at Madge Lake were a surprising number of Bay-breasteds, "many more than I have seen anywhere before or since" (BL). One seen in 1976 (JAS). A presumed transient seen, 21 Aug 1989 (TH). Rare transient, one seen at Selwood campground on 15 Jun 1981 (Hatch and Weidl), and another on 25 May 1993 (DH).

BLACKPOLL WARBLER *Dendroica striata*

Fairly common transient some years.

ESA(14): 10 May 1959 at GSL (WA) and 12 May 1945 and 1946 at Sheho (WN), Av 18 May. **LSD**(3): 29 May 1954. **EFA**: 1 and 3 Sep 2002 at Ketchen (DF) and 3 Sep 1961 at GSL (WA). **LFD**(6): 16 Sep 1956, Av 5 Sep.

Banded 1, mistnetted, 102 Logan Crescent, Yorkton, 18 Sep 1958 (CSH).

DUCK MOUNTAIN: SMNH 1926. SMNH 1951. One seen 21 May 1989 (TH). Uncommon transient, one seen near Batka Lake on 9 Sep 1981 (Hatch and Weidl).

BLACK-AND-WHITE WARBLER *Mniotilta varia*

Fairly common transient. No breeding records.

Abundant in Duck Mountains, 10 June 1884 (Seton 1891).

At Sheho, 29 May 1927 (Niven 1973). At GSL on 6 Aug 1938 (IMP). One killed when it flew into a picture window in Yorkton, 17 May 1977 (LM). One seen at Hyas on 25 May 1984 (CP). Seen near Preeceville, 24 Jun 1991 (MIC,MIH,CSH).

ESA(23): 3 May 1959, Av 13 May. **LSD**: 3 Jun 1956, Av 19 May. **EFA**(13): 7 Jul 2000, Av 19 Aug. **LFD**(9): 17 Sep 1982, Av 5 Sep.

Banded 1, mistnetted 30 Sep 1959 (CSH).

Langenburg: One singing male in coulee running into the Assiniboine River valley, 1990 (TH,BL).

DUCK MOUNTAIN: Prefers deciduous trees in ravines and draws, with birch and willow in the mix (RK). SMNH 1926. SMNH 1951. SNHS 1956. Seen in 1974, 1976 (JAS). Fairly common in deciduous woodland (Hatch and Weidl 1981). May 1986 (MDG,SJS).

Point counts: Rank #33, 5 individuals at 5 stops (ARS).

Aspen count totals: OA 3, YA 1, MA 1 (TH).

AMERICAN REDSTART *Setophaga ruticilla*

Fairly common transient and summer resident in northern parts of YDM area. Breeds.

Nest with eggs, 8 June 1884, Duck Mountains (Seton 1891). Was more common west of Saltcoats until about 1935 (Frank Baines).

Summer resident in heavy aspen woods, including GSL in 1959, 1962, 1963, 1980, 1983, 1986, and consistently each year from 1988 through 2002 (WA). One singing in coulee leading down to Lake of the Prairies, north of Highway 10, 21 Jun 1989 (CSH,MIH).

ESA(20): 17 May 1959 and 1981, Av 23 May. **LFD**(19): 21 Sep 1992, Av 6 Sep.

Breeding attempt: female beginning nest with bits of down, spider webbing, and one feather, in heavy willows in deep aspen woods, 29 May 1960; completed with fine inner bark and poplar fluff and feathers on 5 Jun 1960; nest then deserted (WA).

BBS: Theodore, 6 years, including 3 in 1994, mean 0.2.

Banded 7 (Smith 7, at DMPP) + pre-1955, 1 (CSH).

Langenburg: Four singing males in wooded valleys, 1990 (TH,BL).

DUCK MOUNTAIN: Found in a wide variety of young, mature, and old deciduous tree habitat (RK). SMNH 1926: 5 eggs on 8 Jun, group, #1870; study skin female, #1923; two sets of 4 eggs, #1958 and #1959, on 13 Jun. SMNH 1951: 4 eggs on 26 Jun, #5502. SNHS 1956. Seen in 1974, 1975, 1977; nest 20 feet above ground in birch crotch, four young almost fledged, 1976 (JAS). Common in deciduous woodland throughout the park, one of two commonest warblers; 15 were heard during a hike along Pelly Point Nature Trail on 15 Jul 1981 (Hatch and Weidl). May 1986 (MDG,SJS). Fifteen heard, SNHS 1989 (JFR). Commonest warbler in north fringe of park and near Pelly, 24 Jun 1991 (MIC,CSH,MIH).

Point counts: Rank #3, 68 individuals at 39 stops (ARS).

Aspen count totals: OA 23, YA 9, MA 22 (TH).

OVENBIRD *Seiurus aurocapilla*

Uncommon summer resident in aspen bush near GSL, common resident at Duck Mountain. Breeds.

Only Sheho observation, 22 May 1959 (WN). Two heard at GSL on 11 Jun 1977 (WCH).

ESA(45): 8 May 1988, Av 21 May. **LFD**(5): 22 Sep 1931 at Melville (LTMcK), and 12 Sep 1956 at GSL (WA).

Breeding: Ralph Jowsey found a nest 7 miles north of Saltcoats about 1941 (JRJ). A partly completed nest found at GSL on 23 May 1951 had only 1 egg hatch (Anaka 1951). Nest, one young and two cowbird young, 22 Jul 1959; nest, five young, 29 Jun 1961 (WA).

BBS: overall mean, 0.1. Churchbridge, 1. Theodore, 1 each in 1969, 1987, 1990, 1995.

Banded 1 (CSH) + pre-1955, 1 (CSH, 3 June 1945).

Langenburg: Four singing males in wooded valleys, 1990 (TH,BL).

DUCK MOUNTAIN: Requires a mature deciduous forest with an open floor and a low understory of Sarsaparilla; sometimes in mixed woods if floor is open (RK). SMNH 1926. SMNH 1951: nest but no eggs on 1 Jul, #5512. SNHS 1956. Seen in 1974, 1975, 1976 (JAS). Common in mixed woods throughout park (Hatch and Weidl 1981). May 1986 (MDG,SJS). Twenty-five heard, SNHS 1989 (JFR). Present in north fringe of park and near Pelly, 24 Jun 1991 (MIC,CSH,MIH).

Point counts: Rank #2, 70 individuals at 61 stops (ARS).

Aspen count totals: OA 5, YA 6, MA 12 (TH).

NORTHERN WATERTHRUSH *Seiurus noveboracensis*

Fairly common transient. Uncommon summer resident at Duck Mountain. No breeding records.

May breed in valleys. At least six identified along Whitesand River, 11 Aug 1988 (DH).

ESA(6): 4 May 1964 at Sheho (WN) and GSL (WA), Av 14 May. **LFD**(17): 24 Sep 1980, Av 5 Sep.

Banded 4 (CSH) + pre-1955, 2 (CSH).

Langenburg: Two singing males along Cutarm Creek, 1990 (TH,BL).

DUCK MOUNTAIN: In sphagnum bogs, and near streams or springs (RK). SMNH 1926. SMNH 1951. SNHS 1956. Uncommon in deciduous woodland, along west boundary road, Midway subdivision, and Little Boggy Creek (Hatch and Weidl 1981). May 1986 (MDG,SJS).

One heard, SNHS 1989 (JFR).

Point counts: Rank #11, 22 individuals at 19 stops (ARS).

Aspen count total: OA 5 (TH).

CONNECTICUT WARBLER *Oporornis agilis*

Rare transient and summer resident. Breeds at Duck Mountain.

One or two pairs in Tamarack swamp [near present Runnymede], 14 Jun 1884 (Seton 1884, Seton 1891). [Note: a year later, Seton found the world's first nest of this species, near Carberry, Manitoba—CSH].

An adult female, beating against a store window in Yorkton, 2 Jun 1945, was caught by Ed Lawrence and banded by CSH (Priestly 1945b). Another individual was seen four days later by A. McPherson. For three successive years, 1971-72-73, two or three were seen daily during the last week in May, feeding on snowberry shrubs on the east side of the Ravine Reserve near Yorkton (BL).

In mid-June 1973, I heard and saw two singing males in dry aspen bluffs on the upper east slopes of the Assiniboine River, south of the Cote Reserve and north of Kamsack. The understory had been burned out in a fire the previous year. The other habitat type of this warbler is spruce bogs, and that day I found another male in such habitat south of Sturgis (BL).

Spring dates at GSL: 3 Jun 1956; 3 and 6 Jun 1981, 5 Jul 1992, 17 Jun 2001 (WA). Two present on 20 acres near study plots at GSL, 1969 (Flack 1976).

Banded pre-1955, 1 (CSH, see above).

DUCK MOUNTAIN: SMNH 1926: mounted male, #1883. SMNH 1951: two nests on 29 Jun, #11765 and #11667, with 3 eggs and 1 cowbird egg, #5511, and with four young (FWL). SNHS 1956. Rare in coniferous woodland; singles seen along Whitethroat Nature Trail, 4 Jun, and along the Pelly Point Nature Trail, 20 Jul (Hatch and Weidl 1981, Weidl 1983). May 1986 (MDG,SJS). One well seen on 29 May 1986 (JFR et al.). One seen among young aspen in Little Boggy Creek valley, 18 Jun 1988 (TH) and 27 May 1989 (RK,CP,RE,TH). One seen, SNHS 1989 (JFR).

ESA: 22 May 1991 (ARS).

MOURNING WARBLER *Oporornis philadelphia*

Uncommon transient. Common summer resident at Duck Mountain. No breeding records.

Spring dates: singles seen 29 May 1927 at Sheho (WN), 4 Jun 1945 at Yorkton (A. McPherson), 24 May 1967 at GSL (WA), and 29 May 2002 at Ketchen (DF).

Fall dates: Sep 1941 and 13 Sep 1942 (IMP), 5 Aug 1994 and 17 Aug 1995 at GSL (WA).

Two in one of two mature aspen study plots at GSL, 1969 (Flack 1976).

LFD: mistnetted 19 Sep 1954 (CSH).

Banded pre-1955, 1 (CSH).

Langenburg: One singing male in a coulee running into the Assiniboine River valley, 1990 (TH,BL).

DUCK MOUNTAIN: Prefers mixed or deciduous woods, with brushy thickets of willow and hazel in the understory, often with deadfalls (RK). SMNH 1926. SMNH 1951. SNHS 1956. Seen in 1974, 1975, 1976 (JAS). Common throughout park (Hatch and Weidl 1981). May 1986 (MDG,SJS). Ten heard, SNHS 1989 (JFR). Present in north fringe of park and near both Preeceville and Pelly, 24 Jun 1991 (MIC,CSH,MIH).

Point counts: Rank #16, 13 individuals at 13 stops (ARS).

Aspen count totals: OA 2, YA 5, MA 4 (TH).

COMMON YELLOWTHROAT *Geothlypis trichas*

Common summer resident locally in most larger marshes. Breeds.

ESA: 8 May 1938 at Yorkton (IMP) and 13 May 1981 at GSL (WA), Av 28 May. **LFD**(16): at Yorkton, mistnetted 30 Sep 1958 (CSH) and 24 Sep 1990 at GSL (WA), Av 5 Sep.

Breeding: Nest, 4 eggs, Rokeby marsh, 7 Jun 1953 (J.B. Belcher and CSH). Singing males also present at Crescent marsh, Leech and Horseshoe lakes, and both Rousay lakes.

BBS: overall mean, 2.8. Theodore, mean 3.8 (1970s, 3.2; 1980s, 4.4; 1990s, 4.2; n=33); highest, 13, 1986.

Banded 1 (CSH).

Langenburg: Fairly common in marshy areas with sufficient cover, 1990 (TH,BL).

DUCK MOUNTAIN: SMNH 1951. SNHS 1956. Seen in 1974,

1975, 1976, 1977 (JAS). Common in shallow wetlands; one of two commonest warblers in park; nest, 4 eggs, in Black Spruce bog near southeast corner of park, only metres from Bonaparte's Gull nest; family group near Batka Lake on 13 Aug 1981 (Hatch and Weidl). May 1986 (MDG,SJS). Twenty heard, SNHS 1989 (JFR). Present in north fringe of park, and near Pelly, 24 Jun 1991 (MIC,CSH,MIH).

Point counts: Rank #13, 18 individuals at 16 stops (ARS).

WILSON'S WARBLER *Wilsonia pusilla*

Fairly common transient at Sheho, less common at Yorkton.

Seen near Pelly, 24 Jun 1991 (MIC,MIH,CSH).

ESA(4): 12 May 1941 at Yorkton (IMP) and 17 May 1963 and 1981 at GSL (WA). Third spring sighting at GSL was on 22 May 1954 (WA). **LFD**(5): 24 Sep 2000. One fall date at Hyas, 8 Sep 1985 (CP).

DUCK MOUNTAIN: SMNH 1926. Rare in shallow wetlands; one adult near Kipp Lake on 12 Aug 1981 (Hatch and Weidl).

CANADA WARBLER *Wilsonia canadensis*

Rare transient.

Seen on 5 Jun 1945 at Yorkton (A. McPherson) and 17 Aug 1998 at GSL (WA), and 25 May 2000 south of Ketchen (DF).

DUCK MOUNTAIN: Prefers hillsides with low areas and springs, with Beaked Hazelnut or Mountain Maple in the understory (RK). SMNH 1926: study skin male, #1920. SMNH 1951. Uncommon in deciduous woodland; two near Kipp Lake on 3 Jun, one along Pelly Point Nature Trail on 23-24 Jun, and male heard and seen on road to Bear Head Lake on 12 Aug 1981 (Hatch and Weidl, Weidl 1983). May 1986 (MDG,SJS). Found regularly in dense Beaked Hazelnut understory: along Red Squirrel Trail, Pelly Point Trail, and road beyond Batka Lake on the rim of Little Boggy Creek valley, in 1989 and 1990 (TH). Spring date of 22 May 1991 (ARS).

Point counts: Rank #48, 2 individuals at 2 stops (ARS).

Aspen count totals: OA 2, MA 2 (TH).

YELLOW-BREASTED CHAT *Icteria virens*

Vagrant.

Two sightings at GSL: 8-10 Jun 1962 (D.E. Wade and R.W. Nero), and one in Anakas' yard, 26 Jul 1980 (WA).

SUMMER TANAGER *Piranga rubra*

Vagrant.

One present at Dallas Fairburn's feeder at Ketchen from 7 through 9 November 1999; a diagnostic videotape taken by Dallas was viewed by Frank Roy, Stan Shadick, and Bob Luterbach. This is Saskatchewan's third confirmed record of this species (BL).

SCARLET TANAGER *Piranga olivacea*

Occasional visitor.

Reported from Assiniboine valley near Togo in 1945 (Priestly 1945c). One male at York Lake Golf Course, 21-22 May 1950 (Nancy Morrison, CSH,MIH; Houston 1950; Gammell and Gammell 1950b). One seen at Canora by V. Tataryn, 11-12 May 1993 (Koes and Taylor 1993c).

DUCK MOUNTAIN: Bright-plumaged male singing sporadically at ski hill area, 18 Jun 2000 (DV,PSa,BL, Bain and Shanahan 2000b). An adult male singing along the ski hill road, 3 Jun 2001 (RK, J. Pollock, Bain 2001).

WESTERN TANAGER *Piranga ludoviciana*

Occasional visitor.

A male seen 5 miles east of Saltcoats, 10 Jun 1945 (RB). An unmistakable male seen at Stornoway on 20 Apr 1960 by Stanley and Richard Zazelenchuk (Gammell 1960). One near Togo, 25 Jun 1991 (MIC,MIH,CSH). One male at Preeceville in June 1997 (LP).

DUCK MOUNTAIN: One singing at Kamsack Beach, 7 Jun 1989 (TH, Bennett 1989). One vocalizing 20 Jun 1993, almost at Manitoba boundary (RK, Koes and Taylor 1993d).

SPOTTED TOWHEE *Pipilo maculatus*

Occasional visitor.

Four eggs, Yorkton, 1 Jun 1902, W. Raine, AMNH, may apply to the next species.

Between 1954 and 1995 the two towhees were lumped together by the AOU as one species, so few birders attempted to differentiate between them. As a result, we have only two reasonably certain records of the Spotted Towhee.

One on 26 Jun 1938 (IMP). At GSL, on 21 May 2000, there was one positive identification (WA).

EASTERN TOWHEE *Pipilo erythrophthalmus*

Uncommon summer resident in the sand dunes at GSL. No breeding records.

ESA(11): 8 May 1995, Av 17 May. **LFD**(5): 21 September 2001. A very late fall date at Preeceville from 29 November to 2 December 1994 (LP, Koes and Taylor 1995a).

Suspected breeding at GSL: On 11 July 1999 a young Eastern Towhee, recently out of the nest but able to fly, was "begging" from an adult in our yard. Along 1 mile of beachfront dunes on 29 May 2002 we heard and saw four males singing "on territory"(WA).

BBS: Theodore, one in 1996 (WA,FAS) and one in 2000 (Lionel Bonneville), mean 0.06.

Langenburg: Two males singing along coulee and valley slopes, 1990 (TH,BL).

DUCK MOUNTAIN: Rare in deciduous woodland; one heard along Little Boggy Creek, 18 Jun 1981 (Hatch and Weidl). One near south end of park, 10 Jul 1990 (TH).

AMERICAN TREE SPARROW *Spizella arborea*

Common transient.

Numbers decreased between 1980 and 1990.

ESA(45): 12 Mar 1974 at Sheho (WN) and 13 Mar 1973 at GSL (WA), Av 3 Apr. **LSD**(45): 10 May 1966 and 1967, Av 27 Apr. **EFA**(38): 19 Sep 1957 (CSH) and 20 Sep 1937 at Yorkton (IMP), and 25 Sep 1960 and 1966 at GSL (WA), Av 7 Oct. **LFD**(41): 12 Dec 1998 at Ketchen (DF) and 24 Nov 1966 at GSL (WA), Av 28 Oct.

CBC: Rank #38. 2 counts, 21 birds. Highest: 20, Kamsack 2000.
Banded 66 (CSH 52, WA 14) + pre-1955, 10 (Wotherspoon).

CHIPPING SPARROW *Spizella passerina*

Common summer resident. Breeds.

ESA(42): 19 Apr 1942 (WN) and 27 Apr 1987 (WA), Av 9 May. **LFD** at GSL(7): 12 Sep 1956, Av 19 Aug (WA). **LFD** at Ketchen: 11 Oct 2000, also 30 Sep 1999, 13 Sep 2001, 2 Oct 2002 (DF).

Breeding: Typically nests in small cultivated conifers. Nest, 1 egg and 1 cowbird egg, 3 Jun 1952, hatched 4 Jun 1952 (Lionel Coleman).

PNRS: at Saltcoats with 4 eggs on 28 Jun 1959 (BH), four young and 1 cowbird egg on 20 Jun 1970 (JRJ); at GSL, 4 eggs on 8 Jun 1960 (JG). Nests regularly in Anakas' yard at GSL; completed nest, 26 May 1986, had 5 eggs on 31 May, five young on 10 Jun and four fledged by 13 Jun; 3 eggs and 1 cowbird egg, 27 May 1989 (WA). Four young near Togo on 20 Jun 1988 (HWRC).

BBS: overall mean, 3.9. Theodore, mean 5.5; increasing (1970s, 2.6; 1980s, 6.8; 1990s, 7.1; n=33).

Banded 1 (CSH) + pre-1955, 27 (CSH 18, Wickencamp 5, Wotherspoon 4).

Recoveries, banded in area, 1 [3.6%](SK 1, it drowned the day after it was banded).

Return: 3, one bird returned in two successive years and one at one year (CSH).

Langenburg: Common, 1990 (TH,BL).

DUCK MOUNTAIN: SMNH 1926. SMNH 1951. SNHS 1956. Present 1974, 1976, 1977; nest with three young, 1975 (JAS). Common throughout the park; a fledged young was caught near Little Boggy Creek on 18 Jun 1981 (Hatch and Weidl). SNHS 1989.

PNRS: at Madge Lake, four young on 19 Jun 1988 and 3 eggs on 26 Jun 1989 (HWRC).

Point counts: Rank #4, 54 individuals at 48 stops (ARS).

Aspen count total: OA 2 (TH).

CLAY-COLORED SPARROW *Spizella pallida*

Abundant summer resident. Breeds.

Very abundant. Easy to find three or four nests in an hour or two (Seton 1891).

Four in one of two mature aspen study plots at GSL, 1969 (Flack 1976).

ESA(47): 30 Apr 1955, Av 10 May. **LFD**(24): 8 Oct 1999, 29 Sep 2000 and 28 Sep 2001 at Ketchen (DF); 26 Sep 1943 at Yorkton (CSH); 19 Sep 1954 at GSL (WA), Av 2 Sep.

Breeding: Nest with 2 eggs on 19 Jul 1942; another with 3 eggs and 1 cowbird egg on 18 Jul 1946 (CSH). Early nest, 1 egg and 1 cowbird egg, 26 May 1957. Of eight nests, 1964-1967, five had 1 cowbird egg each (WA).

PNRS: at Melville, seven nests with 1, 1, 2, 3, 3, and 4 eggs and one with two young; five of these nests had a single cowbird egg (GAA,VS); at Saltcoats, 12 nests with 1, 3, 3, 3, 3, 3, 4, 4, 4, 4, and 4 eggs, and one with four young, the latter on 12 Jun 1958; one nest had 1 cowbird egg (BH); at Kamsack, 4 eggs on 23 May and four young on 27 May (JHJ); at Stornoway, 3, 3, 4, 4, and 5 eggs, and two young (SZ,RD); along Sorefoot Creek near Beaver Dale, 2 eggs with 3 cowbird eggs (Doug Graham).

BBS: overall mean, 43.8. Theodore, mean 43.4; increasing (1970s, 38.0; 1980s, 36.3; 1990s, 58.7; n=33).

Banded 64 (CSH) + pre-1955, 2 (CSH 1, Wotherspoon 1).

Return, 3 (all at one year, CSH).

Langenburg: Common, 1990 (TH,BL).

DUCK MOUNTAIN: SNHS 1956. Seen in 1974, 1976 (JAS). Fairly common; recorded at west boundary, Bear Head Lake, Kamsack subdivision, Spruce Island, and Little Boggy Creek (Hatch and Weidl 1981). SNHS 1989.

Point counts: 1 individual at 1 stop (ARS).

VESPER SPARROW *Pooecetes gramineus*

Abundant summer resident. Commonest native sparrow at GSL. Breeds.

Five eggs near Melville, 10 Jun 1920 (H.C. Grose, SMNH). One egg collected north of Saltcoats (NJ,PMA).

ESA(48): 17 Apr 1946 (WN) and 19 Apr 1974 (WA), Av 29 Apr. **LFD**(32): 8 Oct 2001, Av 24 Sep.

Breeding: A few nests were noteworthy, since they contained no cowbird eggs when found. Near Yorkton, nests with 4 eggs on 23 May 1946, with 3 eggs on 7 Jul 1946, and with two young, 8 Jun 1943; also a nest with 5 eggs near Runnymede on 25 Jun 1991 (CSH). At GSL, only three of 11 nests, 1963-66, had cowbird eggs (WA). Late nest, two young, 25 Aug 1962.

The remainder showed parasitism, often extreme. Most nests at Sheho contained more cowbird eggs than sparrow eggs (WN). At GSL, 1 egg and 3 cowbird eggs, 9 Jun 1984, with hatching of the three cowbirds by 13 Jun (WA). Horseman's only nest at Saltcoats had 2 sparrow eggs and 1 cowbird egg on 26 May 1958 (BH). A Vesper Sparrow nest on the edge of the Assiniboine valley north of Marchwell, 1 Jun 1976, contained 2 cowbird eggs (DH).

PNRS: At Melville, two of four nests contained cowbird eggs: 3 sparrow and 2 cowbird eggs on 9 Jun 1959 (GAA), and 2 sparrow and 3 cowbird eggs on 12 Jun 1961 (VS). Nests without cowbird eggs were as follows: at Saltcoats, two young ready to fledge on 21 Jun 1970, with a presumably addled egg (JRJ); at Stornoway and Wroxton, 17 nests including one with three young and an unhatched egg on 28 May 1962 and four young the next day; other nests had four young on 16 Jun 1960, three young on 4 Jun 1961, five young on 4 Aug 1961, and 1, 2, 2, 3, 3, 3, 3, 3, 4, 4, 4, 4, and 5 eggs (RD,SZ); at McOuat Lake on 8 Jun 1981, 4 eggs (Doug Graham).

Also breeds at Togo, Langenburg (ARS, *Atlas*).

BBS: overall mean, 25.4. Theodore, mean 29.4 (1970s, 30.3; 1980s, 30.5; 1990s, 30.1; n=33).

CBC: 1 count, 1 bird. Saltcoats 1952.

Banded pre-1955, 22 (CSH 1, Blakeley 7, Wickencamp 11, Wotherspoon 3).

Langenburg: Abundant, 1990. Along roadsides and open areas, its dominant song could be heard at almost every stop beside pastures and farmland. It was more numerous by far than the Horned Lark (TH,BL).

DUCK MOUNTAIN: Fairly common in grassy areas near Little Boggy Creek in June, July, and August (Hatch and Weidl 1981).

LARK SPARROW *Chondestes grammacus*

Rare summer resident. Breeds.

Noted at Sheho only once, at a cattle feed trough on 14 and 15 May 1966 (Niven 1968). At GSL on 6, 20, and 23 Jun 1967 (Anaka 1967). It was next seen at GSL in 1992 and in six of the 10 years since (WA).

Photographed in Fairburn's yard at Ketchen, 11 May 2002 (DF).

ESA(7): 5 May 2001, Av 20 May.

Breeding: An adult carrying food near Preeceville on 12 Jun 1987 was presumptive evidence of breeding (D.R.M. Hatch). Nest, 5 eggs on 7 Jun 1992 in crown of alfalfa plant on road allowance, 2 miles north of junction between Highways 47 and 229, west of south end of GSL; only 3 eggs remained on 20 Jun 1992 (WA).

BBS: Theodore, 1 in 1999.

Langenburg: A pair in the Assiniboine River valley opposite Shellmouth, 31 May 1976 (DH). Surprisingly common on valley slopes; eight birds seen, at least five of which were singing males, 1990 (TH,BL).

DUCK MOUNTAIN: Rare; one seen in grassy area near Little Boggy Creek on 18 Jun 1981 (Hatch and Weidl) and again on 8 Sep (Weidl 1983).

LARK BUNTING *Calamospiza melanocorys*

Occasional visitor, just beyond the extreme northern limit of its range.

L.T. McKim had three records at Melville: 3 Jun 1934, 12 Jun 1935, and 28 May 1944. One at GSL, 2 Jun 1967 (Anaka 1967), and another on 14 May 1991 (WA,JA, Koes and Taylor 1991b).

SAVANNAH SPARROW *Passerculus sandwichensis*

Very common summer resident, particularly in moist, thick grass, often near marshes. Breeds.

ESA(42): 19 Apr 1976, Av 4 May. **LFD**(16): 27 Sep 1971, Av 24 Aug.

Breeding: Early nest, 4 eggs, 25 May 1958 (WA). Nest, 6 eggs, Crescent marsh, 22 Jun 1948 (CSH). Three nests at Horseshoe Lake, Jun 1975, including one with 5 eggs (DH), and one with 1 cowbird egg (DH). Nest with 1 cowbird egg, south shore Little Quill Lake, 27 May 1994 (DH).

PNRS: at Saltcoats, 2, 4, 4, and 5 eggs from 22 May to 21 Jun 1958 (BH); one young and four cowbirds at Melville, 5 Jul 1961 (VS); at GSL, five young on 11 Jun 1958 and 5 eggs on 5 Jun 1960 (WA); at Kuroki Lake, nest with two young and one cowbird young on 12 Jun 1981 (Doug Graham).

Also breeds at Margo (ARS, *Atlas*).

BBS: overall mean, 18.7. Theodore, mean 25.6 (1970s, 24.9; 1980s, 24.5; 1990s, 29.6; n=33).

Banded 6 (Smith at Little Quill Lake) + pre-1955, 1 (Wickencamp).

Langenburg: Fairly common, 1990 (TH,BL).

DUCK MOUNTAIN: Four seen on shore of Bear Head Lake on 10 Sep 1981 (Hatch and Weidl).

GRASSHOPPER SPARROW *Ammodramus savannarum*

Rare summer resident.

One probable sighting, singing on fence post along Highway 9, 4 miles south of Yorkton, 13 Jun 1954 (CSH).

BBS: Theodore, one in 1976 and one in 1977, mean 0.06.

BAIRD'S SPARROW *Ammodramus bairdii*

Common summer resident in pastures, including the Good Spirit Lake Community Pasture until 1988, but not since. Declining. Breeds.

Numerous in dry alkali bottoms, Shell River country, 22 June 1884. (Seton 1891). Nest, 8 June 1901, in a tuft of grass, a few inches above the ground; of grass, lined with hair (Walter Raine, in Reed 1904:255).

First identified, presumably nesting, on ungrazed upland meadows immediately north of Salt Lake near Sheho on 28 May 1944. Also seen and heard in 1945, 1946, 1948, 1949, and 1950, but not again until 1973 (Niven 1973, Houston 1973c), and 1977 (WN). At GSL, heard during 19 of 29 years since 1974 (WA).

At Upper Rousay Lake, one found dead, 17 Aug 1944 (CSH), and one singing, 13 Jun 1989 (CSH,MIH). Present and possibly breeding near Foam Lake and Preeceville (Smith 1996).

ESA: 28 May 1944 and 8 Jun 1945 at Sheho (Niven 1968).

Breeding: An immature, learning to fly, banded at Upper Rousay Lake, 24 Jul 1945.

BBS: overall mean, 1.6. Churchbridge, 2. Willowbrook, 16. Theodore, 11 years, including 13 in 1976 and 29 in 1988, but not recorded since 1988. This decline coincided with changes in agricultural practices.

Banded pre-1955, 1 (CSH; see above).

DUCK MOUNTAIN: Present in north fringe of park, 24 Jun 1991 (MIC,CSH,MIH).

LE CONTE'S SPARROW *Ammodramus leconteii*

Common summer resident locally in grass surrounding sloughs and marshes. Breeds.

Abundant, breeding on west slope of Duck Mountains, Jun 1884 (Seton 1891). Nest, Crescent Lake, 10 Jun 1901, nest built in a tuft of prairie grass 4 inches above the ground; made of grass, lined with finer [grass] and a few hairs (Walter Raine in Reed 1904:256). Three eggs, Crescent Lake, 23 Jun 1897, F. Baines, ROMZ #7419. Five eggs, Little Quill Lake, 3 Jun 1908 (Rev. C.J. Young, WFVZ).

Regularly found around Upper Rousay, York, Leech, and Horseshoe lakes, and Rokeby marsh.

An increase was evident at GSL in 1971 (WA, Houston 1971c). Five singing males at Horseshoe Lake, 4 Jun 1975 (DH). Twenty counted at Little Quill Lake on 10 Jul 1976 (DH). Two near GSL on 11 Jun 1977 (WCH,SL).

ESA(20): 25 Apr 1976, Av 11 May. **LFD**(8): 6 Sep 1998, Av 19 Aug.

Breeding: Near Sheho, one nest found on 26 Jun 1951 (Niven 1951), and two in Jun 1953 (WN). Took to nesting in long, tangled brome grass near Sheho in the late 1960s; the young of the first nest were barely able to get out of the way of the machinery in mid-July haying, and the second nest was usually destroyed during haying, the nest being found only after the hay was cut and raked (Niven 1968). Nesting occurred again in 1973 (WN, Serr 1975b).

BBS: overall mean, 2.1. Calder, 1. Willowbrook, 7. Theodore, mean 3.2 (1970s, 4.6; 1980s, 2.2; 1990s, 3.4; n=27).

Banded pre-1955, 9 (Wotherspoon).

Langenburg: one singing on 31 May 1976 (DH).

DUCK MOUNTAIN: SMNH 1926: study skin female, #1905. Seen in 1975, 1976 (JAS). Uncommon, heard in sedge meadows near Bear Head Lake on 1 and 25 Jun, and at Roy Island on 16 Jul 1981 (Hatch and Weidl). Present in north fringe of park and near Pelly, 24 Jun 1991 (MIC,CSH,MIH).

NELSON'S SHARP-TAILED SPARROW *Ammodramus nelsoni*

Fairly common summer resident. Breeds.

Its numbers are grossly underestimated because it is inconspicuous and secretive, and its weak song travels only a short distance. Can be heard near the margins of Upper and Lower Rousay, Leech, and

Horseshoe lakes, and Rokeby marsh. One near south shore of Little Quill Lake on 1 Jun 1990 (DH).

ESA: 5 May 1945 at Ravine Reserve (A. McPherson). Recorded at GSL, 11-12 Jun 1977 (MB), and seen there in marsh-meadow areas south of Anakas' yard and at Horseshoe Lake in June and July of 14 years (WA).

Breeding: Two family groups at Rokeby marsh, 19 Jul 1946; one immature just able to fly; adult collected (Farley Mowat).

BBS: overall mean, 0.4. Willowbrook, 2. Theodore, 9 years, including 10 in 1971. Surprisingly, their location is not consistent; they have been heard over the years at 19 of the 50 different stops, but never at the stop of any previous year.

Banded 3 (Smith at Little Quill Lake) + pre-1955, 1 (Wotherspoon).

Langenburg: two heard on 31 May 1976 (DH).

DUCK MOUNTAIN: Rare; one heard in bulrushes at Ranger Bay on 25 Jun 1981 (Hatch and Weidl). One heard each evening in marsh behind cabin at Kamsack Beach during week of 5-11 Aug 1990 (RK).

FOX SPARROW *Passerella iliaca*

Fairly common transient some years, but numbers are declining.

Quite common, evidently breeding, Duck Mountains [near present Runnymede], 19 Jun 1884 (Seton 1891).

Once was common at Sheho in migration, in the 1920s and 1930s. Usually only one to three are seen, such as three on 30 Apr 1976 at Horseshoe Lake (DH). None was seen at Yorkton between 1944 and 1958 (CSH, Krause 1959a).

ESA(37): 6 Apr 1989, Av 20 Apr. **LSD**(29): 26 May 1961, Av 2 May. **EFA**(12): 15 Sep 1957, Av 23 Sep. **LFD**(13): 3 Nov 2002 at Ketchen (DF) and 23 Oct 1979 at GSL (WA), Av 8 Oct.

CBC: 1 count, 1 bird. Kamsack 1993.

Banded 23 (CSH 2, these two provided at the time Yorkton's earliest [24 Sep] and latest [11 Oct] fall dates, both in 1958; WA 21).

SONG SPARROW *Melospiza melodia*

Abundant summer resident, but less common at Sheho. Breeds.

Breeding, west side of Duck Mountains, 1884 (Seton 1891). Five fresh eggs in side of haystack near Melville on 5 Jun 1920 (H.C. Grose, SMNH).

Decreased in numbers north of Sheho by the 1970s and 1980s as farming became more mechanized; none seen or heard in 1984 until 7 Jun (WN).

ESA(47): 3 Apr 1991, Av 17 Apr. **LFD**(32): 31 Oct 1983, Av 5 Oct.

Breeding: Early nest, 3 eggs, 17 May 1964 (WA). Nest, 4 eggs, 27 May 1946, and another nest, 3 eggs on 27 May 1948 (CSH). At GSL, nests without cowbird eggs: 6 eggs, 3 Jun 1958; 2 eggs, 4 Jun 1958; 5 eggs, 24 May 1987. The remainder contained cowbird eggs: 3 eggs and 1 cowbird egg, 28 Jun 1959; 4 eggs and 2 cowbird eggs on 10 Jun 1963; 3 eggs and 2 cowbird eggs, 17 May 1964; 4 eggs and 2 cowbird eggs on 10 and 19 Jun 1963; 2 eggs and 2 cowbird eggs on 24 May 1964; a nest with 2 sparrow eggs on 28 May 1967 had 5 sparrow eggs and 2 cowbird eggs seven days later (WA). Early hatching, two nests, with three and four young, respectively, 16 Jun 1946 (CSH).

PNRS: Non-parasitized nests contained: at Melville, 2, 5, and 5 Song Sparrow eggs; at Saltcoats, 4 and 5 eggs, and two nests with four young each; at Stornoway, 16 nests, with 1, 3, 3, 3, 4, 5, 5, 5, and 6 eggs, and three, three, four, four, four, four, and five young (SZ,RD). Either there was a regional difference at Stornoway, or Zazelenchuk and Derkatch failed to mention any cowbird egg that might have been present.

At Melville, 11 of 14 nests were parasitized, 1959-1961. One had 5 cowbird eggs and one had 1 cowbird egg, both without any Song Sparrow eggs remaining. The remaining combinations were: 4 cowbird and 2 sparrow eggs; 3 cowbird and 1 sparrow egg, twice; 2 cowbird and 5 sparrow eggs; 2 cowbird and 3 sparrow eggs; 1 cowbird and 5 sparrow eggs; 1 cowbird and 2 sparrow eggs, twice; and three cowbird young and one sparrow young (GAA).

At Saltcoats, three parasitized nests had 1 sparrow egg and 2 cowbird eggs, 4 eggs and 1 cowbird egg and five young and one cowbird young (BH); at GSL, a nest had only two sparrow young on 17 and 21 Jun 1958, no doubt because of the three cowbird nestmates; another had 5 eggs and 4 cowbird eggs on 7 Jun 1960 (JG).

BBS: overall mean, 24.1. Theodore, mean 29.9 (1970s, 31.7; 1980s, 24.2; 1990s, 32.1; n=33).

CBC: 1 count, 1 bird. GSL 1949.

Winter: A late arrival on 8 Nov 1959, the first since 14 Sep, attempted to winter at Joyce Gunn's feeding tray at GSL, but was found dead under the tray on 5 Jan 1960 after a night of -30° F.

Banded 16 (CSH 9, WA 7) + pre-1955, 32 (Blakeley 4, Wotherspoon 28).

Langenburg: Three singing males in valleys and near large marshes, 1990 (TH,BL).

DUCK MOUNTAIN: SMNH 1926. SMNH 1951: nest and 5 eggs on 1 Jun, #5505 (FWL). SNHS 1956. Seen in 1974 and 1977; nest with five young, 1975; nest with 4 eggs, 28 Jun 1976 (JAS,SMNH). Common throughout the park, including Green Lake, Batka Lake, Spruce and Roy islands, Pickerel Point, Selwood campground, and Little Boggy Creek (Hatch and Weidl 1981). SNHS 1989.

PNRS: Nest with 3 eggs and 1 cowbird egg, 27 Jun 1989 (HWRC).

Point counts: Rank #29, 6 individuals at 6 stops (ARS).

LINCOLN'S SPARROW *Melospiza lincolnii*

Common but inconspicuous transient. Breeds at Duck Mountain.

At Yorkton, early and late dates were determined when migrants entered banding traps.

ESA(27): 27 Apr 1954 and 1979, Av 6 May. **LSD**(9): 22 May 1967, Av 16 May. **EFA**(6): 28 Aug 1958, Av 7 Sep. **LFD**(7): 6 Oct 2000 at Ketchen (DF) and 5 Oct 1957 at GSL (WA), Av 25 Sep (WA).

An immature bird, barely able to fly, at Preeceville on 8 Jul 1988 offered suggestive evidence of breeding (R.R. Hooper).

Breeding: Duck Mountain.

Banded 70 (CSH 67, WA 3) + pre-1955, 12 (CSH 8, Wickencamp 4).

Recoveries, banded outside area, 2 (MO 1, ND 1; both indicated the speed of spring migration: banded at Independence, MO, 28 Apr 1983 by M.L. Meyers, found dead at Langenburg 14 days later; banded 12 May 1929 at Jamestown, ND, by M. Guest, found dead north of Foam Lake 12 days later).

DUCK MOUNTAIN: Uncommon; scolding pairs were heard near the boat lagoon on 4 Jun, at Little Boggy Creek on 18 Jun, and one was seen along road to Batka Lake on 11 Sep 1981 (Hatch and Weidl). Nest and young, 21 May 1988 (CP).

SWAMP SPARROW *Melospiza georgiana*

Uncommon transient. Breeds at Duck Mountain.

Territorial male at Loch Lomond near Ketchen, 2000 through 2002 (DF).

ESA(10): 24 Apr 1997, Av 5 May. **LSD**(3): 17 May 1983. **EFA:** 18

Sep 1942 and 22 Sep 1944 (IMP). **LFD**(5): 13 Oct 1986.

Breeding: DMPP.

Langenburg: One singing male in a scrubby, willow brush habitat along Cutarm Creek, 1990 (TH,BL).

DUCK MOUNTAIN: SMNH 1926: Breeding; nest, #1988, with 2 eggs on 3 Jun, #1987a; male, #1878. SMNH 1951: nest, #1779, 4 eggs on 4 Jun, #5504 (FWL). SNHS 1956. Nest, 4 eggs, near shore of west bay of Madge Lake, June 1973 (BL). Seen in 1977 (JAS). Heard and seen near Batka Lake, Bear Head Lake, east park boundary, and Spruce Island in Madge Lake; one heard at Little Boggy Creek, 17 Jul; one observed at Batka Lake, 9-11 Sep; a single near Bear Head Lake on 10 Sep 1981 (Hatch and Weidl, Weidl 1983). 1986 (JFR et al.). SNHS 1989. Present in north fringe of park and near Pelly, 24 Jun 1991 (MIC,CSH,MIH).

Point counts: Rank #43, 3 individuals at 3 stops (ARS).

WHITE-THROATED SPARROW *Zonotrichia albicollis*

Abundant transient and fairly common summer resident near the north end of the YDM area. Common at Duck Mountain. Breeds.

Large waves in spring migration can peak at 200, as on 11-17 May 1983 (WA, Harris 1983).

ESA(47): 23 Apr 2000 at Ketchen (DF), also 23 Apr 1954, 1989, 1994, 2002 at GSL (WA), Av 30 Apr. **LSD**(20): 31 May 1968, Av 22 May. **EFA**(36): 18 Aug 1984, Av 10 Sep. **LFD**(44): 12 Nov 1992, Av 11 Oct.

Summer: Present during summers of 1939 and 1940 at Yorkton (IMP). Singing male on territory through June and early July, 1947, north of Sheho (WN). At GSL, a territorial songster was present during the summers of 1955 through 1961 inclusive, and 1982, 1983, 1985 to 1989 inclusive, and in 11 sporadic years for a total of 25 years (WA).

Breeding: Nest with 3 eggs, on south edge of Yorkton in Jun 1961; the young disappeared on 21 Jun (Jeff and Ruth Smith).

BBS: overall mean, 0.3. Willowbrook, 1. Theodore, 11 years, but annually since 1994; highest, 3, 1998.

Langenburg: one in migration on 10 May 1976 (DH).

CBC: 1 count, 1 bird. Kamsack 1959.

Banded 349 (CSH 196, WA 153) + pre-1955, 56 (CSH 28, Blakeley 3, Wotherspoon 25).

Return, 1, banded 15 Sep 1957 and retrapped 28 Sep 1958 (WA).

DUCK MOUNTAIN: SMNH 1926. SMNH 1951. SNHS 1956. Seen in 1974, 1975, 1976, 1977 (JAS). Common throughout park. Pairs with one fledged young each were seen along Pelly Point Nature Trail on 23 Jun and the Red Squirrel Nature Trail on 15 Jul 1981 (Hatch and Weidl). SNHS 1989.

Point counts: Rank #5, 41 individuals at 37 stops (ARS).

Aspen count totals: OA 14, YA 6, MA 4 (TH).

HARRIS'S SPARROW *Zonotrichia querula*

Fairly common transient. In flocks, Fort Pelly, 1881 (Macoun in Seton 1891).

Large numbers passed through in the fall of 1942.

ESA(34): 30 Apr 2001 at Ketchen (DF), 1 May 1938 and 1946 at Yorkton (IMP), and 4 May 1964 at GSL (WA), Av 12 May. **LSD**(23): 19 Jun 2000 at Ketchen (DF) and 27 May 1974 at GSL (WA), Av 20 May. **EFA**(29): 10 Sep 1955 and 1986, Av 25 Sep. **LFD**(27): 8 Dec 1974 at Sheho (WN) and 18 Nov 1965 at GSL (WA), Av 17 Oct.

CBC: Rank #41. 7 counts, 13 birds. Highest: 3, Bangor 1999.

Winter: One present at Sheho, 8-11 Dec 1974 (WN, Serr 1975a).

Banded 159 (CSH 74, WA 85) + pre-1955, 3 (CSH)

Recoveries, banded in area, 1 [0.6%] (OK 1, banded at GSL on 30 Sep 1955, found dead at Lake Overholser, Bethany, Oklahoma, on 1 Mar 1957, WA).

WHITE-CROWNED SPARROW *Zonotrichia leucophrys*

Common transient most years.

Common in spring of 1938 and much less so in spring of 1939 (IMP). Migrant numbers decreased by 1960 and have remained low since (WA).

ESA(33): 28 Apr 1946 at GSL (WA) and 1999 and 2001 at Ketchen (DF), Av 7 May. **LSD**(20): 26 May 2002 at Ketchen (DF), and 22 May 1967 and 1994 at GSL (WA), Av 18 May.

EFA(18): 8 Sep 2001 at Ketchen (DF) and 9 Sep 1954 at GSL (WA), Av 22 Sep. **LFD**(11): 28 Oct 2002, 19 Oct 1999 and 2001 at Ketchen (DF) and 17 Oct 1998 at GSL (WA), Av 9 Oct.

Banded 61 (CSH 20, WA 41 [including 40 Gambel's and 1 Eastern], + pre-1955, 11 (CSH 9, Wickencamp 2).

DUCK MOUNTAIN: Seen in 1977 (JAS).

DARK-EYED (SLATE-COLORED) JUNCO *Junco hyemalis hyemalis*

Abundant transient. Fairly common summer resident at Duck Mountain. Breeds.

Assiniboine River, July 1881, probably breeding (Macoun in Seton 1891). Duck Mountains, 1884 (Seton 1891).

Noted in smaller numbers at Sheho, 1960 through 1989 (WN).

ESA(46): 5 Mar 2000 at Ketchen (DF) and 11 Mar 1981 at GSL (WA), Av 29 Mar. **LSD**(48): 12 Jun 1954, Av 10 May. **EFA**(43): 30 Aug 1955, Av 15 Sep. **LFD**(46): 19 Dec 1990, Av 7 Nov.

Breeding: Nest, 1 egg and 3 cowbird eggs, 31 May 1953; removed the cowbird eggs on 5 Jun only to have another cowbird egg present on 7 Jun; on 21 Jun there was only one junco and one cowbird young in the nest (WA). Nest, GSL, 1971 (BL).

BBS: Theodore, 1 in 1999.

CBC: Rank #19. 32 counts, 441 birds. Highest: 132, Kamsack 1999.

Winter: at GSL, 1 Dec 1992, 2 Dec 1979, 19 Dec 1990, 23 Dec 1980 and 1981. One overwintered in Dallas Fairburn's yard at Ketchen, 2002-2003 (DF).

Banded 1,106 (CSH 829, WA 277) + pre-1955, 477 (CSH 95, Wickencamp 5, Wotherspoon 377).

Recoveries, banded in area, 4 [0.3%] (SK 2, AK 1, AR 1). Both distant recoveries were over a year later, one to Doddridge, AR, and the other to Soldotna, AK (Houston 1959c).

Recoveries, banded outside area, 2 (MB 1 [banded in Winnipeg by A. Haak on 27 Apr 1937, caught by cat at Norquay on 18 Apr 1938]; ND 1 [banded at Park River on 23 Mar 1938, found dead at Buchanan on 7 May 1943]).

DUCK MOUNTAIN: SMNH 1926, study skin male, #1906. SMNH 1951. SNHS 1956. Seen in 1974, 1975, 1976 (JAS). Fairly common, especially along the nature trails (Hatch and Weidl 1981). SNHS 1989. One nest found (ARS).

Point counts: Rank #16, 13 individuals at 12 stops (ARS).

"Oregon Junco" ***Junco hyemalis oreganus***

This subspecies is occasionally seen in migration. A typical high-plumage male was seen at 102 Logan Crescent on 29 Sep 1959 (MIH).

McCOWN'S LONGSPUR *Calcarius mccownii*

Vagrant.

One sighted just west of Melville, 25 May 1935 (LTMcK).

LAPLAND LONGSPUR *Calcarius lapponicus*

Abundant transient, often in flocks of untold thousands.

Very common on the plain at Fort Pelly, September 1881 (Macoun in Seton 1891).

For a few days in early spring and late fall, this may be numerically the commonest bird species in the YDM area.

ESA(12): 22 Mar 1968 at Sheho (WN), 29 Mar 1940 at Yorkton (IMP), and 7 Apr 1968 and 1984 at GSL (WA), Av 4 May. **LSD**(7): 29 May 1957, Av 23 May. **EFA**: 10 Sep 1960 (WA) and 21 Sep 1944 (IMP). **LFD**: 9 Oct 1938 (IMP).

Banded pre-1955, 4 (Wickencamp 3, Wotherspoon 1).

DUCK MOUNTAIN: Uncommon transient; a few in a flock of Snow Buntings over Ranger Bay on 5 Nov 1981 (Hatch and Weidl).

SMITH'S LONGSPUR *Calcarius pictus*

Unpredictable transient, probably declining.

Not seen at all most years near Yorkton. Seen more commonly at Sheho between 1944 and 1955, when it was recorded every year except for 1953 (in six of these years both spring and fall), though not seen in any year before or since! (Niven 1973).

North of Sheho: in 1944, flocks of 200 to 300 were noted from 24 Sep to 14 Oct, perhaps the largest and most prolonged concentration in

> Winter: On 27 Oct 1987, Joyce Anaka watched a flock of approximately 100 Snow Buntings over the open water of GSL. "There was an offshore wind creating a slightly wavy surface, and the birds were very low over the water, apparently feeding on insect life and behaving very much like swallows. They dipped down to the surface repeatedly and one actually dipped into the water momentarily and resumed flight. The flock worked farther on, staying about 150 m from shore and gradually moving east." (Harris 1987).

Saskatchewan's history; in 1946, a flock of 50 to 100 birds arrived on 9 Sep and stayed through the end of the month (WN). In 1949, 200 to 300 passed through Sheho from 11 Apr through 1 May 1949 (WN, Gammell and Gammell 1949b); in the fall of 1950, Sheho was the only locality in the Northern Great Plains reporting this species (Gammell and Gammell 1951).

From one to five were present at GSL from 15 to 28 Sep 1955 (WA).

ESA: 27 Mar 1946 at Sheho (WN). A flock of 35 or 40 just north of Melville on 1 May 1959 (GAA). **LSD:** 4 May 1945 at Yorkton (CSH) and 9 May 1957 at GSL (WA). **EFA:** 1 Sep 1945 (WN, Priestly 1945c).

Langenburg: A flock of between 20 and 30 was sighted on overgrazed prairie about 10 miles north of Langenburg on 6 Oct 1975 (DH).

DUCK MOUNTAIN: "A company" seen at west end of park, 20 Sep 1945 (JDS).

CHESTNUT-COLLARED LONGSPUR *Calcarius ornatus*

Rare summer resident on larger tracts of remaining grassland, particularly near lakes, but numbers have decreased drastically since 1989. Breeds.

At Sheho, at the northern edge of their range, Chestnut-collareds were first seen by Niven on 31 May 1942. There were two small nesting colonies on grassy upland meadows, one immediately north of Salt Lake and the other just east of Silver Lake, with 10 to 20 pairs in each colony, in 1932, 1944, 1945, and 1949. One pair was presumed to be nesting near Salt Lake on 6 Jul 1971 (Niven 1968, 1973). In 1943-1948, there were several pairs on territory each year near Upper Rousay Lake. In 1989, seven were singing over pastures near Crescent Lake (CSH,MIH).

ESA: 10 May 1945 (WN), Av at Sheho 16 May.

Breeding: Small nesting colony present at Upper Rousay Lake in 1944 (CSH). Nest, 3 eggs, 6 Jun 1973, in pasture beside Lower Rousay Lake (AS).

BBS: no sightings!

SNOW BUNTING *Plectrophenax nivalis*

Common to abundant in fall and winter.

Somehow they appear near roads whenever a snowfall is beginning. At Sheho, flocks decreased in size after the 1960s; none seen during winter of 1980-81 (WN). A large late flock of 200, west of Saltcoats on

10 May 1976 (DH).

EFA(46): 28 Sep 1942 at Yorkton (IMP) and 1 Oct 1961 at GSL (WA), Av 19 Oct. **LSD**(46): 18 May 1956 and 1967 at GSL (WA) and 19 May 1940 at Yorkton (IMP), Av 20 Apr.

CBC: Rank #2. 169 counts, 28,135 birds. Highest: 2,428, Saltcoats 1954. The Yorkton total of 520 individuals in 1952 was the top number for North America that year.

Banded 33 (WA) + pre-1955, 1 (Wotherspoon).

DUCK MOUNTAIN: SMNH 1926. Fairly common in winter; a flock of 25 over Ranger Bay on 4 Nov 1981 (Hatch and Weidl).

NORTHERN CARDINAL *Cardinalis cardinalis*

Occasional visitor.

A pair seen by Mrs P. Fraser and one by P. Fraser in the Assiniboine valley, east of Calder, just inside Saskatchewan in May 1895 (Fraser 1961, published 66 years after the observation!).

One observed at close range at the edge of Preeceville, in summer of 1954 (LP). One visited Harry and June Crossley's feeder, 225 Fifth Avenue South, Yorkton, from 4 to 23 Apr 1985. One was seen at Pelly in January and February 1990 (C. Dobby in Smith 1996).

CBC: 1 count, 2 birds. Kamsack 1997.

ROSE-BREASTED GROSBEAK *Pheucticus ludovicianus*

Common to fairly common summer resident in remaining heavy stands of aspen, especially near lakes or creeks. Breeds.

Frank Baines found a nest near Crescent Lake about 1918.

On his farm 5 miles east of Saltcoats, Robert Barnhart had sightings only on 15 May 1946, 12 Jun 1949 and 14 May 1953. Six in one of two mature aspen study plots at GSL, 1969 (Flack 1976). Single males singing near Runnymede on 10 Jun 1989, at Crescent Lake on 20 Jun, in coulee leading down to Lake of the Prairies on 21 Jun, and one at York Lake on 22 Jun (CSH,MIH).

ESA(46): 10 May 1953 and 1983, Av 17 May. **LFD**(31): 9 Sep 2002 at Ketchen (DF) and 6 Sep 1988 at GSL (WA), Av 24 Aug.

Breeding: A flimsy nest, where one could see the 3 eggs by looking up through it from below, 7 feet up in a clump of willows, at York Lake on 14 Jun 1953, had two young and 1 egg two days later (PPa, Houston

1953b). Present on territory at GSL; four young in juvenile plumage in yard from 3 to 8 Aug 1991 may have been raised locally (WA). A nest with 3 eggs on 15 Jun 2002, 7 feet up in 9-foot spruce, 3 miles east of Parkerview, had two young and an unhatched egg on 23 Jun (AS).

BBS: overall mean, 1.5. Calder, 3. Willowbrook, 17. Theodore, mean 1.8; increasing (1970s, 0.9; 1980s, 1.5; 1990s, 3.5; n=22).

Banded 2 (CSH) + pre-1955 2 (Wotherspoon).

Langenburg: Seven singing males and one pair seen in coulees, 1990 (TH,BL).

DUCK MOUNTAIN: SMNH 1926, group, #1869 including 4 eggs; study skin male, #1907; female, #1908. SMNH 1951. SNHS 1956. Seen in 1975, 1976, 1977 (JAS). Fairly common in deciduous woodlands, particularly along east and west boundary roads and along Pelly Point Nature Trail (Hatch and Weidl 1981). SNHS 1989.

Point counts: Rank #37, 4 individuals at 4 stops (ARS).

Aspen count totals: OA 6, YA 4, MA 3 (TH).

BLACK-HEADED GROSBEAK *Pheucticus melanocephalus*

Vagrant.

One male sighted at GSL on 12 May 1984 (WA,JA, Gollop 1984).

LAZULI BUNTING *Passerina amoena*

Vagrant.

One seen at GSL on 1 Jun 1996 (WA, Koes and Taylor 1996).

INDIGO BUNTING *Passerina cyanea*

Occasional visitor, but seen during three of four recent years at Duck Mountain.

At GSL, two sightings: by W.A. Tripp on 27 Apr 1941 (Houston 1949) and a singing male on 11 Jun 1977, joined by a female the next day (WCH,SL, Moray Lewis, B. Godwin, Kathy Meeres, D.W. Robinson; Jensen and Weidl 1977; Serr 1977c). Seen only once at Sheho, in 1984 (WN), but unaccountably there is no date recorded in Niven's notebook. One at GSL, 25 May 1995 (WA,JA, Koes and Taylor 1995b). One stopped for a few minutes in Fairburn's yard at Ketchen, 16 May 2001 (DF).

DUCK MOUNTAIN: Rare. A male was heard and seen near entrance road to Jubilee cottage subdivision, 4 Jun, and two males were seen along trail near Little Boggy Creek, 16 Jun 1981 (Hatch and Weidl, Weidl 1983). Singing males near ski hill area, 18 Jun 1988 (RK,CP,RE,TH); 18 Jun 2000 (PSa,DV,BL, Bain and Shanahan 2000b); and 4 Jun 2001 (Bain 2001).

A hybrid male cross with a Lazuli Bunting, singing more like a Lazuli, was carefully described near the ski hill, 10 Jul 1990 (TH).

DICKCISSEL *Spiza americana*

Vagrant.

One flitted about inside an open garage at Melville on 4 Sep 1942 (LTMcK in Houston 1949).

BOBOLINK *Dolichonyx oryzivorus*

Common summer resident in damp meadows. Breeds.

From 1883 through 1930, it was more common, with nests found, but it decreased noticeably in drought years (Frank Baines).

Numerous at Bangor, 1912 through at least the late 1920s (Thompson 1953). At Sheho, fairly common and nesting in damp meadows in the wet years of the 1920s and seen yearly until 1942; several pairs nested in fields in 1976 (WN). Definite increase noted near Yorkton with the return of wet times in 1942. No records at Sheho from 17 Jun 1942 through 1948, but in 1956 they were common again and stayed until 7 Aug (WN, Gammell 1956b).

Up to three males in damp grass on east side of Foam Lake marsh, 28 May to 24 Jun 1974 (AS). In the dry year of 1989, Bobolinks were sparingly distributed in wet meadows around the Rousay, Leech, and Crescent lakes and Crescent marsh (CSH,MIH). Resident in grassy area

> A male bobolink, with a nest in the hayfield over the fence, took no notice of the many available fence posts. For one entire nesting season he chose for his strutting and singing perch the almost horizontal wooden handle of the yard well pump. This earned him the nickname of "Pump Handle Pete" (WA).

on Jowsey farm, 6 miles north of Saltcoats, in 1987 and 1988 (JRJ,SJ).

ESA(39): 13 May 1988 and 1989, Av 21 May. **LFD**(7): 7 Aug 1965, Av 28 Jul.

Breeding: At GSL, early nest, 5 eggs, 6 Jun 1987. A nest on the ground 1 mile east of Melville contained 1 egg and 2 cowbird eggs on 8 Jun 1957; by 12 Jun it had 4 bobolink eggs and 2 cowbird eggs (GAA). Pairs on territory and believed nesting near Hyas on 23 May 1986 (R. Bellows).

PNRS: At GSL, nest with 5 eggs on 15 Jun 1958 had four young on 20 and 28 Jun (JG); a nest with 1 egg and three young on 14 Jul 1959 was predated by 20 Jul (WA).

Also breeds at Togo (ARS, *Atlas*).

BBS: overall mean, 5.0. Theodore, mean 6.3; possibly decreasing (1970s, 7.5; 1980s, 7.2; 1990s, 5.4; n=30).

Langenburg: Three males singing together in a moist meadow, 1990 (TH,BL).

DUCK MOUNTAIN: Rare in open areas immediately outside the park near Kamsack airport on 3 Jun 1981 (Hatch and Weidl).

RED-WINGED BLACKBIRD *Agelaius phoeniceus*

Abundant summer resident, chiefly in and near sloughs and lakes, but also in deserted farmsteads. Breeds.

Four eggs, Crescent Lake, 5 Jun 1900, F Baines for W Raine, WFVZ. One egg collected north of Saltcoats (NJ,PMA).

ESA(48): 24 Mar 1945 at Saltcoats and Yorkton (RB,CSH) and 29 March 1981 at GSL (WA), Av 12 Apr. **LFD**(12): 21 Oct 1989, Av 30 Sep.

Breeding: Early nest, 4 eggs, 27 May 1973 at Lower Rousay Lake (AS), and 5 eggs, 1 Jun 1946 and 1948. Between 1 Jun and 1 Jul 1946, 53 nests were recorded at Upper and Lower Rousay lakes, with earliest hatching, 12 Jun (Houston 1946b).

At GSL in 1955, nine nests on 29 May, five with 4 eggs, one with 2 eggs, one with 1 egg, two with no eggs; on 6 Jun there were three more nests and on 26 Jun, a nest with 3 eggs and 1 cowbird egg. In 1958, there were 38 nests with 94 eggs on 1 Jun, and on a return visit on 8 Jun there were 59 nests with 194 eggs and 31 young. A nest with 3 eggs and 1 cowbird egg, 26 Jun 1955 (WA).

Oology: one egg near Canora, 1965-66 (H. Cymbalisty, SMNH).

In June 2002 a heavy infestation of Forest Tent Caterpillars stripped the Anaka yard of most of its foliage. Caterpillars formed cocoons everywhere–on buildings, in the windsock, in spruce, in the crevices of poplars and willows, and in the leaves of the few species that they only partially devoured (Balsam Poplar, Red Osier Dogwood, Highbush Cranberry, and some willow species). On 3 July, male Redwings moved into the yard and for nearly a week checked every leaf bundle and crevice they could reach. They did not eat the caterpillars, but as they pupated, they tore open the cocoons for the chrysalis inside. This action was imitated by an occasional chickadee, robin, and nuthatch (WA).

PNRS: at Saltcoats, 18 nests (BH,LM); at Stornoway, 25 nests, including a nest with 3 blackbird eggs and 2 cowbird eggs on 5 Jun 1963 (SZ,RD); at GSL, 141 nests in 1958, 81 in 1960, two in 1961 (WA,JG); at Kamsack, one nest (JHJ); at Melville 20 nests [VS, 2; GAA, 18], one of the latter with 1 blackbird egg and 1 cowbird egg on 30 May 1959.

BBS: overall mean, 115.9. Theodore, mean 99.0; decreasing (1970s, 136.4; 1980s, 91.0; 1990s, 77.8; n=33).

CBC: Rank #50. 3 counts, 6 birds. Highest: 4, Yorkton 1951, where the only other sighting in first 19 years was in 1954.

Banded 6 (CSH 1, WA 5) + pre-1955, 13 (CSH).

Recoveries, banded outside area, 1 (AR, banded east of Forrest City, 23 Feb 1962, found injured at Canora on 4 Jul 1962).

A male wintered near an open spring at Ravine Reserve, winter of 1940-41, and again in 1944-45; he appeared healthy on 22 and 31 Dec 1944, 27 Jan and through March 1945, when he began calling continuously from the top of a tall cottonwood.

Langenburg: Very common, 1990 (TH,BL).

DUCK MOUNTAIN: SMNH 1926. SMNH 1951. SNHS 1956. Seen in 1974, 1975, 1977; barely flying young, 1976 (JAS). Fairly common in shallow wetlands (Hatch and Weidl 1981). SNHS 1989.

Point counts: Rank #15, 15 individuals at 13 stops (ARS).

WESTERN MEADOWLARK *Sturnella neglecta*

Very common summer resident of pastures and fields, declining in numbers since about 1990. Breeds.

1 egg collected north of Saltcoats (NJ,PMA).

For most residents this is the true harbinger of spring, singing what has been rendered as "Oh, dear, my feet are cold." Decreased in numbers at Sheho from the early 1960s as pastures were overgrazed and provided less nesting cover.

ESA(48): 21 Mar 1945 at Saltcoats (RB), 1946 at Yorkton (CSH), and 25 Mar 1968 at GSL (WA), Av 7 Apr. **LFD**(36): 3 Nov 1956, Av 11 Oct.

Breeding: Nest, 4 eggs, 20 May 1946 (CSH). Nests found at Sheho (WN). Five eggs at GSL, 11 Jun 1965, with three young hatched the next day (WA).

Oology: 1 egg near Canora, 1965-66 (H. Cymbalisty, SMNH).

PNRS: at Saltcoats, nests with 6 eggs on 3 Jun 1958 and on 27 May 1959, and another with four young on Jun 20 1959 (BH); at Stornoway, 3 eggs on 9 Jul 1962; at Mikado, 4 eggs on 14 Jun and four young on 28 Jun 1966, in spite of the presence of 2 cowbird eggs (EK); at GSL, 2 eggs on 28 Jun 1960 (JG); near Buchanan, 1 egg on 13 May, 3 eggs on 15 May and 5 eggs on 17 May 1959 (WC); at Melville, two nests with 5 eggs each on 25 May and 10 Jun 1959 (GAA); at Duff, nests with five young each on 21 and 24 May 1968 (MM).

BBS: overall mean, 21.2. Theodore, mean 16.7; possibly decreasing (1970s, 19.7; 1980s, 19.8; 1990s, 11.8; n=33).

Winter: at Sheho, waste grain around strawstacks was a particular attraction in fall (Niven 1973). One bird attempted to winter on a neighbour's farm north of Sheho, late in 1942, until it disappeared about 1 Jan, possibly killed by a cat.

Banded pre-1955, 5 (CSH 2, Wotherspoon 3).

Langenburg: Fairly common, but not nearly as numerous as on the Regina plains, 1990 (TH,BL).

DUCK MOUNTAIN: one seen on a small island near Spruce Island on 18 Jun 1981 (Hatch and Weidl).

YELLOW-HEADED BLACKBIRD *Xanthocephalus xanthocephalus*

Abundant summer resident of larger marshes. Breeds.

Four eggs, 10 Jun 1911 near Foam Lake (H.C. Grose, SMNH). One egg collected north of Saltcoats (NJ,PMA).

Three times as common near Crescent Lake until the drought of the 1930s, and similarly decreased sharply at Sheho about 1933. At Yorkton, numbers increased with the wet seasons beginning in 1942.

ESA(46): 22 Apr 1986, Av 4 May. **LFD**(17): 10 Sep 1943 at Yorkton (CSH) and 5 Sep 1954 at GSL (WA), Av 25 Aug.

Breeding: Early nests, 4 eggs, 27 May 1973 at Lower Rousay Lake (AS), and 5 eggs, 29 May 1958 at Melville (GAA). Nest, 4 eggs, 1 Jun 1946, near Yorkton (CSH).

PNRS: at GSL, colony of 19 nests with 30 eggs and 42 young on 8 Jun 1958 (WA); north of Gerald, four nests on 24 May 1959 (A. Hruska); at Melville in 1959, 15 nests on 30 May and 6 Jun, and north of Melville on 28 May and 1 and 6 Jun, 97 nests, all with 1 to 4 eggs; at Saltcoats, two nests (BH); at Stornoway, 1960-63, 22 nests with 1, 1, 2, 2, 2, 3, 3, 3, 3, 3, 3, 4, 4, 4, 4, 4, 4, 4, 4, 4, 4, and 4 eggs (SZ,RD).

BBS: overall mean, 31.4. Theodore, mean 31.2; decreasing (1970s, 49.3; 1980s, 31.3; 1990s, 17.9; n=32).

Banded 1 (Smith at Little Quill) + pre-1955, 3 (CSH).

Recoveries, banded outside area, 1 (banded as an adult male at Kenmare, ND, by Ann Gammell, shot 2 Jun 1970 at Goodeve).

Langenburg: Abundant at any body of water with bulrush or cattails; several colonies of 30 to 50 birds. Much more numerous than the Red-winged Blackbird, 1990 (TH,BL).

DUCK MOUNTAIN: SMNH 1926. SMNH 1951. Uncommon in shallow wetlands, including the slough along the road to Bear Head Lake on 1 Jun 1981 (Hatch and Weidl). SNHS 1989.

RUSTY BLACKBIRD *Euphagus carolinus*

Uncommon spring transient, common in fall in most years through the 1970s. Greatly decreased since about 1974, now uncommon even in fall.

Very abundant on Swan River in September 1881 (Macoun in Seton 1891).

A flock of over 1,000 from 13-17 Apr 1958 (WA, Krause 1958), another of 200 at GSL, 15-30 Apr 1977 (WA, Serr 1977b), and 300 at Horseshoe Lake on 7 Oct 1993 (DH).

Perhaps one of the first local warnings of impending declines was William Niven's observation of their unaccountable absence—for the first time—at Sheho in the fall of 1974 (WN, Houston and Houston 1975). The last migrants were noted at Sheho on 9 Oct 1976 and 7 Nov 1979 (WN). Numbers in migration at GSL have dropped off dramatically since about 1980, in keeping with the general decline evident throughout most of North America (Greenberg and Droege 1999).

ESA(15): 3 Apr 1955, Av 16 Apr. **LSD**(6): 7 May 1955, Av 1 May. **EFA**(23): 11 Sep 1955, Av 3 Oct. **LFD**(23): 26 Nov 1966 (WA, Hatch 1967a), Av 5 Nov.

BBS: overall mean, 0.2. Churchbridge, an unexpected group of 8 in 1973 (JRJ).

CBC: Rank #47. 3 counts, 11 birds. Highest: 6, Yorkton 1999.

Banded 4 (WA) + pre-1955, 10 (Wotherspoon).

Recoveries, banded in area, 2 [14.3%] (SK 1, Wotherspoon bird found dead locally five days after 12 Nov 1933 banding at Hyas; SD 1, banded at GSL on 30 Oct 1953 and found dead at Hecla, SD, on 4 Feb 1957).

DUCK MOUNTAIN: Uncommon transient; two at Little Boggy Creek on 11 Sep 1981 (Hatch and Weidl). On 22 Apr 1989, 15 were feeding among the wrack of a wooded marsh (TH).

BREWER'S BLACKBIRD *Euphagus cyanocephalus*

Very common summer resident in farming areas. Breeds.

Half a dozen nests around a single small lake in Duck Mountains, June 1884 (Seton 1891). One egg collected north of Saltcoats (NJ,PMA).

For many years, nested in small colonies around almost every farm near Sheho, but less common since the 1950s (Niven 1973). An albino was noted by Mrs J. R. Foreman near Wroxton on 8 Aug 1942 among a flock of normally coloured Brewer's Blackbirds. A month later, on 10 Sep, Mrs J. Pearce of Tonkin saw the same or another albino closer to Yorkton, about 20 miles west of the first sighting.

ESA(27): 23 Mar 1945 (WN) and 28 Mar 1958, Av 27 Apr. **LFD**(8): 11 Nov 1942 at Yorkton (CSH) and 21 Sep 1953 at GSL (WA), Av 1 Sep.

Breeding: At GSL, colony of 12 to 15 pairs, three nests with eggs and young and some young already flying, 23 Jun 1962. Nest, 5 eggs, 8 Jun 1963, hatched by 16 Jun and fledged by 28 Jun. A nest on a road shoulder contained 1 egg and a rock twice as large as the egg, on 16 Jun 1991 (WA).

PNRS: at Mikado, 4 blackbird eggs and 2 cowbird eggs on 16 Jun 1966; nevertheless the four blackbird young all hatched by 24 Jun (EK); west of Canora, 1 egg on 31 May 1959, 4 eggs on 3 Jun, first hatched by 21 Jun and 2 more by 24 Jun (WC); at Stornoway, 3, 4, 5, and 5 eggs (SZ,RD); at Saltcoats, 2, 3, 5, 5, 6, 6, and 6 eggs, and three young (BH); at Duff, 2 eggs on 17 Jun and 4 on 28 Jun 1968, with three hatched on 12 Jul, while the next year 5 eggs on 15 Jun were deserted and cold two days later (MM); at Melville, 5 eggs on 16 Jun 1960 (VS); nest, 2 eggs, on ground, 27 May 1958 (GAA).

BBS: overall mean, 17.6. Theodore, mean 15.3; increasing (1970s, 9.8; 1980s, 17.4; 1990s, 16.8; n=30).

CBC: Rank #31. 8 counts, 36 birds. Highest: 20, Yorkton 1996. Seven were sighted during the 1954 CBC at Yorkton (Gammell et al. 1955).

Winter: 26 Dec 1973 and 18 Nov 1996 at GSL (WA).

Banded pre-1955, 22 (Briggs 7 [misidentified as nestling Rusty Blackbirds], Wickencamp 11, Wotherspoon 4).

Recoveries, banded in area, 5 [23.8%] (SK 4, SD 1, Wotherspoon's was shot the following year; three Wickencamp bands were found locally at Stenen; the Briggs bird was shot at Lake City, SD, the same fall).

Langenburg: Present, but much less numerous than the Brown-headed Cowbird, 1990 (TH,BL).

DUCK MOUNTAIN: Recorded by R.J. Long, July-August 1970. SNHS 1989 (CSH).

COMMON GRACKLE *Quiscalus quiscula*

Common summer resident in villages and cities. Breeds.

Three eggs collected north of Saltcoats (NJ,PMA).

For many years, nested in small colonies around most Sheho farms, but since the 1950s only a transient (Niven 1973).

ESA(31): 1 Apr 1945 at Yorkton (IMP) and 6 Apr 1963 at GSL (WA), Av 20 Apr. **LFD**(10): 26 Oct 1972, Av 8 Oct. A flock of 150 was seen at GSL on 9 Oct 1966 (WA, Hatch 1967a).

Breeding: Oology: one egg collected near Canora, 1965-66 (H. Cymbalisty, SMNH #11195).

PNRS: at Melville, 11 nests, two of which were destroyed by property owners after the first egg was laid, but the remainder had 1, 1, 3, 3, 4, 4, 4, and 5 eggs, and four young (GAA,VS); at Saltcoats, 9 nests with 3, 3,

4, 4, 5, 5, 6, and 6 eggs, and one with three young (BH); at Stornoway, 2 eggs on 7 Jun 1959 and a third egg in the same nest on 28 Jun (SZ).

BBS: overall mean, 4.0. Theodore, mean 5.1; decreasing (1970s, 8.7; 1980s, 3.9; 1990s, 2.0; n=28).

CBC: Rank #56. 4 counts, 4 birds. Kamsack 1991 and 1998, Yorkton 1996, Preeceville 2001.

Winter: One present in Christmas week of 1961 at Melville (GAA).

Banded 6 (CSH 5, WA 1) + pre-1955, 320 (CSH 313, Wotherspoon 7).

Recoveries, banded in area, 17 [5.2%] (SK 13, MB 1, LA 1, MN 1, MS 1). A Wotherspoon grackle was caught in a trap at Hyas the following year. Recoveries from Ebenezer, Springside, Bangor, Lestock, and Langenburg, Saskatchewan; Birtle, Manitoba; Euclid, Minnesota; Webb, Mississippi; and Baton Rouge, Louisiana (CSH, Houston 1968c).

Recoveries, banded outside area, 8 (SK 1, AR 3, LA 1, ND 1, TN 2).

Returns, 9, including a one-legged adult female, banded on its remaining leg on 4 Sep 1944, returned to parade around the City Hall lawn as a local favourite for four summers before it succumbed and was found dead on 14 Sep 1948. Two others were retrapped at Yorkton at four years, three at two years and three at one year (CSH).

DUCK MOUNTAIN: SMNH 1926: 5 eggs on 3 Jun. SMNH 1951. SNHS 1956. Seen in 1976, 1977; grown, barely flying young in 1975 (JAS). Uncommon in shallow wetlands and clearings around cottage areas and park core (Hatch and Weidl 1981).

BROWN-HEADED COWBIRD *Molothrus ater*

Abundant summer resident. Breeds.

Three eggs collected north of Saltcoats (NJ,PMA).

At Sheho, they increased after the early years of settlement, becoming an increasing menace to the nests of small birds (Niven 1973). Four and six in the two mature aspen study plots at GSL, 1969 (Flack 1976).

ESA(48): 25 Apr 1974 at Sheho (WN) and 30 Apr 1988 at Sheho (WN) and GSL (WA), Av 5 May. **LFD**(28): 10 Sep 1944 at Yorkton (CSH) and 1 Sep 1955 at GSL (WA), Av 22 Aug.

Breeding: Of the 51 species in Saskatchewan known to have been parasitized by the cowbird, only 19 have been documented in the YDM area to date. See accounts of Least Flycatcher, Warbling Vireo, Red-eyed

Vireo, Horned Lark, Veery, Yellow Warbler, Yellow-rumped Warbler, Connecticut Warbler, Ovenbird, Chipping Sparrow, Clay-colored Sparrow, Vesper Sparrow, Savannah Sparrow, Song Sparrow, Dark-eyed Junco, Bobolink, Red-winged Blackbird, Western Meadowlark, and Brewer's Blackbird.

PNRS: Anaka 14, Horseman 7, Anweiler 4, Schmidt 1, Jowsey 1, Gunn 1, Zazelenchuk 2, Kozak 2.

Also breeds at Margo, Canora, Yorkton, Langenburg (ARS, *Atlas*).

BBS: overall mean, 25.3. Theodore, mean 27.7; possibly decreasing (1970s, 34.3; 1980s, 30.6; 1990s, 21.2; n=32).

Banded pre-1955, 20 (Patrick).

Recoveries, banded outside area, 1 (SD, banded at Mellette by Dr J. F. Brenckle on 24 May 1938, band reported from Buchanan in Jul 1940).

Langenburg: Disturbingly abundant in all habitats: farmland, pastures, aspen bluffs, valleys, and beside sloughs, 1990 (TH,BL).

DUCK MOUNTAIN: SMNH 1926. SNHS 1956. Seen in 1974, 1975, 1976 (JAS). Common in open areas throughout park (Hatch and Weidl 1981). SNHS 1989.

Point counts: Rank #18, 12 individuals at 11 stops (ARS).

Aspen count totals: OA 7, YA 12, MA 1 (TH).

BULLOCK'S ORIOLE *Icterus bullockii*

Vagrant.

A pair spent much of the summer of 1940 around J.R. Foreman's home on Wallace Avenue, Yorkton. No nest was found. Another was seen 5 miles west of Saltcoats in 1950 (RB).

BALTIMORE ORIOLE *Icterus galbula*

Common summer resident. Breeds.

Upper Assiniboine, 1884 (Seton 1891).

Numbers declined near Sheho between 1950 and the 1970s, perhaps related to the decline in woodland habitat due to bulldozers. Ten in one of two mature aspen study plots at GSL, 1969 (Flack 1976).

A female oriole fed at the feeding tray of Mrs A. Kryski in Yorkton for three weeks, then moved half a city block to be banded on 17 Dec; it was not seen after 20 Dec (Houston 1954).

ESA(48): 9 May 1985, Av 17 May. **LFD**(29): 6 Sep 1939 (IMP) and

Nests are built annually in Anakas' yard at GSL; the oriole's favourite nesting material is the strips from disintegrating blue woven poly sheets; the two mostly blue nests from 2001 and 2002 are only inches apart, hanging over the driveway (WA). Used nests from the previous year become readily visible in the early spring before trees come into leaf (AS).

31 Aug 1967 (WA), Av 21 Aug.

Breeding: Nest, four young, Saltcoats, 27 Jun 1957 (BH). Nest, 15 feet above ground in aspen at GSL, 1 Jun 1969 (JADF). Female collecting nest material, 19-21 May 1989 (WA). Adult with two young in Fairburn's yard at Ketchen, on 10 Aug 2002 (DF).

PNRS: at Melville, 2 eggs and two young on 11 Jul 1960, 1 egg and two young on 24 Jun 1969, 3 eggs on 28 Jun 1969 (VS,GAA); at Stornoway, two young on 7 Jul 1961; at Kamsack, 6 eggs on 13 Jun 1961 (RD); at Duff, 5 eggs on 15 Jun 1969 and five young on 26 Jun but destroyed by 30 Jun (MM).

BBS: overall mean, 6.6. Theodore, mean 6.5 (1970s, 4.2; 1980s, 10.5; 1990s, 6.4; n=32).

Banded 2 (CSH) + pre-1955, 4 (CSH).

Return, 1, retrapped two years later (CSH).

Langenburg: only five singing males recorded, 1990 (TH,BL).

DUCK MOUNTAIN: SMNH 1926. SMNH 1951. SNHS 1956. Seen in 1974, 1976; immature in 1975 (JAS). Uncommon; heard along Bear Head Lake road, west boundary road, and at Green Lake (Hatch and Weidl 1981). One on 13-15 Jun 1990 (MB).

GRAY-CROWNED ROSY-FINCH *Leucosticte tephrocotis*

Vagrant.

One over-wintered at GSL from 25 Dec 1994 through 3 Mar 1995 (WA, Koes and Taylor 1995a); it fed on sunflower seeds but preferred canola seed. Two at bird feeder on farm 4 miles north of Togo, February 1974 (May 1974).

PINE GROSBEAK *Pinicola enucleator*

Common but unpredictable in winter.

Very common at Yorkton in 1939 and 1941 but scarce in winter of 1943-44 (IMP). Common from 18 to 28 Feb 1940 at Yorkton and from January and February through 8 Mar 1950 at Sheho (WN, Gammell and Gammell 1950a). On 1 Feb 1952, large flocks fed in unharvested flax fields, together with redpolls, near Sheho (WN, Gammell and Gammell 1952b). Until 1973, more regular in aspen bluffs at Sheho than either the Bohemian Waxwing or Evening Grosbeak. A flock on 30 Nov 1989 was the first at Sheho since 1980 (WN). At Bangor, Pine Grosbeaks preferred sunflower seeds to flax seed (Hilton 1986).

> At GSL, this species feeds on sunflower seeds but prefers canola. Pine Grosbeaks treat redpolls amicably except on the rare occasion when a grosbeak steps on a redpoll; on hearing the alarm call from the redpoll, the grosbeak steps off and feeding continues. Grosbeak-to-grosbeak conflict is settled with a short beak-to-beak vertical flight accompanied by hissing; feeding soon resumes. During the winter of 1994-95, we fed five bushels of canola and 220 pounds of sunflower seeds (WA).

EFA(46): 30 Sep 1942 at Yorkton (CSH) and 9 Oct 2000 at GSL (WA), Av 25 Oct. **LSD**(36): 24 Apr 1972, Av 18 Mar.

CBC: Rank #7. 156 counts, 3,915 birds. Highest: 291, Kamsack 2000.

DUCK MOUNTAIN: Uncommon in winter (Hatch and Weidl 1981).

PURPLE FINCH *Carpodacus purpureus*

Uncommon summer resident. Common at Duck Mountain. Breeds.

This "sparrow dipped in raspberry juice" is common in migration, but a few remain in summer in thick woods. In the spring of 1965, 50 males were present at one time at a feeder on a farm near Arran (Collard 1965). Four in one of two mature aspen study plots at GSL, 1969; unexpected, but some planted spruce were nearby (Flack 1976). One at GSL, 25 Jun 1988 (TH).

ESA(45): 28 Mar 1995, Av 18 Apr. **LFD**(31): 14 Nov 2001 at

Ketchen (DF) and 6 Nov 1995 at GSL (WA), Av 12 Oct.

Breeding: Adult pair feeding recently fledged young in Gunn yard at GSL on 19 Jun 1957. Two pairs singing territorially in town of Foam Lake, 1979 (WN). Present each summer, with young, in Fairburn's yard at Ketchen (DF).

BBS: Theodore, one in 1970 and another in 1971, mean 0.06.

CBC: Rank #35. 5 counts, 29 birds. Highest: 18, Kamsack 1991.

Winter: One at GSL, 16 Nov to 26 Dec 1993 (WA).

Banded 66 (CSH) + pre-1955, 125 (CSH 11, Wotherspoon 114).

Recoveries, banded in area, 3 [1.6%] (SK 3, all found freshly dead in Yorkton, one the month after banding and two in the subsequent spring; one hit a power line and one a picture window (Houston 1961, Morgotch 1962).

Recoveries, banded outside area, 3 (MN 1, NE 1, WV 1; all three banded in winter, and recovered dead during the subsequent northward migration at Canora, Yorkton, and Hyas, respectively).

DUCK MOUNTAIN: SMNH 1926. SMNH 1951. SNHS 1956. Common in mixed woods; individuals seen at Kamsack subdivision, Selwood campground, Pelly Bay, southeast corner of park, and Little Boggy Creek in June, and near Ranger Bay and Rundle Lake in July (Hatch and Weidl 1981). A total of 150 at feeders throughout the Kamsack beach subdivision on 22 Apr 1989 (TH, Gollop 1989a).

Point counts: Rank #43, 3 individuals at 3 stops (ARS).

HOUSE FINCH *Carpodacus mexicanus*

Rare year-round resident. No breeding records.

The first observation was just outside the southwest corner of the area at Abernethy on 3 Jun 1990 (Bittner 1995a). One appeared at a Preeceville feeder in Jun 1996 (LP). Six to eight were present at Regina Aye's feeder at Springside, 26-28 Sep 1997 (RK,BL, Bain and Holder 1997d). The first sightings within Yorkton were two birds, 31 Dec 1998 (WA, Bain and Shanahan 1999a).

ESA at Ketchen: 7 Apr 1999, 29 Mar 2000, 7 Apr 2001, 25 Mar 2002, 12 Apr 2003. **LFD** at Ketchen: 11 Nov 1998, 4 Nov 1999, 14 Nov 2000, 11 Nov 2001, 20 Oct 2002 (DF).

CBC: 1, due to an error. Yorkton's first and only House Finch on a CBC was on count day in 1998. It was seen by WA,JA, Julia Wiwchar and Ray Riesz in a backyard on Maple Avenue, but the compiler failed to include it in the submitted report! House Finches were also seen during count week in Melville in 1993.

RED CROSSBILL *Loxia curvirostra*

Occasional visitor at almost any time of year.

Small flocks seen around Yorkton in summer, 21-29 Jun 1951, with two specimens (SMNH 5437) found dead (Houston 1951). Present at GSL during only one year, August and 1-2 Oct 1960 (Gunn 1960), and never again. Another specimen was found by a pupil of Miss Pauline Summers, 24 Oct 1960. Two in Warren Hjertaas' yard in Yorkton on 19 May 1997 (DH).

CBC: Rank #33. 2 counts, 30 birds. Highest: 15, Kamsack 1986, Melville 1998.

DUCK MOUNTAIN: SMNH 1926. SMNH 1951. Twenty on 12 Jul 1994 (RK, Koes and Taylor 1994).

WHITE-WINGED CROSSBILL *Loxia leucoptera*

Occasional visitor, most often seen in winter.

One picked up dead, Yorkton, 3 Apr 1938. One caught in William Niven's barn, Sheho, winter of 1939. An adult female picked up dead in Yorkton, 7 Jun 1947. One found dead by Ron Coghill in 1947 or 1948 and four seen, 6 Feb 1949 (Summers 1949). One found injured by Paul Shordee at Ebenezer, 30 Oct 1954, was brought to CSH for identification.

Seen at GSL in 14 years and in every month but May and August, including one on 20 Mar 1959 (WA, Krause 1959b). Small flock feeding on sunflower heads in garden near Ituna, late Aug 1963 (Brennan 1965a).

EFA(10): 9 Sep 1996, Av 17 Oct. **LSD**(3): 25 May 2002 at Ketchen (DF) and 15 Apr 1988 at GSL (WA).

Summer: 28 Jun and 23 Jul 1987 (WA).

CBC: Rank #25. 12 counts, 187 birds. Highest: 70, Yorkton 1995.

Winter: Seen in 10 of 48 years (WA).

DUCK MOUNTAIN: Uncommon in fall and winter in coniferous woods; nine along Red Squirrel Nature Trail on 9 Sep, and six near Benito cottage subdivision, 5 Nov 1981 (Hatch and Weidl). Five on ski trail behind golf course, 27 Dec 1989 (TH). Fifteen on 23 Feb 1997 (DH).

COMMON REDPOLL *Carduelis flammea*

Common transient in late fall and early spring, and throughout the winter when snowfall amounts are not excessive.

On 1 Feb 1952, redpolls and Pine Grosbeaks appeared in large flocks, feeding on unharvested flax in fields near Sheho (WN, Gammell and Gammell 1952b).

EFA: 29 Sep 1997 at a Springside feeder (G. and R. Wapple, Bain and Holder 1997d) and 14 Oct 1977 and 1993 at GSL (WA), Av 25 Oct. **LSD**(43): 14 May 1939 (IMP) and 4 May 1956 (WA), Av 16 Apr.

CBC: Rank #4. 139 counts, 9,203 birds. Highest: 870, Yorkton 1978. The unprecedented province-wide invasion of redpolls in 1999-2000 extended throughout the Yorkton area, with above average sightings of 563 at Kamsack-Madge Lake, 322 at Yorkton, 226 at Bangor, and 186 at GSL, but unexceptional numbers of 30 each at Togo and Melville (Harris 2000; Bain and Shanahan 1999f).

Banded 280 (CSH 156, WA 124) + pre-1955, 9 (CSH 1, Wickencamp 2, Wotherspoon 6).

Recoveries, banded in area, 1 [0.3%] (SK 1, found dead locally on 24 May after late banding on 30 Apr, CSH).

Recoveries, banded outside area, 1 (ND, Bismarck, shot north of Foam Lake the next year).

DUCK MOUNTAIN: Common in winter (Hatch and Weidl 1981).

HOARY REDPOLL *Carduelis hornemanni*

Fairly common in some winters, absent in others.

About one percent of redpolls, on average, appear whitish, as though they had been romping in white flour.

Twenty Hoary Redpolls appeared in a single, pure flock at GSL on 1 Jan 1993—although basically redpolls were scarce that winter (WA, Koes and Taylor 1993b). Five Hoaries were present on 24 Feb 1997 with Common Redpolls (WA, Bain and Holder 1997b).

EFA(13): 6 Nov 1980, Av 6 Dec. **LSD**(17): 2 May 1997, Av 3 Apr.

CBC: Rank #22. 28 counts, 256 birds. Highest: 54, Kamsack 1989.

Banded 11 (Anaka 7, CSH 4 unequivocal Hoary Redpolls, plus 28 pale birds with immaculate white undertail coverts that, in retrospect, should have been registered as Hoaries [Houston 1963a]).

DUCK MOUNTAIN: Six on ski trail behind golf course, 27 Dec 1989 (TH).

PINE SISKIN *Carduelis pinus*

Fairly common year-round resident, increasing. Breeds.

These irregular little nomads appear opportunistically in confiding flocks, eating dandelion seeds on lawns, most often in July and August; some years they are common. They certainly increased as farming practices changed, as trees and shrubs were planted, and as bird feeding stations came into being.

Robert Barnhart saw none in the 1930s or 1940s, but a small flock appeared on his farm 5 miles east of Saltcoats on 20 Apr 1954. Not observed at Sheho until 14 Oct 1944; subsequently seen erratically in spring, summer, or fall, in most of the succeeding 20 years (Niven 1973). They were especially common in Yorkton on 3 Aug 1947, with 35 on the Houston lawn, and again during the first three weeks of Jul 1951 (Gammell and Gammell 1952a).

At GSL, there was only one record until 1975, hence none was banded; since then there has been a striking increase in numbers. Between 28 Apr and 11 Nov 1987, there were 40 sightings and in Jun-Jul 1990 and Jul-Aug 1991, siskins were present through much of the summer.

A partial albino with a largely white head, visited the feeder in Ketchen on 20 May 1999 (DF).

ESA(18): 21 Mar 1994, Av 2 May. **LSD**: 12 Jun 1990, Av 28 May. **EFA**: 11 Aug 1985 (WA). **LFD**: 21 Nov 1999 and 1 Dec 2000 at Ketchen (DF) and 20 Nov 1989 at GSL (WA).

Breeding: A pair was observed feeding young in mature spruce in the town of Theodore, during the BBS on 29 Jun 1986 and 3 Jul 1987 (FAS,WA).

Summer: seen on 4 Jul 1945 at Yorkton (CSH) and in 14 years at GSL (WA).

BBS: overall mean, 0.3. Theodore, 8 years; highest, 9, 2001.

CBC: Rank #17. 20 counts, 628 birds. Highest: 130, Kamsack 2001.

Winter: 4 to 10 Dec 1993, 22 Dec 1978, and 31 Dec 2000 at GSL (WA), and 7-8-9 Jan 2001 at Ketchen (DF).

Recoveries, banded outside area, 1 (NY, near Endicott on 11 Apr 1963, found dead at Stenen before 12 Jul).

DUCK MOUNTAIN: SMNH 1926. Common during most summers. SMNH 1951. SNHS 1956. Seen in 1974, 1975, 1977 (JAS). Common summer resident in mixed woods; 20 eating sunflower seeds at Pickerel Point Beach on 1 Oct 1981 (Hatch and Weidl). Four present at a feeder, 27 Dec 1989 (TH). One on 13-15 Jun 1990 (MB).

Point counts: Rank #37, 4 individuals at 2 stops (ARS).
Aspen count totals: YA 5, MA 4 (TH).

AMERICAN GOLDFINCH *Carduelis tristis*

Very common summer resident. Breeds.

Ten and two in the two mature aspen study plots at GSL, 1969 (Flack 1976). An adult female feeding a full-grown immature, able to fly, on 3 Oct 1956 (WA).

ESA(48): 18 Apr 2003 at Ketchen (DF) and 1 May 1994 at GSL (WA), Av 19 May. **LFD**(41): 10 Nov 2002 at Ketchen (DF) and 1 Nov 1996 at GSL (WA), Av 1 Oct.

Breeding: One deserted egg in a nest at Stornoway on 8 Jul 1962 (RD).

BBS: overall mean, 12.3. Theodore, mean 12.8; possibly increasing (1970s, 10.7; 1980s, 12.4; 1990s, 15.9; n=33).

CBC: 1 count, 1 bird. Yorkton 1999.

Banded 1 (CSH) + pre-1955, 24 (Bill and Lindsay Wotherspoon at Hyas used hemp seeds as bait; the goldfinches, after eating a marijuana equivalent, appeared to lose all fear and some could be picked up by hand. One lay on Lindsay Wotherspoon's hand for 25 minutes, then rolled on the ground and lay there another 30 minutes (Wotherspoon 1976).

Recoveries, banded outside area, 1 (SD, Brookings, on 17 Mar 1982, found dead at Invermay, 30 Sep 1982).

Langenburg: Very common, especially in pasture land with intermittent poplar bluffs and Caragana hedges, 1990 (TH,BL).

DUCK MOUNTAIN: SMNH 1926. SMNH 1951. SNHS 1956. Seen in 1974, 1975, 1976, 1977 (JAS). Common throughout park (Hatch and Weidl 1981). SNHS 1989.

Point counts: Rank #43, 3 individuals at 3 stops (ARS).
Aspen count total: YA 1 (TH).

EVENING GROSBEAK *Coccothraustes vespertinus*

Unpredictable in winter in parkland. Year-round resident at Duck Mountain. Breeds.

Common winter transient at Yorkton in some years, including 18 Feb through 14 Apr 1937 (IMP), and the winters of 1946-47 (CCS) and 1958-59 (CSH, Krause 1959b), and almost absent in others. None stopped over at Sheho until March 1939 and March 1943, when cultivated Manitoba Maples were large enough to bear seeds. Apart from the winter of 1970-71, when only a single individual was seen, small flocks occurred each winter through 1972 (Niven 1973).

In Yorkton, largest flock was 200 on 8 April 1950 (Shaw 1950). At Ituna, present at Brennan's farm from late November for three weeks, and then from late March through at least 28 April 1976 (Brennan 1976). At GSL, this is the dominant species at feeders, not allowing other species to feed whilst they are on the table.

EFA(44): 28 Aug 2001, Av 11 Oct. **LSD**(46): 23 May 1973, Av 29 Apr.

Breeding: Adults brought young to feeder at Ketchen on 28 Jul 1999 and 31 Jul 2000 (DF).

Summer: Recorded during 10 of 48 summers at GSL: recorded 15 times between 20 Jul and 23 Aug 1987; four times between 7 and 26 Jul 1988; three times between 19 Jun and 28 Aug 1990; 12 times between 21 Jul and 9 Aug 1991 (WA); 18 Jul 1999 (B. Korol, in Bain and Shanahan 1999d).

CBC: Rank #5. 109 counts, 8,108 birds. Highest: 685, Kamsack 1997.

Winter: The largest flocks in the area visit feeders at Togo, as shown by CBC numbers and by the consumption of 700 pounds of sunflower seeds at one feeder in Togo during the winter of 1964-65 (Krupp 1965). In the winter of 1972-73, farmers near Togo fed two tons of sunflower seeds to flocks of 70 in February that increased to 200 by 21 Mar (Houston and Shadick 1974).

Banded 19 (CSH 7, WA 12) + pre-1955, 9 (CSH).

Recoveries, banded outside area, 9 (SK 1, MB 2, ON 1, MI 3, MN 1, NY 1). These demonstrate the west-east and east-west migration of this species: St Vital, MB, to Churchbridge, where it was caught by a dog two winters later, and another to Theodore, where it was found dead one April later; from Thornloe in extreme eastern Ontario to Togo two winters later; from Marquette, MI, caught by cat at Canora two winters later; from Imlay City, MI, caught by cat at Gorlitz three Februarys later; from Midland, MI, killed by weather at Sheho two winters later; from Walker, MN, to Pelly two winters later; from Watertown, NY, to south of Pelly five Aprils later.

DUCK MOUNTAIN: At Kamsack, 10 in 1959 CBC. Uncommon year-round in mixed woods (Hatch and Weidl 1981). Four on 20 May 1989 (MDG,SJS). About 200 present at feeders at Kamsack Beach, 24

Mar 1989, including one group of 100 at one feeder in March, and a few in late May (TH). SNHS 1989. Small numbers seen almost daily through June and July, with larger flocks by mid- August. Young barely able to fly, photographed about 1991 (RK). Family with newly fledged young near ski hill, 18 Jun 2000 (PSa,DV,BL).

Point counts: Rank #43, 6 individuals at 6 stops (ARS). Aspen count totals: OA 1 , MA 1 (TH).

HOUSE SPARROW *Passer domesticus*

Abundant year-round resident in the vicinity of every human habitation. Breeds.

Arrived about 1901, the year a nest was found at Yorkton by Walter Raine (Houston 1978a). Three eggs collected north of Saltcoats (NJ,PMA).

> In Yorkton in the 1930s, my parents owned an electric refrigerator, so ice was delivered daily in summer to our neighbours' iceboxes but not to us. But milk was also brought by horse-drawn carts. The seeds and fauna in horse droppings were a favourite nourishment for the immigrant House Sparrows, whose numbers dropped as both farm machinery and city delivery systems were motorized (CSH).

An isabelline (cream-coloured) House Sparrow was seen in Melville in late June and early July 1933 (LTMcK, Lawrence 1933c).

Breeding: Nest, 6 eggs, 18 Jun 1959 (JG). Numerous nests of this weaver-finch have been found: in boxes put up for swallows and bluebirds, in barns, and beneath nests of Swainson's and Red-tailed hawks. Common along the nest box trail until each box was relocated to a minimum of 0.6 miles from nearest farmyard (WA).

BBS: overall mean, 58.5. Theodore, mean 69.6; decreasing (1970s, 115.2; 1980s, 57.9; 1990s, 35.1; n=33).

CBC: Rank #1. 209 counts, 29,704 birds. Highest: 1,156, Yorkton 1959.

Langenburg: Locally common in Churchbridge, Langenburg, and Marchwell, and at some farms, 1990 (TH,BL).

DUCK MOUNTAIN: SMNH 1951. SNHS 1956.

APPENDICES

Appendix 1 - Bird Species in the Yorkton-Duck Mountain area

Lower case is before about 1970 (i.e., formerly)
B - breeding
b- formerly bred
H - hypothetical for Saskatchewan
I - Introduced
O -occasional visitor
T - transient
V - vagrant
W - winter
X - present

	Melville S'coats Yorkton	Sheho	GSL Theodore	Wadena Little Quill Foam L	Norquay Hyas P'ville Ketchen	Langenb	DMPP	Other
Pacific Loon			V					
Common Loon	B	x	X	X	B		B	
Pied-billed Grebe	B	B	B	X	B	B	B	
Horned Grebe	B	B	B	X	B	X	X	
Red-necked Grebe	B	B	B	X	B	B	B	
Eared Grebe	B	X	B	X	B	B	X	
Western Grebe	X	X	B		O		X	
American White Pelican	X	T	X	B	X	X	X	
Double-crested Cormorant	B	T	X	X	X		X	
American Bittern	B	X	B	X	B	X	B	
Great Blue Heron	B	X	B	B	X	B	B	
Great Egret	O		O				O	
Snowy Egret			V	V				
Little Blue Heron	O		O					
Cattle Egret				V				
Green Heron	V							
Black-crowned Night-Heron	B	X	X	B		X		
Yellow-crowned Night-Heron	V							
White-faced Ibis				V				
Turkey Vulture	X	X	X		B	X	B	
Greater White-fronted Goose	T	T	T	T	T	T	T	
Snow Goose	T	T	T	T	T	T	T	
Ross's Goose	T		T	T	T	T	T	
Canada Goose	B	B	B	B	B	B	X	
Trumpeter Swan								Beaver Dale B?
Tundra Swan	T	T	T	T	T	T		
Wood Duck	X		X	X	X	X	X	
Gadwall	B	B	B	B	B	X	B	
American Wigeon	B	B	B	B	B	X	B	
American Black Duck	X	X	X				X	
Mallard	B	B	B	B	B	B	B	
Blue-winged Teal	B	B	B	B	B	B	B	
Cinnamon Teal	X			X				
Northern Shoveler	B	B	B	B	B	B	B	
Northern Pintail	B	B	B	B	B	B	B	
Green-winged Teal	B	B	B	B	B	B	B	
Canvasback	B	B	B	B	B	B	B	
Redhead	B	B	B	B	B	X	X	
Ring-necked Duck	B		T	X	B	X	B	
Greater Scaup	T		T				T	
Lesser Scaup	B	B	B	B	B	B	B	
Surf Scoter	T						T	
White-winged Scoter	B	x	B	X	T		B	
Black Scoter							V	
Long-tailed Duck				V	V			

Bufflehead	B		B	X	B	X	B	
Common Goldeneye	B	X	B	X	B	X	B	
Barrow's Goldeneye				V				
Hooded Merganser	T		T	T	T		T	
Common Merganser	T	T	B	T	T	T	B	
Red-breasted Merganser	T		T	T	O		X	
Ruddy Duck	B	B	B	B	B	X	X	
Osprey	T	T	X		B	X	X	
Bald Eagle	B	T	T	T	T	T	B	
Northern Harrier	B	B	B	B	B	X	X	
Sharp-shinned Hawk	T	T	T	T	T		X	
Cooper's Hawk	B	B	B	X	B	X	X	
Northern Goshawk	B	W	W	W	W		X	
Red-shouldered Hawk (H)								Bankend X?
Broad-winged Hawk	T		B		B		B	
Swainson's Hawk	B	X	b	B	X	B	X	
Red-tailed Hawk	B	B	B	B	B	B	X	
Ferruginous Hawk	b							
Rough-legged Hawk	T	T	T	T	T	T	T	
Golden Eagle	T	T	T	T			T	
American Kestrel	B	B	X	B	B	X	X	
Merlin	B	B	B	B	B	B	X	
Gyrfalcon	W		W					
Peregrine Falcon	T	T	T	T	T	T		
Prairie Falcon	T				T			
Chukar	Ib							
Gray Partridge	IB	IB	IB	IB	IB	X	X	
Ring-necked Pheasant	Ib		x					
Ruffed Grouse	B	B	B	B	B	X	B	
Spruce Grouse			x				X	
Willow Ptarmigan		w			w			
Sharp-tailed Grouse	B	B	B	B	B	B	X	
Greater Prairie-Chicken	b	b	x					
Wild Turkey							IB	
Yellow Rail	b	X	X				O	
Virginia Rail	X		B	B			X	
Sora	B	B	B		B	X	B	
American Coot	B	B	B	B	B	B	X	
Sandhill Crane	B	b	b		b		T	
Whooping Crane	b		T		T			
Black-bellied Plover	T	T	T		T	T	T	
American Golden-Plover	T	T	T		T			
Semipalmated Plover	T	T	T	T	T		T	
Piping Plover	T	T	b	T				
Killdeer	B	B	B	B	B	X	B	
American Avocet	B	X	B	B	X	X	X	
Greater Yellowlegs	T	T	T	T	T	T	T	
Lesser Yellowlegs	T	T	T	T	T		T	
Solitary Sandpiper	T	T	T		T		X	
Willet	B	X	B		X	X	X	
Spotted Sandpiper	B	X	B	B	B	X	B	
Upland Sandpiper	B	X	B		O			
Whimbrel	V							
Long-billed Curlew	V			V				
Hudsonian Godwit	T		O	T	T			
Marbled Godwit	B	B	B		X		O	
Ruddy Turnstone	T	T	T	T			X	

Surfbird (H)				V?				
Red Knot				T				
Sanderling	T		T	T	T		T	
Semipalmated Sandpiper	T	T	T	T	T		T	
Western Sandpiper				T				
Least Sandpiper	T	T	T	T	T		T	
White-rumped Sandpiper	T		T	T	T		T	
Baird's Sandpiper	T		T	T	T		T	
Pectoral Sandpiper	T	T	T	T			T	
Dunlin	T	T	T	T				
Stilt Sandpiper	T		T	T	T			
Buff-breasted Sandpiper		T	T	T				
Short-billed Dowitcher	T		T	T	T		T	
Long-billed Dowitcher	T		T	T	T			
Wilson's Snipe	B	X	B	T	X	X	X	
American Woodcock (H)					V?		V?	
Wilson's Phalarope	B	X	B	X	B	X	X	
Red-necked Phalarope	T		T	T	T		T	
Red Phalarope				V				
Franklin's Gull	X	X	X	B	X		X	
Bonaparte's Gull	X		X	X	T		B	
Ring-billed Gull	b	T	X	B	T	X	B	
California Gull	X	T	X	B	T		X	
Herring Gull	X		X	X	T	X	X	
Sabine's Gull							V	
Caspian Tern			X				X	
Common Tern	B	T	B	X	T	X	B	
Forster's Tern	X		X	X	T	B	X	
Black Tern	B	X	B		B	X	B	
Rock Dove	B	B	B	X	B	X	X	
Band-tailed Pigeon			V		V			
Mourning Dove	B	B	B	X	B	X	B	
Passenger Pigeon							x	
Black-billed Cuckoo	B	X	X		O		X	
Barn Owl					V			
Eastern Screech-Owl	X		X			X	X	
Great Horned Owl	B	B	B	B	B	B	B	
Snowy Owl	W	W	W	W	W	W	W	
Northern Hawk Owl	b	W	W	W	W			
Burrowing Owl	b		b		b			
Barred Owl							B	
Great Gray Owl	W		W		W		X	
Long-eared Owl	B	B	B	B	B	X	X	
Short-eared Owl	b	B	X	B	O		X	
Boreal Owl	W	W	W		W			
Northern Saw-whet Owl	X	X	X		X		X	
Common Nighthawk	B	X	X	X	X	X	X	
Whip-poor-will	x		x				X	
Chimney Swift	O					O		
Ruby-throated Hummingbird	B	B	B		B	X	B	
Anna's Hummingbird			V					
Rufous Hummingbird								Bangor V
Belted Kingfisher	X	b	X		B		B	
Lewis's Woodpecker	v							
Red-headed Woodpecker	X	X	X		O		?	
Red-bellied Woodpecker	V							
Williamson's Sapsucker	V							

Yellow-bellied Sapsucker	B	X	B		B	X	B	
Downy Woodpecker	B	B	B	X	B	X	X	
Hairy Woodpecker	X	B	B	X	B	B	X	
Three-toed Woodpecker	W		W				B	
Black-backed Woodpecker			W				W	
Northern Flicker	B	B	B	B	B	X	X	
Pileated Woodpecker	x		B	W	B	B	B	
Olive-sided Flycatcher	X		x		X		B	
Western Wood-Pewee	X	b	X		X	X	B	Parkerview B
Eastern Wood-Pewee			X				X	
Yellow-bellied Flycatcher							X	
Alder Flycatcher	X		X		X		B	
Least Flycatcher	B	B	B	B	B	B	B	
Eastern Phoebe	B	B	B		B	X	B	
Say's Phoebe		T						
Great Crested Flycatcher	X	B	B	X	B	X	X	
Western Kingbird	B	B	X	X	X		X	
Eastern Kingbird	B	B	B	X	B	X	X	
Loggerhead Shrike	B	X	X	X	O	X	X	
Northern Shrike	W	W	W		W			
Yellow-throated Vireo	x					X	X	
Blue-headed Vireo	X		B		X		B	
Warbling Vireo	B	X	B		X	X	X	
Philadelphia Vireo	X		X		X		b	
Red-eyed Vireo	B	X	B	B	B	X	B	
Gray Jay	W	W	W		bW		B?	
Steller's Jay	V							
Blue Jay	B	X	B	X	B		X	
Black-billed Magpie	B	B	B	B	B	X	X	
American Crow	B	B	B	B	B	X	B	
Common Raven	B	X	B	X	B	X	X	
Horned Lark	B	B	B	X	B	X	X	
Purple Martin	B	B	B	B	B	X	B	
Tree Swallow	B	B	B	B	B	B	B	
Northern Rough-winged Swallow			X					
Bank Swallow	B	B	B	B	B	B	B	
Cliff Swallow	B	B	B		B	B	B	
Barn Swallow	B	B	B	B	B	X	B	
Black-capped Chickadee	B		B	X	B	X	B?	
Boreal Chickadee	W		W		W		B	
Red-breasted Nuthatch	X	X	X	X	X	X	X	
White-breasted Nuthatch	B		B		B		B	
Brown Creeper	X	T	X				B	
House Wren	B	B	B	B	B	B	B	
Winter Wren							B?	
Sedge Wren	B	X	B?	B	B		B?	
Marsh Wren	B	X	B	X		X	X	
Golden-crowned Kinglet	X		X				X	
Ruby-crowned Kinglet	X		B		X		X	
Eastern Bluebird	B	X	B		B	B	B	
Mountain Bluebird	B	X	B	B	B	B	B	
Townsend's Solitaire	X		O					
Veery	B		B		X	X	B	
Gray-cheeked Thrush	T		T		T		T	
Swainson's Thrush	X		X		T		B	
Hermit Thrush	X	X	X		X		X	
Wood Thrush			V					

American Robin	B	B	B	B	B	B	X	
Varied Thrush	V							Jasmin V
Gray Catbird	B	B	B	X	B	X	X	
Brown Thrasher	B	X	B	X	O	X	X	
European Starling	B	B	B	B	B	B	X	
American Pipit	T	T	T					
Sprague's Pipit	b		X	X				
Bohemian Waxwing	W	W	W	W	W	W	W	
Cedar Waxwing	B	X	B	X	B	X	X	
Golden-winged Warbler	X						X	
Tennessee Warbler	X	T	X		X		X	
Orange-crowned Warbler	X		X		T		X	
Nashville Warbler	T		T		T		X	
Northern Parula	V						V	
Yellow Warbler	B	B	B	B	B	X	X	
Chestnut-sided Warbler	T	T	T				X	
Magnolia Warbler	T	T	T		T		X	
Cape May Warbler	T	T	T		T		T	
Black-throated Blue Warbler			T					
Yellow-rumped Warbler	T	T	B		X	T	X	
Black-throated Green Warbler	T	T	T		T		X	
Blackburnian Warbler			T		T		X	
Palm Warbler	T	T	T	T	T		T	
Bay-breasted Warbler	T	T	T		T		T	
Blackpoll Warbler	T	T	T		T		T	
Black-and-white Warbler	T	T	T		T	T	T	
American Redstart	X	T	B?		X	X	B	
Ovenbird	B	X	B		B	X	B	
Northern Waterthrush	T		T		X	X	X	
Connecticut Warbler	T		T				B	
Mourning Warbler	T	T	T		T	X	X	
Common Yellowthroat	B	X	B		X	X	B	
Wilson's Warbler	T	T	T		T		T	
Canada Warbler	T				T		X	
Yellow-breasted Chat			V					
Summer Tanager					V			
Scarlet Tanager	O						X	
Western Tanager	O						X	
Spotted Towhee	O		O					
Eastern Towhee			B?			X	X	
American Tree Sparrow	T	T	T	T	T	T		
Chipping Sparrow	B	B	B	X	B	X	B	
Clay-colored Sparrow	B	X	B	X	B	X	X	
Vesper Sparrow	B	B	B	X	B	X	X	
Lark Sparrow		X	B		O	X	X	
Lark Bunting	o		O					
Savannah Sparrow	B	B	B		B	X	X	
Grasshopper Sparrow	X		X					
Baird's Sparrow	B	X	X	X			X	
Le Conte's Sparrow	b	B	B		X	X	X	
Nelson's Sharp-tailed Sparrow	B		X	X		X	X	
Fox Sparrow	T	T	T		T		T	
Song Sparrow	B	X	B		X	X	B	
Lincoln's Sparrow	T		T		X		B	
Swamp Sparrow	T		T		X	X	B	
White-throated Sparrow	B	X	X	T	B	X	B?	
Harris' Sparrow	T	T	T	T	T	T		

White-crowned Sparrow	T		T	T	T	T	T	
Dark-eyed Junco	T	T	B	T	T	T	B	
McCown's Longspur	v							
Lapland Longspur	T	T	T	T	T	T	T	
Smith's Longspur	T	T	T			T	T	
Chestnut-collared Longspur	b	b						
Snow Bunting	W	W	W	W	W	W	W	
Northern Cardinal	O				O			Togo O
Rose-breasted Grosbeak	B	T	B		B	X	B	
Black-headed Grosbeak			V					
Lazuli Bunting			V					
Indigo Bunting		O	O		O		X	
Dickcissel	v							
Bobolink	B	b	B	X	B	X	X	
Red-winged Blackbird	B	B	B	B	B	X	B?	
Western Meadowlark	B	B	B	X	B	X	X	
Yellow-headed Blackbird	B	B	B	B	B	B	X	
Rusty Blackbird	T	T	T		T		T	
Brewer's Blackbird	B	B	B	B	B	X	X	
Common Grackle	B	X	X	X	B		B?	
Brown-headed Cowbird	B	B	B	B	B	B	X	
Bullocks Oriole	v							
Baltimore Oriole	B	B	B	X	B	X	X	
Gray-crowned Rosy-Finch			V					Togo W
Pine Grosbeak	W	W	W	W	W	W	W	
Purple Finch	X	B	B	X	B	X	X	
House Finch	X				X			
Red Crossbill	O		O				O	
White-winged Crossbill	X	W	X		X		X	
Common Redpoll	W	W	W	W	W	W	W	
Hoary Redpoll	W	W	W		W		W	
Pine Siskin	X	X	B	X	X	X	X	
American Goldfinch	X	X	B	X	B	X	X	Stornoway B
Evening Grosbeak	W	W	W	W	B	W	B?	
House Sparrow	B	B	B	B	B	X	X	

A challenge: We hope, over the years, readers will exert effort to fill in the blanks in the columns above, and to change a whole bunch of X's to B's .

Appendix 2 - Breeding Bird Surveys (53) in the Yorkton-Duck Mountain Area

	33 yr Theodore	9 yr Chrchbrdg	7 yr W'brook	3 yr Ituna	1989 Calder	Total individuals	Mean per 53 counts
Common Loon	6	1				7	0.1
Pied-billed Grebe	101	3	34	2		140	2.6
Horned Grebe	63	19	17	3		102	1.9
Red-necked Grebe	36		3		1	40	0.8
Eared Grebe	55	33				88	1.7
Western Grebe	5	2				7	0.1
American White Pelican	3					3	0.1
Double-crested Cormorant	23	2	3			28	0.5
Am. Bittern	80	46	47	3	2	178	3.4
Great Blue Heron	112	2	3		4	121	2.3
Black-crowned Night Heron	1					1	0.0
Canada Goose	337		242			579	10.9
Gadwall	42	6	39	1		88	1.7
American Wigeon	50	9	26	8	1	94	1.8
Mallard	1070	250	345	35	27	1727	32.6
Blue-winged Teal	198	59	82	34	4	377	7.1
Northern Shoveler	44	16	46	13		119	2.2
Northern Pintail	67	48	30	11		156	2.9
Green-winged Teal	19	9	15	2		45	0.8
Canvasback	76	13	34	20		143	2.7
Redhead	62		34			96	1.8
Ring-necked Duck	21		13	2		36	0.7
Lesser Scaup	460	7	97	21	5	590	11.1
White-winged Scoter	2					2	0.0
Bufflehead	35		2			37	0.7
Common Goldeneye	2					2	0.0
Ruddy Duck	112	1	64	1		178	3.4
Northern Harrier	48	18	16	4		86	1.6
Cooper's Hawk	11		2			13	0.2
Broad-winged Hawk	3					3	0.1
Swainson's Hawk	7	14	1	2		24	0.5
Red-tailed Hawk	148	18	52	11		229	4.3
American Kestrel	11	3				14	0.3
Merlin	13		4			17	0.3
Gray Partridge	7	6	3			16	0.3
Ring-necked Pheasant	1					1	0.0
Ruffed Grouse	1	3	4			8	0.2
Sharp-tailed Grouse	18	32	87			137	2.6
Virginia Rail	2					2	0.0
Sora	412	45	109		2	568	10.7
American Coot	625	81	96	33	6	841	15.9
Sandhill Crane	1					1	0.0
Killdeer	437	123	95	23	6	684	12.9
American Avocet	1		6			7	0.1
Greater Yellowlegs	1					1	0.0
Lesser Yellowlegs	2		3			5	0.1
Willet	27	12	15			54	1.0
Spotted Sandpiper	15	1	8		3	27	0.5
Upland Sandpiper	32	4	1			37	0.7
Marbled Godwit	30		4			34	0.6
Wilson's Snipe	174	34	32	7	11	258	4.9
Wilson's Phalarope	9	3	4			16	0.3
Franklin's Gull	152					152	2.9
Bonaparte's Gull	1					1	0.0
Ring-billed Gull	66	1	11	1		79	1.5
Common Tern			5	1		6	0.1

Black Tern	775	93	107	45	2	1022	19.3
Rock Dove	72	48	75	9		204	3.8
Mourning Dove	476	99	181	9	9	774	14.6
Black-billed Cuckoo	127	21	15	1		164	3.1
Great Horned Owl	13	21	11	1	2	48	0.9
Short-eared Owl		1		2		3	0.1
Northern Saw-whet Owl	1					1	0.0
Ruby-throated Hummingbird	1	1	1			3	0.1
Belted Kingfisher	4		1			5	0.1
Yellow-bellied Sapsucker	**51**	2	7	1		61	1.2
Downy Woodpecker	20	1	3	2		26	0.5
Hairy Woodpecker	26		1			27	0.5
Northern Flicker	166	51	20	19	2	258	4.9
Pileated Woodpecker	2					2	0.0
Western Wood-Pewee	7					7	0.1
Eastern Wood-Pewee	**4**					4	0.1
Alder Flycatcher	25	8	33	1	7	74	1.4
Least Flycatcher	537	13	174	1	10	735	13.9
Eastern Phoebe	44	4	7			55	1.0
Great Crested Flycatcher	126		26			152	2.9
Western Kingbird	31	3	3	1	1	39	0.7
Eastern Kingbird	156	64	70	18	6	314	5.9
Loggerhead Shrike	1	1	1	1		4	0.1
Warbling Vireo	**454**	37	132		9	632	11.9
Philadelphia Vireo	8	1				9	0.2
Red-eyed Vireo	317	3	21		8	349	6.6
Blue Jay	**13**		2			15	0.3
Black-billed Magpie	714	82	87	12	14	909	17.2
American Crow	2392	612	426	265	38	3733	70.4
Common Raven	45		3		1	49	0.9
Horned Lark	1826	542	355	84	12	2819	53.2
Purple Martin	82	15		1		98	1.8
Tree Swallow	528	96	37	30	5	696	13.1
N Rough-winged Swallow				1		1	0.0
Bank Swallow	**202**	103			4	309	5.8
Cliff Swallow	5701		329	1	10	6041	114.0
Barn Swallow	764	171	128	46	30	1139	21.5
Black-capped Chickadee	104	4	2	3		113	2.1
Red-breasted Nuthatch	2					2	0.0
House Wren	1292	158	280	17	20	1767	33.3
Sedge Wren	43	1				44	0.8
Marsh Wren	28		8			36	0.7
Ruby-crowned Kinglet	2	1		3		6	0.1
Eastern Bluebird	1					1	0.0
Mountain Bluebird	198	30	15	5		248	4.7
Veery	217	7	6		1	231	4.4
Swainson's Thrush	1					1	0.0
Hermit Thrush	3					3	0.1
American Robin	939	82	122	33	13	1189	22.4
Gray Catbird	166	62	70	5	4	307	5.8
Brown Thrasher	33	24	17	1	1	76	1.4
European Starling	1044	29	138	12		1223	23.1
Sprague's Pipit	61	2			1	64	1.2
Cedar Waxwing	187	8	15	6	5	221	4.2
Tennessee Warbler	11			2		13	0.2
Orange-crowned Warbler	7					7	**0.1**
Nashville Warbler	4					4	0.1
Yellow Warbler	439	59	132	30	19	679	12.8
Yellow-rumped Warbler	2					2	0.0
American Redstart	8					8	0.2

Ovenbird	4	1				5	0.1
Common Yellowthroat	124	10	15		1	150	2.8
Spotted Towhee	1					1	0.0
Eastern Towhee	2					2	0.0
Chipping Sparrow	180	11	6	4	4	205	3.9
Clay-colored Sparrow	1433	312	359	173	43	2320	43.8
Vesper Sparrow	969	115	238		23	1345	25.4
Lark Sparrow	1					1	0.0
Savannah Sparrow	846	39	88	7	9	989	18.7
Grasshopper Sparrow	**2**					2	0.0
Baird's Sparrow	68	2	16			86	1.6
Le Conte's Sparrow	105		7		1	113	2.1
Nelson's Sharp-tailed Sparr	21		2			23	0.4
Song Sparrow	986	102	115	70	6	1279	24.1
White-throated Sparrow	14		1			15	0.3
Slate-colored Junco	1					1	0.0
Rose-breasted Grosbeak	60		17		3	80	1.5
Bobolink	207	16	41		3	267	5.0
Red-winged Blackbird	3267	1554	801	457	64	6143	115.9
Western Meadowlark	552	351	135	78	5	1121	21.2
Yellow-headed Blackbird	1029	80	534	15	5	1663	31.4
Rusty Blackbird		8				8	0.2
Brewer's Blackbird	504	318	71	38	4	935	17.6
Common Grackle	167	25	12	5	5	214	4.0
Brown-headed Cowbird	914	119	279	8	21	1341	25.3
Baltimore Oriole	214	30	87	8	12	351	6.6
Purple Finch	2					2	0.0
Pine Siskin	18					18	0.3
American Goldfinch	421	82	118	13	16	650	12.3
House Sparrow	2297	508	116	176	4	3101	58.5
Total species for that route	**142**	**92**	**101**	**68**	**56**	**146**	

Mean for each species (last column) is total for that species divided by 53

Appendix 3 - Species banded in the Yorkton-Duck Mountain Area

174 species

Eight species in **boldface** were NOT banded in YDM area

	pre-1955 Houston	Patrick	Blakeley	Wickencamp	Wotherspoon	Bard Briggs Wilson	Post-1955 N of 51	S of 51	**Total Banded**	**Recoveries Banded in area** n of 51	s of 51	**Total Recoveries** Banded in area	Percent of banded	**Recoveries** Banded outside YDM area
Common Loon							1		1			0		
Pied-billed Grebe	2				1		1		4			0		
Horned Grebe	2						1		3			0		
Red-necked Grebe	2								2			0		
Western Grebe	1								1			0		
American White Pelican		38					157		195	14		14	7.2%	6
Double-crested Cormorant		60							60	4		4	6.7%	11
Am. Bittern	8					1	24		33	2		2	6.1%	
Great Blue Heron	8	11					95		114	7		7	6.1%	1
Black-crowned Night-Heron	88	68					24		180	6		6	3.3%	
Gr White-fronted Goose														10
Snow Goose														112
[Greater Snow Goose]														1
[Blue Goose]														44
Ross's Goose														45
[Small Canada Goose]														112
Canada Goose		(14)					4309	254	4563	1411	61	1472	32.3%	745
Tundra Swan														3
Wood Duck							161		161	3		3	1.9%	
Gadwall							820	8	828	82		82	9.9%	4
American Wigeon							1538	45	1583	179	5	184	11.6%	13
Mallard/Black Duck hybrid							66		66	9		9	13.6%	1
Am Black Duck							219		219	39		39	17.8%	9
Mallard		(68)	(1)	(5)			109691	257	109948	17874	42	17916	16.3%	1916
[Mottled Duck]														1
Blue-winged Teal		(21)					105295	576	105871	3625	27	3652	3.4%	233
Cinnamon Teal							3		3	1		1	33.3%	
Northern Shoveler		(19)					640	36	676	41		41	6.1%	5
Northern Pintail		(11)					12065	32	12097	881	2	883	7.3%	130
Green-winged Teal							2366	42	2408	116		116	4.8%	4
Canvasback		(5)					2158	6	2164	292	4	296	13.7%	33
Redhead		(6)					6869		6869	646		646	9.4%	29
Ring-necked Duck							109		109	14		14	12.8%	1
Greater Scaup							2		2			0		1
Lesser Scaup							598	13	611	49	4	53	8.7%	19
White-winged Scoter							16		16			0		
Bufflehead							82		82	12		12	14.6%	1
Common Goldeneye							21		21	4		4	19.0%	
Hooded Merganser							4		4	1		1	25.0%	
Common Merganser							1		1			0		
Red-breasted Merganser							2		2			0		
Ruddy Duck							70		70	4		4	5.7%	
Northern Harrier	13				20		53		86	12		12	14.0%	
Cooper's Hawk	5						47		52	1		1	1.9%	
Northern Goshawk							1		1			0		4
Broad-winged Hawk							13		13			0		
Swainson's Hawk	12						4		16	3		3	18.8%	
Red-tailed Hawk	7		3	2	2		119	34	167	10	2	12	7.2%	3
Golden Eagle														1
American Kestrel						1	28	1	30	1		1	3.3%	
Merlin							288	135	423	4	4	8	1.9%	1
Peregrine Falcon														1
Ruffed Grouse							1		1					
Sharp-tailed Grouse	3								3			0		
Sora	3			1					4			0		
American Coot		(5)	(1)	(1)			2202	32	2234	82	2	84	3.8%	8
Sandhill Crane							2		2			0		1
Whooping Crane														1
Semipalmated Plover							55		55			0		
Piping Plover							3		3			0		
Killdeer	10		6	7			21		44			0		
Greater Yellowlegs							5		5			0		
Lesser Yellowlegs	6						365		371	1		1	0.3%	
Solitary Sandpiper			2				2		4			0		
Willet	6						37		43			0		
Spotted Sandpiper				2			4		6			0		
Upland Sandpiper	1								1			0		
Hudsonian Godwit							110		110			0		
Marbled Godwit	1					1	29		31	2		2	6.5%	
Red Knot							2		2			0		
Semipalmated Sandpiper	4						3449		3453	9		9	0.3%	1
Western Sandpiper							2		2			0		
Least Sandpiper	4		1				452		457			0		
Baird's Sandpiper							6		6			0		
Pectoral Sandpiper			1				99		100			0		
Stilt Sandpiper	1						1454		1455	2		2	0.1%	
Short-billed Dowitcher							102		102			0		
Long-billed Dowitcher							99		99			0		
Wilson's Snipe	4		2				12		18			0		
Wilson's Phalarope							127		127			0		

	pre-1955 Houston	Patrick	Blakeley	Wickencamp	Wotherspoon	Bard Briggs Wilson	Post-1955 N of 51	S of 51	Total Banded	Recoveries Banded in area n of 51	s of 51	Total Recoveries Banded in area	percent of banded	Recoveries Banded outside YDM area
Red-necked Phalarope							42		42			0		
Ring-billed Gull							236		236	9		9	3.8%	10
California Gull							2		2			0		
Herring Gull														1
Common Tern	24						1		25			0		
Black Tern			1				3		4			0		
Mourning Dove	1						7		8			0		
Eastern Screech-Owl	1								1			0		
Great Horned Owl	9			5	3		861	91	969	70	11	81	8.4%	6
Snowy Owl	1								1			0		
Burrowing Owl	4			4					8			0		
Long-eared Owl	6						97	1	104	2		2	1.9%	
Short-eared Owl							51	5	56			0		
Boreal Owl							15	9	24	1	2	3	12.5%	
Northern Saw-whet Owl							2	1	3			0		
Common Nighthawk	3								3			0		
Ruby-throated Hummingbird	3								3			0		
Yellow-bellied Sapsucker	1				2		4		7			0		
Downy Woodpecker				1	1		25	1	28	1		1	3.6%	1
Hairy Woodpecker					2		21		23			0		1
Northern Flicker	10	9	5	25	6	63	2		120		1	1	0.8%	
Western Wood-Pewee					19				19			0		
Least Flycatcher	1						8		9			0		
Eastern Phoebe			2	1			1		4			0		
Western Kingbird						7			7					
Eastern Kingbird	1				2	6			9			0		
Loggerhead Shrike			5		20	6			31			0		
Red-eyed Vireo							1		1			0		
Gray Jay	1						5		6	1		1	16.7%	
Blue Jay				1			15		16			0		1
Black-billed Magpie	6					11	166	15	198	8	1	9	4.5%	1
American Crow	8				23	13	219		263	32	1	33	12.5%	4
Common Raven							10		10			0		
Horned Lark			1	23			1		25			0		
Purple Martin	1				3	1	8		13			0		
Tree Swallow	2			4		7	568	223	804		2	2	0.2%	
Bank Swallow	12				9				21			0		
Cliff Swallow							1		1			0		
Barn Swallow	10		1	19		1	36	1	68			0		
Black-capped Chickadee	18			5	14		260		297	4		4	1.3%	
Boreal Chickadee	1								1			0		
Red-breasted Nuthatch				1			1		2			0		
White-breasted Nuthatch							2		2			0		
Brown Creeper	1						2	1	4			0		
House Wren	5		19	36	67		4		131			0		
Marsh Wren							1		1			0		
Ruby-crowned Kinglet							1		1			0		
Eastern Bluebird							1		1			0		
Mountain Bluebird							273	169	442		1	1	0.2%	1
Veery							6		6			0		
Gray-cheeked Thrush							11		11			0		
Swainson's Thrush							17		17			0		
Hermit Thrush					1		9		10			0		
American Robin	127		17	27	35	15	64		285	10		10	3.5%	1
Gray Catbird	4			1	10	2	11		28			0		1
Brown Thrasher							5		5			0		
European Starling								5	5			0		
Sprague's Pipit					1				1			0		
Bohemian Waxwing	26						7		33			0		
Cedar Waxwing							9		9			0		1
Tennessee Warbler							13		13			0		
Orange-crowned Warbler							4		4			0		
Nashville Warbler							1		1			0		
Yellow Warbler	3				5		8		16			0		
Magnolia Warbler							1		1			0		
Myrtle Warbler	3						35		38			0		
Palm Warbler							3		3			0		
Blackpoll Warbler							1		1			0		
Black-and-white Warbler							1		1			0		
American Redstart	1						7		8			0		
Ovenbird	1						1		2			0		
Northern Waterthrush	2						4		6			0		
Connecticut Warbler	1								1			0		
Mourning Warbler	1								1			0		
Common Yellowthroat							1		1			0		
American Tree Sparrow					10		66		76			0		
Chipping Sparrow	18			5	4		1		28	1		1	3.6%	
Clay-colored Sparrow	1				1		64		66			0		
Vesper Sparrow	1		7	11	3				22			0		
Savannah Sparrow				1			6		7			0		

	pre-1955 Houston	Patrick	Blakeley	Wickencamp	Wotherspoon	Bard Briggs Wilson	Post-1955 N of 51	S of 51	**Total Banded**	**Recoveries Banded in area** n of 51	s of 51	**Total Recoveries** Banded in area	percent of banded	**Recoveries** Banded outside YDM area
Baird's Sparrow	1								1			0		
Le Conte's Sparrow					9				9			0		
Nelson's Sharp-tailed Sparrow					1		3		4			0		
Fox Sparrow							23		23			0		
Song Sparrow			4		28		16		48			0		
Lincoln's Sparrow	8			4			70		82			0		2
White-throated Sparrow	28		3		25		349		405			0		
Harris's Sparrow	3						159		162	1		1	0.6%	
White-crowned Sparrow							1		1			0		
Gambel's	9			2			60		71			0		
Slate-colored Junco	95			5	377		1106		1583	4		4	0.3%	2
Lapland Longspur				3	1				4			0		
Snow Bunting					1		33		34			0		
Rose-breasted Grosbeak					2		2		4					
Red-winged Blackbird	13						6		19			0		1
Western Meadowlark	2				3				5			0		
Yellow-headed Blackbird	3						1		4			0		1
Rusty Blackbird					10		4		14	2		2	14.3%	
Brewer's Blackbird				11	4	7			22	4	1	5	22.7%	
Common Grackle	313				7		6		326	17		17	5.2%	8
Brown-headed Cowbird		20							20			0		1
Baltimore Oriole	4						2		6			0		
Purple Finch	11				114		66		191	3		3	1.6%	3
Common Redpoll	1			2	6		280		289	1		1	0.3%	1
Hoary Redpoll							11		11			0		
American Goldfinch					24		1		25			0		1
Evening Grosbeak	9						19		28			0		9
Total, banded YDM area	1010	206	80	209	876	142	262128	1993	266644	25614	173	25787	9.7%	3578
Number of species	80	16	19	32	45	27			174					
Total, banded outside area												3578		
TOTAL RECOVERIES												29365		

Note that numbers of ducks and coots banded before 1955 were computerized and hence are not treated separately
Patrick also banded 563 Am White Pelicans and 416 D-c Cormorants at Big Quill Lake, outside of the YDM area
In addition another 3,578 recoveries of birds banded outside the Yorkton area were recovered within it. Total 29,365 recoveries
Two banders are listed in "other" column: Fred G. Bard of the Sask. Museum banded 11 magpies and 13 crows as part of the anti-crow campaign; J.A. Briggs came out from Regina and banded 117 individuals of 12 species.

Appendix 4 - Waterfowl banders in the Yorkton-Duck Mountain area

	CAGO	WODU	GADW	AMWI	hybrid	ABDU	MALL	BWTE	CITE	NOSH	NOPI	AGWT	CANV	REDH
Bard, FG	5			9										
Benning, D			16		3	7	3595	2425			472	144	9	7
Blazevic, VA							65	1065			4	16	2	5
Brazda, R							387	762			14	11		
Brevig, R			15	94			871	542		51	16	16	15	11
Buelna, E			7	1	4	4	2079	2600			34	15		38
Buller, RJ							34	13			1		2	
Carter, ET				1			1				4			
Clark, RG			14	2			78	50		5		1		
Colls, DG	7													
Ducks Unlimited	325						442							
England, J					1		5585							
Ferguson, E		1		4		8	6694	199			193		11	18
Goldsberry, J			4	15		7	4026	8508		1	213	107	11	422
Glover, FA							5							
Gruener, CJ				3			316	4			21			
Hague, JA							16	1		3	1			
Halstead, WW				14			790	130		26	64		1	1
Houston, CS	1		1	71		2	1969	1404		15	476	70	92	376
Kennedy, WH			11	16			253	927		4	32	95	144	128
Kniffin, FC			28	120			380	438		163	56	29	363	189
Krieble, DW							8	2						
Martin, LM			2	33			19	146			8	6	4	
Meyerding, RJ				2			98	341		8	10	6		14
Nelson, JL			23	7			66	932			194	236		1
Nieman, D	3858	139	7	30	1	3	895	1463		2	87	53	18	149
Niemeyer, AE							25				1			
Norman KD			1		3	2	807	1499		1	262	15	7	28
Novara, A		8	334	270	17	57	21312	21951		13	4647	758	196	1063
Patrick, JAM	14						68	21		19	11		5	6
Plant, CD				5			325	54		3	17		2	
Powell, BH		4	12	19	1	8	1347	1844		22	443	21	31	186
Price, WE	9		30	308			997	1789		231	132	202	125	74
Reynolds, Ron					2	4	1979	900			102	4		57
Robinson, AJ				1			14	10		5	4			
Runge, W							1	2108			11	10	1	5
Sanders, WE				29			4	36		17	1	6	116	23
Schroeder, L				4		5	1612	1000		1	273	4	67	69
Seal, D							1							
Sherratt, D	344						252							
Smith, CS			7	5		3	1949	954		1	35	6	8	10
Smith, JD		2		5		1	261	2823			240	11		45
Smith, M			1	27			185	9418		11	399	45	14	257
Tautin, J					2	3	1042	**347**		1	23	13	16	6
Thorpe, P			103	18	3	3	3641	14183	1	18	710	83	3	50
Tice, RC			1	44			23	19						
Voelzer, JF		7	154	295	27	97	42834	22633	1	21	2173	376	892	2975
Walker, LO					2	2	581	555		1	3	5	7	116
Wickencamp, E							1							
Wilson, JH			57	131		2	1452	1775	1	33	710	44	2	540
Wilson, WJ						1	558							
Wotherspoon Bros							5							
TOTAL WATERFOWL	4563	161	828	1583	66	219	109948	105871	3	676	12097	2408	2164	6869
All north of 51 except														
South of 51	254		8	45			257	576		36	32	42	6	

RNDU	GRSC	LESC	WWSC	BUFF	COGO	HOME	COME	RBME	RUDU	AMCO	TOTAL	
											14	Bard, F.G.
		11		1							6690	Benning, D
										25	1182	Blazevic, VA
		1									1175	Brazda, R
11				1					6	2	1651	Brevig, R
		1									4783	Buelna, Eliz
		1								1	52	Buller, RJ
											6	Carter, ET
											150	Clark, RC
											7	Colls, D.G.
											767	Ducks Unlimited
											5586	England
											7128	Ferguson, E
		2		2	1						13319	Goldsberry, J
												Glover, FA
											344	Gruener, CJ
												Hague, JA
										41	1067	Halstead, WW
		11		10	2				1	643	5144	Houston, CS
5		53		1					3	275	1947	Kennedy, WH
29		112	16		9			2	13	359	2306	Kniffin, FC
											10	Krieble, DW
		12								2	232	Martin, L.M.
									1	8	488	Meyerding, RJ
		1								10	1470	Nelson, JL
		12							1		6718	Nieman, D
											26	Niemeyer, AE
		1		2							2628	Norman KD
11		86		5	1	1			21		50751	Novara, A
										5	149	Patrick, JAM
										8	414	Plant, CD
3		5							1		3947	Powell, BH
14		60		4					10	72	4057	Price, WE
											3048	Reynolds, Ron
											34	Robinson, AJ
										28	2164	Runge, W
		24		9							265	Sanders, WE
1											3036	Schroeder, L
											1	Seal, D
											596	Sherratt, D
		7									2985	Smith, CS
		6								10	3404	Smith, JD
		1								351	10709	Smith, M
											1453	Tautin, J
		7		2	2						18827	Thorpe, P
											87	Tice, RC
33		110		21	4	2			13		72668	Voelzer, JF
2	2	17		3							1296	Walker, LO
										1	2	Wickencamp, Emma
		70		21	2	1	1			392	5234	Wilson, JH
											559	Wilson, WJ
										1	6	Wotherspoon Bros
109	2	611	16	82	21	4	1	2	70	2234	250608	TOTAL WATERFOWL
		13								32	1301	South of 51

Appendix 5 - 10-minute blocks (108) in the Yorkton-Duck Mountain Area

MELVILLE MAP SHEET

504	1013	Marchwell
504	1014	West of Marchwell
504	1015	S of Churchbridge
504	1020	Zeneta
504	1021	Atwater
504	1022	Bangor - Dubuc
505	1013	East of Langenburg - P
505	1014	Langenburg
505	1015	Churchbridge
505	1020	Bredenbury
505	1021	West of Bredenbury
505	1022	North of Bangor
505	1023	East of Melville
505	1024	Melville
505	1025	Birmingham
505	1030	Duff
505	1031	Finnie
505	1032	South of Hubbard

YORKTON MAP SHEET (SOUTH)

510	1013	MacNutt - P
510	1014	Calder
510	1015	South of Wroxton
510	1020	Saltcoats
510	1021	West of Saltcoats
510	1022	East of Otthon
510	1023	Otthon
510	1024	South of Willowrook
510	1025	East of Fenwood
510	1030	Fenwood
510	1031	Goodeve
510	1032	Hubbard
510	1033	South of Jasmin
510	1034	South of Kelliher
510	1035	South of Leross
511	1013	West of Roblin - P
511	1014	East of Wroxton
511	1015	Wroxton
511	1020	Stornoway
511	1021	Tonkin
511	1022	Yorkton
511	1023	Rousay Lake
511	1024	Willowbrook
511	1025	W of Willowbrook
511	1030	Jedburgh
511	1031	S of Parkerview
511	1032	Ituna
511	1033	Jasmin
511	1034	Kelliher
511	1035	Leross
512	1013	Togo - P
512	1014	Runnymede
512	1015	South of Kamsack
512	1020	East of Rhein
512	1021	Rhein
512	1022	Gorlitz
512	1023	East of Springside
512	1024	Springside
512	1025	Theodore
512	1030	Insinger
512	1031	Parkerview
512	1032	Tuffnell
512	1033	E of West Bend
512	1034	West Bend
512	1035	South of Wishart

YORKTON MAP SHEET (NORTH)

513	1013	North of Togo - P
513	1014	E of Kamsack
513	1015	Kamsack
513	1020	Veregin
513	1021	Mikado
513	1022	Canora
513	1023	West of Canora
513	1024	S of Buchanan
513	1025	East of Insinger
513	1030	Insinger
513	1031	Sheho
513	1032	Tuffnell
513	1033	Foam Lake
513	1034	East of Wishart
513	1035	Wishart
514	1013	S of Benito [MB] - P
514	1014	South of Arran
514	1015	South of Pelly
514	1020	South of Norquay
514	1021	North of Mikado
514	1022	Amsterdam
514	1023	E of Buchanan
514	1024	Buchanan
514	1025	East of Invermay
514	1030	Invermay
514	1031	North of Sheho
514	1032	South of Kuroki
514	1033	N of Foam Lake
514	1034	Leslie
514	1035	Elfros
515	1013	Benito [MB] - P
515	1014	Arran
515	1015	Pelly
515	1020	Norquay
515	1021	Hyas
515	1022	Stenen
515	1023	Sturgis
515	1024	Preeceville
515	1025	Hazel Dell
515	1030	W of Hazel Dell
515	1031	East of Kuroki
515	1032	Kuroki
515	1033	Kylemore
515	1034	Wadena
515	1035	West of Wadena

P - partial block, a portion of which is in Manitoba

REFERENCES

Adam, C.I.G. 1989. Eastern Screech-Owl in Saskatchewan and adjacent areas. *Blue Jay* 47:164-188.

Allen, R.P. 1952. *The Whooping Crane.* New York: National Audubon Society.

American Ornithologists' Union. 1957. *Check-List of North American Birds, fifth edition.* Baltimore: Lord Baltimore Press.

American Ornithologists' Union. 1998. *Check-List of North American Birds, seventh edition.* Washington: American Ornithologists' Union.

American Ornithologists' Union. 2000. Forty-second supplement to the AOU *Check-List of North American Birds. Auk* 117:847- 858.

American Ornithologists' Union. 2002. Forty-third supplement to the AOU *Check-List of North American Birds. Auk* 119:895-906.

Anaka, J. 1986. First Band-tailed Pigeon in Yorkton area. *Blue Jay* 44:128.

Anaka, J. 1996. *Donald Gunn: Family History.* Published privately.

Anaka, J. 2000. *A Century of Changes: A History of Good Spirit/Devil's Lake.* Published privately.

Anaka, W. 1951. Ovenbirds at Spirit Lake. *Blue Jay* 9(3):4.

Anaka, W. 1952. Winter record of Short-eared Owl. *Blue Jay* 14:52.

Anaka, W. 1954. Unusual Great Horned Owl's nest and unusual flight of buteos. *Blue Jay* 12(2):20.

Anaka, W. 1955. Uncommon birds at Spirit Lake. *Blue Jay* 13(4):8.

Anaka, W. 1956a. The Snowy Egret. *Blue Jay* 14:89-90.

Anaka, W. 1956b. Golden-crowned Kinglet at Spirit Lake. *Blue Jay* 14:119.

Anaka, W. 1958. Whip-poor-will observed at Spirit Lake. *Blue Jay* 16:158.

Anaka, W. 1959. Unusual incubation behaviour of the Ruffed Grouse. *Blue Jay* 17:158.

Anaka, W. 1961a. American Golden Plover in fall migration at Horseshoe Lake. *Blue Jay* 19:164.

Anaka, W. 1961b. Great Gray Owl at Spirit Lake. *Blue Jay* 19:174.

Anaka, W. 1962. Golden-crowned Kinglets. *Blue Jay* 20:8.

Anaka, W. 1967. First record of Lark Bunting and Lark Sparrow at Spirit Lake, Saskatchewan. *Blue Jay* 25:189.

Anaka, W. 1997. Operating a nest box line: a 13-year summary. *Blue Jay* 55:85-89.

Anaka, W., and F. Switzer. 1999. Thirty years on the Theodore, Saskatchewan, route. *BBS Canada* Winter 1999:6-7.

Anderson, C. 1998. Drastic declines in Sharp-tailed Grouse and American Bittern. *Blue Jay* 56:128.

Anweiler, G. 1960. The Boreal Owl influx. *Blue Jay* 18:61-63.

Anweiler, G. 1965. Unusual nest site of a Horned Lark. *Blue Jay* 23:84.

Ashdown, D.N. 1966. Eastern Bluebirds. *Blue Jay* 24:187.

Atkinson, G.E. 1905. *A review-history of the Passenger Pigeon of Manitoba.* Hist. Sci. Soc. Manitoba. Transaction #68.

Bain, M. 2001. Cross Canada Round-up. *Birders Journal* 10:180-181.

Bain, M. 2002a. Cross Canada Round-up. *Birders Journal* 11:167-168.

Bain, M. 2002b. Cross Canada Round-up. *Birders Journal* 11:210-212.

Bain, M. 2003. Cross Canada Round-up. *Birders Journal* 12:7-9.

Bain, M., and M. Holder. 1996. Cross Canada Round-up. *Birders Journal* 5:214-215.

Bain, M., and M. Holder. 1997a. Cross Canada Round-up. *Birders Journal* 6:6-7.

Bain, M., and M. Holder. 1997b. Cross Canada Round-up. *Birders Journal* 6:57-58.

Bain, M., and M. Holder. 1997c. Cross Canada Round-up. *Birders Journal* 6:112-113.

Bain, M., and M. Holder. 1997d. Cross Canada Round-up. *Birders Journal* 6:216-217.

Bain, M., and M. Holder. 1997e. Cross Canada Round-up. *Birders Journal* 6:269-272.

Bain, M., and D. Shanahan. 1998a. Cross Canada Round-up. *Birders Journal* 7:121.

Bain, M., and D. Shanahan. 1998b. Cross Canada Round-up. *Birders Journal* 7:172-173.

Bain, M., and D. Shanahan. 1998c. Cross Canada Round-up. *Birders Journal* 7:224-225.

Bain, M., and D. Shanahan. 1998d. Cross Canada Round-up. *Birders Journal* 7:275-276.

Bain, M., and D. Shanahan. 1999a. Cross Canada Round-up. *Birders Journal* 8:9.

Bain, M., and D. Shanahan. 1999b. Cross Canada Round-up. *Birders Journal* 8:59-60.

Bain, M., and D. Shanahan. 1999c. Cross Canada Round-up. *Birders Journal* 8:112-113.

Bain, M., and D. Shanahan. 1999d. Cross Canada Round-up. *Birders Journal* 8:165-168.

Bain, M., and D. Shanahan. 1999e. Rare birds in Canada in 1998. *Birders Journal* 8:246-251.

Bain, M., and D. Shanahan. 1999f. Cross Canada Round-up. *Birders Journal* 8:268.

Bain, M., and D. Shanahan. 2000a. Cross Canada Round-up. *Birders Journal* 9:113-114.

Bain, M., and D. Shanahan. 2000b. Cross Canada Round-up. *Birders Journal* 9:167-169.

Bain, M., and D. Shanahan. 2000c. Cross Canada Round-up. *Birders Journal* 9:218-219.

Baines, F. 1942. Godwit attacks crow. *Blue Jay* 1:3.

Baines, F. 1950. Birds at Saltcoats. *Blue Jay* 8(1):7.

Baines, K.E. 1956. The ups and downs of game at Crescent Lake. *Blue Jay* 14:65-66.

Bard, F.G. 1952. Annual report of the Provincial Museum, pp. 131-136 *In* Annual Report of the Dept. Of Natural Resources.

Bard, F.G. 1957. A flash from the Wascana Waterfowl Park goose project. *Blue Jay* 15:103.

Bard, F.G., and F.W. Lahrman. 1958. The Wascana Canada Goose project, 1957-1958. *Blue Jay* 16:152.

Beaulieu, R. 1999. The new Porcupine Forest flock of Trumpeter Swans, *Cygnus buccinator*, in Saskatchewan. *Can. Field-Nat.* 113:269-272.

Bechard, M.J., and C.S. Houston. 1984. Probable identity of purported Rough-legged Hawk nests in the western U.S. and Canada. *Condor* 86:348-352.

Belcher, M. 1955. Bird notes from the files. *Blue Jay* 13(2):28- 29.

[Belcher, M.] 1956. Birds of Madge Lake. *Blue Jay* 14:53-54.

Belcher, M. 1959. In memoriam: Clifford Carlyle Shaw. *Blue Jay* 17:116.

Bell, R. 1882. Notes on the birds of Hudson's Bay. *Proceedings Roy. Soc. Canada* 4:49-54.

Bennett, G. 1989. What's happening. *Bird Finding in Canada* 9(6):23-27.

Bittner, R.A. 1995a. House Finch at Abernethy in 1990. *Blue Jay* 53:42-43.

Bittner, R.A. 1995b. Experimental nest boxes for bluebirds: an update. *Blue Jay* 53:200-202.

Bradshaw, F. 1927. *Report of the Chief Game Guardian for 1926-27.*

Brennan, M. 1965a. Birdwatching at Ituna. *Blue Jay* 23:52.

Brennan, M. 1965b. Record of a Wood Duck at Ituna. *Blue Jay* 23:84.

Brennan, M. 1976. Winter birds at Ituna. *Blue Jay* 34:126.

Brennan, M. 1984. Hummingbirds at Ituna, Saskatchewan. *Blue Jay* 42:112-113.

Brewer, D., A. Diamond, E.J. Woodsworth, B.T. Collins, E.H. Dunn. 2000. *Canadian Atlas of Bird Banding*, volume 1. Ottawa: Canadian Wildlife Service.

Bull, J.L. 1958. The changing seasons: a summary of the nesting season. *Audubon Field Notes* 12:391-395.

Burke, A.M. 1996. Habitat use and breeding biology of Sandhill Cranes in the Yorkton region, Saskatchewan, Canada. Report to Canadian Wildlife Service.

Burke, A.M. 2001. Status of Sandhill Cranes in the Yorkton Wetland Complex, Saskatchewan, Canada. M.Sc. thesis. University of Wisconsin, Stevens Point.

Callin, E.M. 1980. *Birds of the Qu'Appelle, 1857-1979.* Sask. Nat. Hist. Soc. Spec. Publ. 13.

Carmichael, L.T. 1954. Protection of hawks and owls. *Blue Jay* 12(2):19.

Cartwright, B.W., and A.H. Shortt. 1956. *Waterfowl banding, 1939-1954.* Winnipeg: Ducks Unlimited.

Christy, R.M. 1885. Notes on the birds of Manitoba. *The Zoologist*, third series, 9:121-133.

Chytyk, P., K. Scalise, and R. Tillie. 2002. Robert Harry Kreba, 1954-2001. *Blue Jay* 60:124-125.

Clapp, R.B., M.K. Klimkiewicz, and A.G. Futcher. 1983. Longevity records of North American birds, *Columbidae* through *Paridae. J. Field Ornithol.* 54:123-137.

Collard, Mrs J.E. 1965. Bird notes from Arran. *Blue Jay* 23:151-152.

Copland, H.W.R. 1969. A Sabine's Gull in southern Saskatchewan. *Blue Jay* 27:146-147.

Cross, N. 1995. A chickadee's larder. *Blue Jay* 53:116.

Cumming, E.E. 1988. Northern Wheatears sighted in Duck Mountain Provincial Park, Manitoba. *Blue Jay* 46:84-85.

Cumming, E.E. 1998. *Distribution and habitat associations of Golden-winged Warblers in Duck Mountain, Riding Mountain, and Porcupine Hills, Manitoba, 1998.* Saskatoon: Canadian Wildlife Service and Manitoba Mixed-Wood Forest Research. 46 pp.

Deighton, H. 1951. Game birds increasing. *Blue Jay* 9:17.

Deignan, H.G. 1960. Proceedings of the 77th stated meeting of the American Ornithologists' Union. *Auk* 77:56-68.

Dzus, E.H., and R.G. Clark. 1996. Effects of harness-style and abdominally implanted transmitters on survival and return rates of Mallards. *J. Field Ornithol* 67:549-557.

Dzus, E.H., and R.G. Clark. 1998. Brood survival and recruitment of Mallards in relation to wetland density and hatching date. *Auk* 115:311-318.

Elder, W.H., and D. Zimmerman. 1983. A comparison of recapture versus resighting data in a 15-year survivorship study of the Black-capped Chickadee. *Bird Banding* 54:138-145.

Erskine, A.J. 1990. Joint laying in *Bucephala* ducks: "parasitism" or nest-site competition? *Ornis Scandinav.* 21:52-56.

Fast, H.R., and D.A. Neil. 1975. The Yorkton region of Saskatchewan. *Prairie Regional Studies in Geography* #21.

Ferrie, M. 1960. Cooper's Hawk nests. *Blue Jay* 18:175.

Finley, J.K. 2000. The great Bufflehead crash, 1940. *Blue Jay* 58:161-167.

Finley, J.K. 2001. Addendum: the great Bufflehead crash. *Blue Jay* 59:42-43.

Flack, J.A.D. 1976. *Bird populations of aspen forests in western North America.* AOU Ornithological Monographs No. 19.

Forsyth, E.S. 1937. *Report of the Game Commissioner for 1936-37.*

Fox, E. 1959. First annual report on the Prairie Nest Records Scheme. *Blue Jay* 17:12-13.

Fraser, P. 1961. Greater Prairie Chicken, Cardinal. *Blue Jay* 19:142.

Gammell, A. 1956a. *Audubon Field Notes* 10:341-343.

Gammell, A. 1956b. *Audubon Field Notes* 10:390-392.

Gammell, A. 1957a. *Audubon Field Notes* 11:34-35.

Gammell, A. 1957b. *Audubon Field Notes* 11:357-358.

Gammell, A. 1960. *Audubon Field Notes* 14:397-398.

Gammell, A., and H. Huenecke. 1954. Northern Great Plains. *Audubon Field Notes* 8:350-352.

Gammell, A., and H. Huenecke. 1955. Northern Great Plains. *Audubon Field Notes* 9:34-36.

Gammell, R., and A. Gammell. 1949a. Northern Great Plains. *Audubon Field Notes* 3:174-175.

Gammell, R., and A. Gammell. 1949b. Northern Great Plains. *Audubon Field Notes* 3:213-214.

Gammell, R., and A. Gammell. 1950a. *Audubon Field Notes* 4:207-208.

Gammell, R., and A. Gammell. 1950b. *Audubon Field Notes* 4:247-248.

Gammell, R., and A. Gammell. 1951. *Audubon Field Notes* 5:24-25.

Gammell, R., and A. Gammell. 1952a. *Audubon Field Notes* 6:23-25.

Gammell, R., and A. Gammell. 1952b. *Audubon Field Notes* 6:201-202.

Gammell, R., and A. Gammell. 1952c. *Audubon Field Notes* 6:286-288.

Gammell, R., and A. Gammell. 1953. *Audubon Field Notes* 7:23-24.

Gammell, R., A. Gammell and H. Huenecke. 1953. *Audubon Field Notes* 7:313-315.

Gammell, R., A. Gammell and H. Huenecke. 1954a. *Audubon Field Notes* 8:25-27.

Gammell, R., A. Gammell, and H. Huenecke. 1954b. *Audubon Field Notes* 8:319-320.

Gammell, R., A. Gammell, and H. Huenecke. 1955. Audubon Field Notes 9:265-267.

Gilroy, D. 1990. Prairie Wildlife #1337. *Western Producer,* 25 October:14.

Gilroy, D. 1991. Prairie Wildlife #1349. *Western Producer,* 18 April:24.

Gollop, J.B. 1978. White-faced Ibis and Cattle Egret at Little Quill Lake. *Blue Jay 36:122.*

Gollop, J.B. 1979. *Am. Birds* 33:781-782.

Gollop, J.B. 1980a. *Am. Birds* 34:787-789.

Gollop, J.B. 1980b. *Am. Birds* 34:905-906.

Gollop, J.B. 1982. *Am. Birds* 36:304-305.

Gollop, J.B. 1983. *Am. Birds* 37:311-312.

Gollop, J.B. 1984. *Am. Birds* 38:925-927.

Gollop, J.B. 1985. *Am. Birds* 39:315-317.

Gollop, J.B. 1986. *Am. Birds* 40:487-488.

Gollop, J.B. 1988a. *Am. Birds* 42:450-451.

Gollop, J.B. 1988b. *Am. Birds* 42:1304-1305.

Gollop, J.B. 1989a. *Am. Birds* 43:495-497.

Gollop, J.B. 1989b. *Am. Birds* 43:1330-1332.

Gratto-Trevor, C.L., and H.L. Dickson. 1994. Confirmation of elliptical migration in a population of Semipalmated Sandpipers. *Wilson Bull.* 106:78-90.

[Gray, D.] 1969. Saskatchewan Nis'ku project. *Blue Jay* 27:68-70.

Greenberg, R., and S. Droege. 1999. On the decline of the Rusty Blackbird and the use of ornithological literature to document long-term population trends. *Conservation Biology* 13:553-559.

Gunn, D., and C.R. Tuttle. 1880. *History of Manitoba.* Ottawa: Maclean Roger.

Gunn, J. 1954a. Arctic Three-toed Woodpecker. *Blue Jay* 12(2):11.

Gunn, J. 1954b. The return of the hawks. *Blue Jay* 12(2):21,31.

Gunn, J. 1956a. Spruce Grouse and Saw-whet Owl. *Blue Jay* 14:36B.

Gunn, J. 1956b. Myrtle Warbler. *Blue Jay* 14:88.

Gunn, J. 1956c. Natural feeding station for summer birds. *Blue Jay* 14:119.

Gunn, J. 1957a. Myrtle Warblers at Spirit Lake. *Blue Jay* 15:103.

Gunn, J. 1957b. Cliff Swallows use Barn Swallow nest. *Blue Jay* 15:103.

Gunn, J. 1960. Late records of summer birds. *Blue Jay* 18:52.

Gunn, J. 1968. Pileated Woodpecker nests at Good Spirit Lake. *Blue Jay* 26:176.

Hanson, L., and A.R. Smith. 2000. Great Blue Heron colonies in Saskatchewan in 1999, an inventory. *Blue Jay* 58:81-85.

Harmon, D.W. 1820. *A Journal of Voyages and Travels in the Interior of North America.* Andover: Flagg and Gould.

Harris, W.C. 1974. More Hudsonian Godwits in Saskatchewan. *Blue Jay* 32:101-104.

Harris, W.C. 2000. 58th annual Saskatchewan Christmas Bird Count, 1999. *Blue Jay* 58:2-28.

Harris, W.C. 1980. *Am. Birds* 34:172-174.

Harris, W.C. 1981a. *Am. Birds* 35:195-196.

Harris, W.C. 1981b. *Am. Birds* 35:309-311.

Harris, W.C. 1983. *Am. Birds* 37:882-884.

Harris, W.C. 1984. *Am. Birds* 38:214-216.

Harris, W.C. 1985a. *Am. Birds* 39:67-69.

Harris, W.C. 1985b. *Am. Birds* 39:927-928.

Harris, W.C. 1986. *Am. Birds* 40:130-131.

Harris, W.C. 1987. *Am. Birds* 41:104-106.

Harris, W.C., and S.M. Lamont. 1976. *Am. Birds* 30:732-735.

Hatch, D.R.M. 1965. *Audubon Field Notes* 19:554-557.

Hatch, D.R.M. 1966. *Audubon Field Notes* 20:519-522.

Hatch, D.R.M. 1967a. *Audubon Field Notes* 21:47-51.

Hatch, D.R.M. 1967b. *Audubon Field Notes* 21:581-583.

Hatch, D.R.M. 1968a. *Audubon Field Notes* 22:54-57.

Hatch, D.R.M. 1968b. *Audubon Field Notes* 22:448-451.

Hatch, D.R.M. 1968c. *Audubon Field Notes* 22:544-547.

Hatch, D.R.M. 1968d. *Audubon Field Notes* 22:617-620.

Hatch, D.R.M. 1969. *Audubon Field Notes* 23:669-671.

Hawkes, J. 1924. *The Story of Saskatchewan and its People.* Chicago: S.J. Clarke. [L.T. McKim, 3:1494-1495.]

Henny, C.J. 1973. Drought displaced movements of pintails into Siberia. *J. Wildl. Manage.* 37:23-29.

Herriot, T. 1995. Appendix X. "Bird survey methodology, species lists and incidental observations on bird species distribution." *In* Wright et al. 1995.

Hicke, J.M. 1991. Grayson annual report. *Blue Jay* 49:158.

Hilton, J. 1986. Bird feeders. *Blue Jay* 44:140-141.

Hilton, J., and G. Hilton. 1985. Rufous hummingbird sighting. *Blue Jay* 43:54-55.

Hjertaas, D.G. 1975. Horseshoe Lake bird species list, 1-19 July 1975. Regina: Dept. of Natural Resources.

Hjertaas, D.G. 1994. Summer and breeding records of the Whooping Crane in Saskatchewan. *Blue Jay* 52:99-115.

Hjertaas, D.G., D.H. Ellis, B.W. Johns, and S.L. Moon. 2001. Tracking Sandhill Crane migration from Saskatchewan to the Gulf Coast. *Proceedings N. Am. Crane Workshop* 8:57-61.

Hjertaas, W. 1986. Eastern Screech-Owl near Yorkton. *Blue Jay* 44:47.

Hombeck, H. 1993. Great Egret. *Blue Jay* 51:63.

Hooper, D.F. 1992. *Birds of East-Central Saskatchewan, Kelvington to Kelsey Trail.* Sask. Nat. Hist. Soc. Spec. Publ. 18.

Hooper, D.F, and R.R. Hooper. 1953. Surf Scoter at Yorkton. *Blue Jay* 11(4):9.

Hooper, R.R. 1993. Saskatchewan sighting of a Red-shouldered Hawk. *Blue Jay 51:105.*

Hornaday, W.T. 1913. *Our Vanishing Wildlife*. New York: New York Zoological Society.

Houston, C.S. 1945. Saskatchewan bird banders #1. Wotherspoon brothers, of Ilyas. *Blue Jay* 3:16.

Houston, C.S. 1946a. Saskatchewan bird banders #7. J.H. Wilson. *Blue Jay* 4:31.

Houston, C.S. 1946b. Nesting survey, Rousay Lakes, for Ducks Unlimited, Winnipeg. 43 pp. (Mss.).

Houston, C.S. 1949. The birds of the Yorkton district, Saskatchewan. *Can. Field-Nat.* 63:215-241.

Houston, C.S. 1950. Scarlet Tanager. *Blue Jay* 8(3):13.

Houston, C.S. 1951. Yorkton records. Red Crossbills at Yorkton. *Blue Jay* 9(4):12.

Houston, C.S. 1952. Birding at Good Spirit Lake. *Blue Jay* 10(3):9,23.

Houston, C.S. 1953a. Yorkton bird notes. *Blue Jay* 11(1):6.

[Houston, C.S.]. 1953b. 1953 Yorkton records. *Blue Jay* 11(3):7.

Houston, C.S. 1954. Oriole and dove in December. *Blue Jay* 12(1):4.

Houston, C.S. 1955. Sandhill Crane nesting at Rokeby Marsh. *Blue Jay* 13(1):9.

Houston, C.S. 1957a. Golden-crowned Kinglet at Yorkton. *Blue Jay* 15:7.

Houston, C.S. 1957b. Starling records. *Blue Jay* 15:9.

Houston, C.S. 1957c. The Little Blue Heron in Saskatchewan. *Blue Jay* 15:52.

Houston, C.S. 1957d. Cedar Waxwings in March. *Blue Jay* 15:63.

Houston, C.S. 1957e. 1957 summer banding. *Blue Jay* 15:111-112.

Houston, C.S. 1957f. Mallard nesting in barn at Langenburg. *Blue Jay* 15:148.

Houston, C.S. 1957g. The Great Gray Owl in Saskatchewan. *Blue Jay* 15:150-153.

Houston, C.S. 1958. Saskatchewan nesting records of the Cooper's Hawk. *Blue Jay* 16:153-154.

Houston, C.S. 1959a. Northern Three-toed Woodpecker at Yorkton. *Blue Jay* 17:58.

Houston, C.S. 1959b. Hooded Mergansers at Yorkton. *Blue Jay* 17:58.

Houston, C.S. 1959c. Junco recovered in Alaska. *Blue Jay* 17:155.

Houston, C.S. 1960a. Interesting American Wigeon recovery. *Blue Jay* 18:77.

Houston, C.S. 1960b. 1960—The year of the owls. *Blue Jay* 18:105-110.

Houston, C.S., 1961. Right on course—interesting recovery of a banded Purple Finch. *Blue Jay* 19:21.

Houston, C.S. 1963a. Redpoll identification: a problem. *Bird Banding* 34:94-95.

Houston, C.S. 1963b. Far travels of a teal brood. *Blue Jay* 21:101.

Houston, C.S. 1966a. Saskatchewan Long-eared Owl recovered in Mexico. *Blue Jay* 24:64.

Houston, C.S. 1966b. Breeding records of the Giant Canada Goose near Yorkton. *Blue Jay* 24:71-72.

Houston, C.S. 1967a. Recoveries of Red-tailed Hawks banded in Saskatchewan. *Blue Jay* 25:109-111.

Houston, C.S. 1967b. Recoveries of Black-crowned Night Herons banded in Saskatchewan. *Blue Jay* 25:112-113.

Houston, C.S. 1967c. Saskatchewan bird banders #8: Judge J.A.M. Patrick. Blue Jay 25:172-174.

Houston, C.S. 1968a. Recoveries of Marsh Hawks banded in Saskatchewan. *Blue Jay* 26:12-13.

Houston, C.S. 1968b. Recoveries of Swainson's Hawks banded in Saskatchewan. *Blue Jay* 26:86-87.

Houston, C.S. 1968c. Recoveries of Bronzed Grackles banded in Saskatchewan. *Blue Jay* 26:136-138.

Houston, C.S. 1968d. Recoveries of robins banded in Saskatchewan. *Blue Jay* 26:182-184.

Houston, C.S. 1969a. Turkey Vulture breeding records in Saskatchewan. *Blue Jay* 27:37-39.

Houston, C.S. 1969b. Frank Baines, 1877-1968, a tribute. *Blue Jay* 27:62-63.

Houston, C.S. 1969c. Nesting records of the Yellow Rail in Saskatchewan. *Blue Jay* 27:81-82.

Houston, C.S. 1969d. Recoveries of the Common Crow banded in Saskatchewan. *Blue Jay* 27:84-88.

Houston, C.S. 1971a. *Am. Birds* 25:71-74.

Houston, C.S. 1971b. *Am. Birds* 25:593-596.

Houston, C.S. 1971c. *Am. Birds* 25:869-872.

Houston, C.S. 1971d. Brood size of the Great Horned Owl in Saskatchewan. *Bird-Banding 42*:103-105.

Houston, C.S. 1972a. *Am. Birds* 26:869-872.

Houston, C.S. 1972b. The Passenger Pigeon in Saskatchewan. *Blue Jay* 30:77-83.

Houston, C.S. 1972c. Early Whooping Crane nest records near Yorkton, Saskatchewan. *Blue Jay* 30:152-153.

Houston, C.S. 1973a. *Am. Birds* 27:75-78.

Houston, C.S. 1973b. *Am. Birds* 27:630-633.

Houston, C.S. 1973c. *Am. Birds* 27:882-886.

Houston, C.S. 1974. *Am. Birds* 28:653-656.

Houston, C.S. 1977a. How the Blue Jay got its name. *Blue Jay* 35:60-62.

Houston, C.S. 1977b. Changing patterns of *Corvidae* on the prairies. Blue Jay 35:149-156.

Houston, C.S. 1978a. Arrival of the House Sparrow on the prairies. *Blue Jay* 36:99-102.

Houston, C.S. 1978b. Recoveries of Saskatchewan-banded Great Horned Owls. *Can. Field-Nat.* 92:61-66.

Houston, C.S. 1979. The spread of the Western Kingbird across the prairies. *Blue Jay* 37:149-157.

Houston, C.S. 1980. Introduction, pp. v-xv. *In Ernest Thompson Seton in Manitoba, 1882-1892.* Winnipeg: Manitoba Naturalists Society.

Houston, C.S. 1985. Peregrine Falcons harass nesting Great Horned Owls. *Blue Jay* 43:42-43.

Houston, C.S. 1989. Saskatchewan bird banders #14: William Anaka. *Blue Jay* 47:190-192.

Houston, C.S. 1992. Banding recoveries: a 48-year experience, OR: "It's really Mrs Priestly's fault." *Blue Jay* 50:182-194.

Houston, C.S. 1993. The natural history contributions of H.C. Grose. *Blue Jay* 51:239-241.

Houston, C.S. 1997. J. Dewey Soper, pp. 741-742. *In* K.B. Sterling et al., *Biographical Dictionary of American and Canadian Naturalists and Environmentalists.* Westport CT: Greenwood Press.

Houston, C.S. 1998. Northern Goshawks move back south. *Blue Jay* 56:191-192.

Houston, C.S. 2001. William McIntyre Niven, 1906-2001. *Blue Jay* 59:115-116.

Houston, C.S. 2002a. Spread and disappearance of the Greater Prairie-Chicken, *Tympanuchus cupido*, on the Canadian prairies and adjacent areas. *Can. Field-Nat.* 116:1-21.

Houston, C.S. 2002b. Wayne C. Harris, 1951-2002. *Blue Jay* 61:61-63.

Houston, C.S. 2003a. Rural banders in the Yorkton area. *Blue Jay* 61:94-95.

Houston, C.S. 2003b. Ten-year cycles. Appendix E. *In* C.S. Houston, T. Ball and M. Houston. *Eighteenth-century Naturalists of Hudson Bay.* Montreal: McGill-Queen's University Press.

Houston, C.S. 2003c. Saskatchewan bird banders #22: J.A. Briggs of Regina. *Blue Jay* 61:(in press).

Houston, C.S., and M.J. Bechard. 1982. Edward Arnold, enthusiastic oologist. *Blue Jay* 40:184-192.

Houston, C.S., and K.A. Hodson. 1997. Resurgence of breeding Merlins in Canadian grasslands. *Can. Field-Nat.* 111:243-248.

Houston, C.S., and M.I. Houston. 1975. *Am. Birds* 29:74-77.

Houston, C.S., and A. Schmidt. 1981. History of Richardson's Merlin in Saskatchewan. *Blue Jay* 39:30-37.

Houston, C.S., and S.J. Shadick. 1974. *Am. Birds* 28:814-817.

Houston, C.S., M.I. Houston, and H.M. Reeves. 1997. The 19th century trade in swan skins and quills. *Blue Jay* 55:24-34.

Houston, C.S., M.J. Stoffel, and A.R. Smith. 2002. Three Turkey Vulture pairs nest in Saskatoon Bird Area. *Blue Jay* 60:206-209.

Hruska, A.J. 1958. Dry weather affects nesting of Barn Swallows. *Blue Jay* 16:158.

Hruska, A.J. 1997. Cutarm heronry. *Blue Jay* 55:131-132.

Hunt, F.L. 1880. Sketch of the late Hon. Donald Gunn, pp. ix- xxii. *In* Gunn and Tuttle 1880.

Imhof, T. 1962. *Birds of Alabama.* University AL: University of Alabama Press.

James, P.C. 1994. Northern Hawk-owl irruptions in Saskatchewan, 1990-91 and 1991-92. *Blue Jay* 52:25-28.

Jensen, R., and D. Weidl. 1977. Indigo Bunting at Good Spirit Provincial Park, Saskatchewan. *Blue Jay* 35:169-170.

Jmaeff, J.H. 1959. A Gyrfalcon observation. *Blue Jay* 17:77.

Johns, B.W. 1987. Whooping Crane sightings in the Prairie Provinces, 1979-85. *C.W.S. Progress Notes* #169.

Johns, B., D. Collister, R. Bjorge, K. De Smet, W. Harris, E. Wiltse, and S. Barry. 2002. Loggerhead Shrike population trends in western Canada. *C.W.S. Progress Notes* #218.

Johnson, O.W., and P.G. Connors. 1996. American Golden-Plover (*Pluvialis dominica*) *In The Birds of North America*, No. 201 (A. Poole and F. Gill, eds.). Philadelphia: The Academy of Natural Sciences, and Washington: The American Ornithologists' Union.

Johnson, Mrs Gilbert. 1967. Prairie chicken at Marchwell? *Blue Jay* 25:100.

Jowsey, J.R. 1992. Red-bellied Woodpecker sighting at Saltcoats. *Blue Jay* 50:78.

Kehoe, P., and R. Alisauskis. 2001. Declines in White-winged Scoters. Unpublished report, Canadian Wildlife Service.

Koes, R.F., and P. Taylor. 1990a. *Am. Birds* 44:109-111.

Koes, R.F., and P. Taylor. 1990b. *Am. Birds* 44:445-447.

Koes, R.F., and P. Taylor. 1990c. *Am. Birds* 44:1148-1149.

Koes, R.F., and P. Taylor. 1991a. *Am. Birds* 45:118-120.

Koes, R.F., and P. Taylor. 1991b. *Am. Birds* 45:460-462.

Koes, R.F., and P. Taylor. 1991c. *Am. Birds* 45:1128-1130.

Koes, R.F., and P. Taylor. 1992a. *Am. Birds* 46:108-111.

Koes, R.F., and P. Taylor. 1992b. *Am. Birds* 46:279-281.

Koes, R.F., and P. Taylor. 1993a. *Am. Birds* 47:108-110.

Koes, R.F., and P. Taylor. 1993b. *Am. Birds* 47:269-270.

Koes, R.F., and P. Taylor. 1993c. *Am. Birds* 47:423-425.

Koes, R.F., and P. Taylor. 1993d. *Am. Birds* 47:1118-1120.

Koes, R.F., and P. Taylor. 1994. *NAS Field Notes* 48:954-956.

Koes, R.F., and P. Taylor. 1995a. *NAS Field Notes* 49:159-160.

Koes, R.F., and P. Taylor. 1995b. *NAS Field Notes* 49:264-267.

Koes, R.F., and P. Taylor. 1995c. *NAS Field Notes* 49:941-943.

Koes, R.F., and P. Taylor. 1996. *NAS Field Notes* 50:961-963.

Koes, R.F., and P. Taylor. 1997. *NAS Field Notes* 51:761-762.

Koes, R.F., and P. Taylor. 1998a. *ABA Field Notes* 52:76-79.

Koes, R.F., and P. Taylor. 1998b. *ABA Field Notes* 52:468-469.

Koes, R.F., and P. Taylor. 2002a. *North Am. Birds* 56:64-65.

Kornder, M., and K. Belcher. 1996. A multispecies inventory of the prairie parkland in east-central Saskatchewan. *Blue Jay* 54:58-61.

Krause, H. 1958. Northern Great Plains Region, *Audubon Field Notes* 12:362-364.

Krause, H. 1959a. Northern Great Plains Region, *Audubon Field Notes* 13:41-43.

Krause, H. 1959b. Northern Great Plains Region, *Audubon Field Notes* 13:303-304.

Krause, H. 1959c. Northern Great Plains Region, *Audubon Field Notes* 13:380-381.

Krause, H. 1959d. Northern Great Plains Region, *Audubon Field Notes* 13:437-439.

Krause, H. 1960a. Northern Great Plains Region, *Audubon Field Notes* 14:47-50.

Krause, H. 1960b. Northern Great Plains Region, *Audubon Field Notes* 14:318-320.

Krause, H. 1961. Northern Great Plains Region, *Audubon Field Notes* 15:51-54.

Krupp, W.J. 1965. Feeding Evening Grosbeaks. *Blue Jay* 23:190.

Lahrman, F. 1958. Some unusual Saskatchewan bird records. *Blue Jay* 16:109.

Lanyon, W.E. 1997. Great Crested Flycatcher (*Myiarchus crinitus*). *In The Birds of North America*, No. 300 (A. Poole and F. Gill, eds.). Philadelphia: The Academy of Natural Sciences, and Washington: The American Ornithologists' Union.

Larsen, K. 1999. Nesting Common Ravens near Bredenbury eat salamanders. *Blue Jay* 57:189-190.

Lattin, F.H. 1901. Jottings. *Oologist* 18:189.

Lawrence, A.G. 1921. Chickadee Notes #17. *Winnipeg Free Press,* 21 July.

Lawrence, A.G. 1927. Chickadee Notes #327. *Winnipeg Free Press,* 30 June.

Lawrence, A.G. 1931a. Chickadee Notes #512. *Winnipeg Free Press,* 16 January.

Lawrence, A.G. 1931b. Chickadee Notes #520. *Winnipeg Free Press,* 13 March.

Lawrence, A.G. 1932. Chickadee Notes #614. *Winnipeg Free Press,* 30 December.

Lawrence, A.G. 1933a. Chickadee Notes #628. *Winnipeg Free Press,* 7 April.

Lawrence, A.G. 1933b. Chickadee Notes #632. *Winnipeg Free Press,* 5 May.

Lawrence, A.G. 1933c. Chickadee Notes #642. *Winnipeg Free Press,* 13 July.

Lawrence, A.G. 1943. Chickadee Notes #1149. *Winnipeg Free Press,* 29 January.

Ledingham, G. 1956. In memoriam: Judge L.T. McKim. *Blue Jay* 14:38.

Leighton, A.L., J. Hay, C.S. Houston, J.F. Roy, and S. Shadick. 2002. *Birds of the Saskatoon Area.* Sask. Nat. Hist. Soc. Spec. Publ. 23.

Lincoln, F.C. 1939. *The Migration of American Birds.* New York: Doubleday Doran.

Lindgren, M. 1985. Yellow-crowned Night-heron. *Blue Jay* 43:247.

Lister, R. 1965. *Audubon Field Notes* 19:48-53.

Long, Robert J. 1970. Barred Owls at Madge Lake, Saskatchewan. *Blue Jay* 28:163.

Long, Robert J. 1971. Photographing Turkey Vultures at Madge Lake. *Blue Jay* 29:86-87.

Lyon, W. 1987. An owl banding trip in 1986. *Blue Jay* 45:123.

Lyon, W.J., D. Hjertaas, and B.W. Johns. 1995. *Evaluation of three potential reintroduction sites for Whooping Cranes in Canada.* Regina: Canadian Plains Research Centre.

MacFarlane, R.R. 1908. List of birds and eggs, pp. 287-448. *In* C. Mair, *Through the Mackenzie Basin.* Toronto: William Briggs.

Macoun, J. 1883. *Manitoba and the Great North-west.* London: Thomas C. Jack.

Macoun, J., and J. Macoun. 1909. *Catalogue of Canadian Birds.* Ottawa: Government Printing Bureau.

Macoun, J. 1922. *Autobiography of John Macoun, M.A., Canadian Explorer and Naturalist.* Ottawa: Ottawa Field-Naturalists' Club.

May, W.J.C. 1974. Winter bird feeding at Togo, Saskatchewan. *Blue Jay* 32:181.

Mayhew, K. 1946, reprinted 1992. Yorkton mourns death of Mrs Bob Priestly. *Blue Jay* 50:131-136.

McKim, L.T. 1926. Arkansas Kingbird nesting at Melville, Saskatchewan. *Auk* 4:370-371.

Mess, E. 1955. The two bluebirds. *Blue Jay* 13:17.

Mitchell, H.H. 1924. Birds of Saskatchewan. *Can. Field-Nat.* 38:101-118.

Mitchell, H.H. 1927. [Madge Lake collecting trip, pp. 37-38]. *In* Bradshaw 1927.

Morgotch, L. 1961a. Bank Swallows nesting in gravel stockpile near Canora. *Blue Jay* 19:115.

Morgotch, L. 1961b. Watching a heronry at Horseshoe Lake. *Blue Jay* 19:158-159.

Morgotch, L. 1962. Banded Purple Finch recovered at Yorkton. *Blue Jay* 20:9,39.

Morgotch, L. 1966. Partial albino Canada Goose. *Blue Jay* 24:201.

Morgotch, L. 1978. Bonaparte's Gull at McOuat Lake [with photograph]. *Blue Jay* 36:127.

Motherwell, P. 1990. *Ralph Prescott Stueck, 1893-1979, Naturalist and Conservationist.* Privately printed by Wendy Anderson, 16 pp.

Mowat, F.M. 1946. Field notes of Saskatchewan collecting trip [unpublished; in possession of CSH].

Nelson, H.K. 1963. pp. 133-149. *In* Restoration of breeding Canada Goose flocks in the north central states. *Proceedings 28th N. American Wildlife and Natural Resources Conference.*

[Nero, R.W.]. 1960. Second annual report on the Prairie Nest Records Scheme. *Blue Jay* 18:26-27.

Nero, R.W. 1961. *Audubon Field Notes* 15:475-477.

Nero, R.W. 1962a. *Audubon Field Notes* 16:47-50.

Nero, R.W. 1962b. *Audubon Field Notes* 16:423-426.

Nero, R.W. 1962c. *Audubon Field Notes* 16:486-488.

Nero, R.W. 1963. *Audubon Field Notes* 17:40-44.

Ninowski, R. 1962. Crows. *Blue Jay* 20:170.

Ninowski, R. 1963. Ring-necked Pheasant. *Blue Jay* 21:35.

Niven, W. 1946. Phalaropes. *Blue Jay* 4:41.

Niven, W. 1949. Sheho. *Blue Jay* 7(1):8.

Niven, W. 1951. Spring and summer migrations. *Blue Jay* 9(3):6.

Niven, W. 1953. Cuckoo and Fox Sparrow. *Blue Jay* 11(1):7.

Niven, W. 1968. Migrant and resident sparrows at Sheho, Saskatchewan. *Blue Jay* 26:110-111.

Niven, W. 1973. The land, the birds, through 50 years in aspen parkland. *Blue Jay* 31:223-229.

Osborne, Mrs C. 1957. More starling records. *Blue Jay* 15:101.

Padbury, G.A., and D.F. Acton. 1994. Ecoregions of Saskatchewan [poster]. Regina: Central Survey and Mapping Agency.

Pollock, C. 1986a. American Woodcock seen in east-central Saskatchewan. *Blue Jay* 44:180-182.

Pollock, C. 1986b. An unusual bird visitor. *Blue Jay* 44:196

Pollock, C. 1986c. Eastern Saskatchewan bird highlights, 1985. *Blue Jay* 44:197-198.

Pollock, L. 1999. Osprey rescues at Preeceville. *Blue Jay* 57:143-145.

Priestly, I.M. 1942a. *List of Birds Identified in the Yorkton District in Recent Years.* Yorkton: mimeographed by C. Stuart Houston.

Priestly, I.M. 1942b. Local notes. *Blue Jay* 1:3.

Priestly, I.M. 1943a. Information please: Purple Martin. *Blue Jay* 1:9.

Priestly, I.M. 1943b. Bohemian Waxwings. *Blue Jay* 1:19.

Priestly, I.M. 1943c. Bird notes. *Blue Jay* 2:3.

Priestly, I.M. 1944. Y.N.H.S. news. *Blue Jay* 2:20.

Priestly, I.M. 1945a. Information please: Burrowing Owl. *Blue Jay* 3:30.

Priestly, I.M. 1945b. Bird notes. *Blue Jay* 3:34.

Priestly, I.M. 1945c. 1945 bird records of interest in brief. *Blue Jay* 4:5-6.

Raine, W. 1892. *Bird-Nesting in North-West Canada.* Toronto: Hunter Rose.

Rayner, H.M. 1944. Unusual kingbird nest. *Blue Jay* 2:34.

Rayner, H.M. 1951. Starlings. *Blue Jay* 9(3):11.

Reed, C.A. 1904. *North American Birds Eggs.* New York: Doubleday Page.

Renaud, W. 1973. *Am. Birds* 27:785-788.

Renaud, W., G.J. Wapple, and D.W. Edgett. 1979. The Piping Plover in Saskatchewan: a status report. *Blue Jay* 37:90-103.

Richards, W. 1961. Bank Swallows nesting in gravel stockpiles. *Blue Jay* 19:84.

Rooke, R.P. 1947. Bird life in the Yorkton district in the 1890s. *Blue Jay* 5:26.

Roy, J.F. 1996. *Birds of the Elbow.* Sask. Nat. Hist. Soc. Spec. Publ. 21.

Sargeant, A.B., R.J. Greenwood, M.A. Sovada, and T.L. Schaffer. 1993. Distribution and abundance of predators that affect duck population--the prairie potholes region. USF&WS Resource Publ. 194.

Schmidt, A. 1973. An ecological survey of the Yorkton wetland complex. Saskatoon: Dept. Natural Resources Wildlife Research. 105 pp. (mimeographed).

Schmidt, V. 1964. Late nesting Gray Partridge. *Blue Jay* 22:149.

Serr, E.M. 1975a. *Am. Birds* 29:705-707.

Serr, E.M. 1975b. *Am. Birds* 29:995-999.

Serr, E.M. 1976. *Am. Birds* 30:87-90.

Serr, E.M. 1977a. *Am. Birds* 31:190-194.

Serr, E.M. 1977b. *Am. Birds* 31:1013-1016.

Serr, E.M. 1977c. *Am. Birds* 31:1150-1154.

Serr, E.M. 1978a. *Am. Birds* 32:220-223.

Serr, E.M. 1978b. *Am. Birds* 32:1021-1024.

Serr, E.M. 1979. *Am. Birds* 33:188-191.

[Seton], E.E. Thompson. 1884. Nest and habits of the Connecticut warbler. *Auk* 1:192.

[Seton], E.E. Thompson. 1885. Nest and eggs of the Philadelphia Vireo. *Auk* 2:305-306.

[Seton], E.E. Thompson. 1886. The birds of western Manitoba. *Auk* 3:146-156, 319-329, 453.

[Seton], E.E. Thompson. 1891. The birds of Manitoba. *Proceedings of the United States National Museum* 13: 457-643.

[Seton], E.E. Thompson. 1893a. Additions to the list of Manitoba birds. *Auk* 10:49.

[Seton], E.E. Thompson. 1893b. Notes of a trip to Manitoba, part 1. *Forest and Stream* 41:424-425.

Seton, E.T. 1909a. Fauna of Manitoba, pp. 183-227. *In A Handbook to Winnipeg and the Province of Manitoba.* Winnipeg: British Association for the Advancement of Science.

Seton, E.T. 1909b. *Lives of Northern Mammals...The Mammals of Manitoba.* New York: Charles Scribner's Sons. 2 vols.

Shadick, S.J. 1986. Golden-winged Warbler at Duck Mountain, Saskatchewan. *Blue Jay* 44:245-246.

Shantz, B.R. 1986. *Mountain Bluebird Management.* Lacombe, AB: Deer Ridge Consulting.

Shaver, J. 1966. Record of Wood Duck at Little Quill Lake. *Blue Jay* 24:79.

Shaw, C.C. 1946a. Bird notes. *Blue Jay* 4:41.

Shaw, C.C. 1946b. Bird records of interest in brief. *Blue Jay* 4:42.

Shaw, C.C. 1947. Pinnated Grouse. *Blue Jay* 5:32.

Shaw, C.C. 1949. Birds. *Blue Jay* 7(2):3-8.

Shaw, C.C. 1950. Evening Grosbeaks. *Blue Jay* 9(2):16.

Shaw, C.C. 1954. American Screech Owls. *Blue Jay* 9(2):22.

Skeel, M., D.C. Duncan, and E. Wiltse. 1997. Saskatchewan results of the 1996 International Piping Plover Census. *Blue Jay* 55:157-168.

Smith, A.R. 1972. The value of abandoned farmsteads as a wildlife resource. Saskatoon: Department of Natural Resources.

Smith, A.R. 1996. *Atlas of Saskatchewan Birds.* Sask. Nat. Hist. Soc. Spec. Publ. 22.

Soper, J.D. 1936. List of birds observed at Good Spirit Lake, Saskatchewan on 23 June 1936. (Mss.).

Soper, J.D. 1942. List of birds observed at Mery Creek, four miles northwest of Lac la Course, 26-28 May 1942. (Mss.).

Stephen, W.J.D. 1979. Whooping Crane sightings: Prairie Provinces 1977 and 1978. *Blue Jay* 37:163-168.

Street, M.G. 1949. A Cooper's Hawk nest. *Blue Jay* 7(1):9.

Summers, P. 1949. Yorkton birds. *Blue Jay* 7(1):9.

Swallow, H.S. 1941. Rain of ducks at Foam Lake, Saskatchewan. *Can. Field-Nat.* 55:130.

Swallow, H.S. 1946. History of water levels and wildlife in Yorkton. *Yorkton Enterprise,* 12 Sep 1946.

Switzer, F.A. 1970. First Chimney Swift record for Yorkton district. *Blue Jay* 28:122.

Switzer, F.A. 1980. Larry Albert Morgotch: In Memoriam. *Blue Jay* 38:67-68.

Thompson, Mrs Arthur. 1953. A noticeable change. *Blue Jay* 11(1):9.

Thompson, Floyd A. 1954. Saskatchewan observations. *Blue Jay* 12(4):21-22.

Vermeer, K., and G. Anweiler. 1970. Great Blue Heron colonies in Saskatchewan in 1970. *Blue Jay* 28:158-161.

Walkinshaw, L.H. 1948. Mallard nesting in an old magpie nest. *Auk* 65:599.

Walkinshaw, L.H. 1949. Apparent observations of the Whooping Crane in central Saskatchewan. *Can. Field-Nat.* 63:78-80.

Walkinshaw, L.H. 1960. Summer records of the Sandhill Crane in Saskatchewan. *Blue Jay* 18:20-23.

Wapple, R. 1999. *Wildlife Utilization of the Upper Assiniboine River Basin (UARB)*. Saskatoon: Wetlands Committee, Upper Assiniboine River Basin Study. 93 pp.

Weber, W.C. 1974. More Great Crested Flycatchers and Short- billed Marsh Wrens in Alberta. *Blue Jay* 32:232.

Wedgwood, J.A. 1982. *Am. Birds* 36:864-866.

Weidl, D.A. 1983. Avian highlights from Duck Mountain Provincial Park, Saskatchewan, 1981. *Blue Jay* 41:190-193.

Weiman, D. 2001. Suicidal ravens? *Blue Jay* 59:208-209.

Wershler, M. 1949. An unlucky flicker. *Blue Jay* 7(3):14.

Wiley, E. 1955. Flooded out, but survived. *Blue Jay* 13:19-21.

Williams, F. 1969. Southern Great Plains Region. *Am. Birds* 23:492-495.

Wotherspoon, L. 1976. Bird notes from the Manitoba boundary. *Blue Jay* 34:205-207.

Wright, Robert A., J. Vandall, C. Lockerbie, G. Gray, and K. Gianenetta. 1995. *Duck Mountain Provincial Park Vegetation Management Plan.* Ottawa: Canadian Forestry Service Project No. 7002.

ADDENDUM:

American Ornithologists' Union. 2003. Forty-fourth supplement to the *AOU Check-List of North American Birds. Auk* 120:923-931. [As we go to press, this supplement has changed the verncular name of the former Rock Dove to Rock Pigeon and the Three-toed Woodpecker to American Three-toed Woodpecker, and has changed the scientific names of the Eastern Screech-Owl, Snowy Owl, American Three-toed Woodpecker, Black-capped Chickadee, and Ovenbird].